3rd International Conference on Civil Engineering and Materials Science (ICCEMS 2018)

MATEC Web of Conferences Volume 206 (2018)

Chengdu, China
13-15 April 2018

Editors:

C. W. Lim
Xiaohong Zhu

ISBN: 978-1-5108-7179-3

Printed from e-media with permission by:

Curran Associates, Inc.
57 Morehouse Lane
Red Hook, NY 12571

Some format issues inherent in the e-media version may also appear in this print version.

Printed by Curran Associates, Inc. (2018)

For additional information, please contact EDP Sciences – Web of Conferences at the address below.

EDP Sciences – Web of Conferences
17, Avenue du Hoggar
Parc d'Activité de Courtabœuf
BP 112
F-91944 Les Ulis Cedex A
France

Phone: +33 (0) 1 69 18 75 75
Fax: +33 (0) 1 69 28 84 91

contact-edps@webofconferences.org

Additional copies of this publication are available from:

Curran Associates, Inc.
57 Morehouse Lane
Red Hook, NY 12571 USA
Phone: 845-758-0400
Fax: 845-758-2633
Email: curran@proceedings.com
Web: www.proceedings.com

TABLE OF CONTENTS

Preface

2018 the 3rd International Conference on Civil Engineering and Materials Science (ICCEMS 2018) was held in Chengdu, China during April 13-15, 2018.

The event was held with presentations delivered by researchers and scholars from the international community, including keynote speeches and highly selective lectures. It brings together the top researchers from Asian Pacific nations, North America, Europe and around the world to exchange their research results and address open issues in biomaterials, biomedical manufacturing, casting and solidification, characterisation, coatings and surface engineering, composite materials, etc.

The proceedings included papers which were from universities, research institutes and industries. All of the papers were subjected to peer-reviewing by conference committee members and international reviewers. The papers selected for the proceedings depended on their quality and their relevance to conference.

We would like to take this opportunity to thank many people. First and foremost we want to express our deep appreciation to keynote speakers, Prof. Kyoung Sun Moon, Yale University School of Architecture, USA; Prof. C. W. Lim, City University of Hong Kong, Hong Kong; Prof. Xiaohong Zhu, Sichuan University, China.

We would like to thank the committee members for their hard work in making sure a smooth running of this conference. Many thanks to the reviewers for their excellent job to maintain the academic quality of this conference.

Final thanks go to all authors and participants at ICCEMS 2018 for helping to make it a successful event.

With our warmest regards,
ICCEMS 2018 Organizing Committees

Statement of Peer review

In submitting conference proceedings to *Web of Conferences*, the editors of the proceedings certify to the Publisher that

1. They adhere to its **Policy on Publishing Integrity** in order to safeguard good scientific practice in publishing.
2. All articles have been subjected to peer review administered by the proceedings editors.
3. Reviews have been conducted by expert referees, who have been requested to provide unbiased and constructive comments aimed, whenever possible, at improving the work.
4. Proceedings editors have taken all reasonable steps to ensure the quality of the materials they publish and their decision to accept or reject a paper for publication has been based only on the merits of the work and the relevance to the journal.

Title, date and place of the conference

2018 the 3rd International Conference on Civil Engineering and Materials Science (ICCEMS 2018)
Chengdu, China
April 13-15, 2018

Proceedings editor(s):

Prof. C. W. Lim, City University of Hong Kong, Hong Kong

Prof. Xiaohong Zhu, Sichuan University, China

Date and editor's signature

Influence of Dip Angle on Non-Anti-Symmetry Response of Tunnels in Bedding Rock Mass Due to Seismic qSV Wave Propagation

H Y Hu[1, 2], X J Zhou[1, 2], and T Chen[3, 4]

[1]*Key Laboratory of Transportation Tunnel Engineering, Ministry of Education,Southwest Jiaotong University, Chengdu, Sichuan 610031, China;*
[2]*School of Civil Engineering, Southwest Jiaotong University, Chengdu,Sichuan 610031, China;*
[3]*School of Transportation and Logistics, Southwest Jiaotong University, Chengdu,610031, China;*
[4] *National Railway Train Diagram Research and Training Center, Southwest JiaoTong University, Chengdu,610031, China.*

Abstract. By considering the transverse isotropic constitutive law of the bedding rock mass and using the wave method to analyze a high-speed railway tunnel in a bedding rock mass, we studied the influence of dip angle on non-anti-symmetry response of tunnels in bedding rock mass due to seismic qsv wave propagation. The results show that the non-anti-symmetry of tunnel seismic response does not obey linear relationship with the change of rock dip, and the variation law is segmental; the moment of seismic response is anti-symmetric strictly when the rock layer is vertical or horizontal, and it is non-anti-symmetrical when the rock layer is inclined; within each segmented interval, the non-anti-symmetry of seismic response increases first and then decreases with the increase of dip angle. This indicates that the influence of dip angle on the non-anti-symmetry of tunnel seismic response is significant.

1. Introduction

Under the action of seismic qSV wave (i.e. quasi-SV wave), the seismic response of homogeneous rock tunnel is anti-symmetric, but the seismic response of bedding rock tunnel (i.e. the strike line of strata is parallel to the tunnel axis, the dip direction is perpendicular to the tunnel axis) is non-anti-symmetric. The non-anti-symmetric is due to the transverse isotropic constitutive law of bedding rock mass[1]. However, in recent years, scholars' research focused more on homogeneous rock tunnels[2-7], in other words, the current studies considered bedding rock mass as an isotropic medium, this does not reflect the non-anti-symmetry of the seismic response, and the influence of dip angle on the non-anti-symmetry[8-14].

So, the transverse isotropic constitutive law of bedding rock mass was considered, and a high-speed railway tunnel in a bedding rock mass was took as an engineering background, and then the numerical analysis of seismic response was carried out by using ANSYS to study the influence of dip angle on non-anti-symmetry of tunnel seismic response caused by incident qSV wave.

2. Material parameters and calculation model

2.1. Lining parameters

A typical excavation section of high-speed railway deep tunnel is shown as Figure 1, the maximum width of tunnel excavation section is 15m. The parameters of tunnel secondary lining are: thickness is 600mm, C35 concrete, reinforcement ratio is 0.35%, modulus of elasticity is 31.5GPa, Poisson's ratio is 0.2, Bulk weight is 25kN/m^3. In order to observe the characteristics of seismic response of tunnel in bedding rock mass more clearly, the reinforcement layer and initial support are omitted in study.

2.2. Bedding rock mass parameters

Without loss of generality, the dip angles of 0°, 25°, 45°, 70° and 90° of the rock mass and two groups of bedding rock mass parameters D1 and D2 are selected, see Table 1 for details; And five dip angles of 0°, 25°, 45°, 70° and 90° are selected. Therefore, there are 10 combinations of work conditions, as is shown in Table 2.

2.3. Calculation model

The plane strain finite element model of a tunnel in a bedding rock mass takes the tunnel as the centre and intercepts a square area of 1000m×1000m in the infinite rock mass domain. And ANSYS software was used for seismic response analysis, where the rock mass is simulated by element plane42, the lining is simulated by element beam3 and the viscoelastic boundary is simulated by element combin14, and the integral step is 0.01s. The calculation of the viscoelastic boundary spring-damping parameters is detailed in [1], where the normal correction factor α_N and the tangential correction factor α_T are taken as 1.0 and 0.5 respectively [15].

Table 1. Bedding rock mass parameters

Rock mass group	Density (kg·m^{-3})	Elastic constants				
		In-Plane Elastic Modulus (GPa)	Symmetrical Axial Elastic Modulus (GPa)	In-Plane Poisson Ratio	Symmetrical Axial Poisson Ratio	Symmetrical Axial Shear Modulus (GPa)
D1	2700	22.0	17.0	0.24	0.26	7.2
D2	2700	22.0	13.0	0.24	0.26	5.2

In order to reveal the general response characteristics of tunnels in bedding rock mass, the total wave field of the seismic wave only contains the incident wave field and the tunnel reflection wave field, and four viscoelastic boundaries of the finite calculation area were set. And the tunnel reflection wave is naturally absorbed by the viscoelastic boundaries.

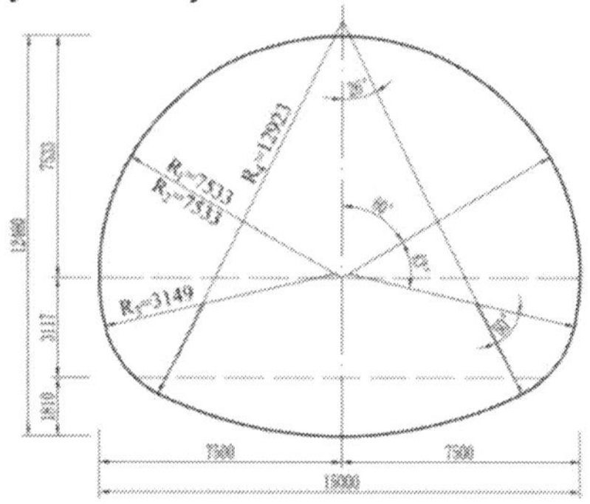

Figure 1 Tunnel excavation section
(Unit: mm)

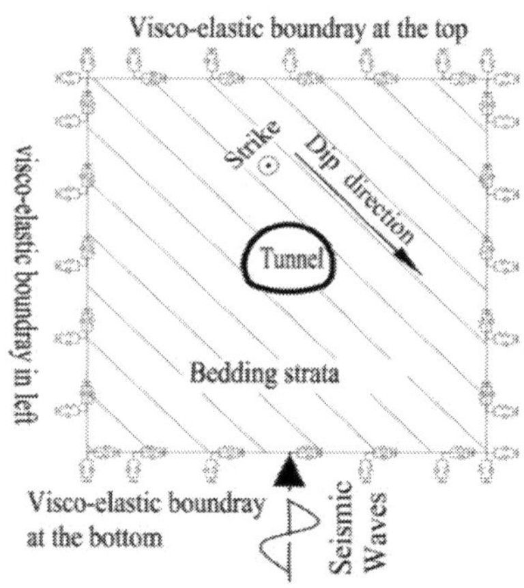

Figure 2. Schematic of calculation model

The incident wave may be located in any direction of the tunnel in actual, however, when the direction of incident seismic wave is on the symmetry axis of the tunnel section, the lining response of non-anti-symmetry caused by rock anisotropy can be presented effectively, for that the transverse component of the qSV is perpendicular to the symmetry axis of the tunnel section and the lining on the axis symmetry. And based on the non-symmetry of the response, the influence of dip angle on the seismic response of the tunnel can be studied. Therefore, the direction of the incident seismic wave was set on the symmetry axis of the tunnel section.

The resulting calculation model is shown as Figure 2.

3. Incident qSV Wave

The Wolong wave in Wenchuan earthquake was selected, and the most violent of the 10~21s band was extracted, filtered and baseline corrected, and the resulting displacement time and velocity time history are shown in Figure 3 and Figure 4.

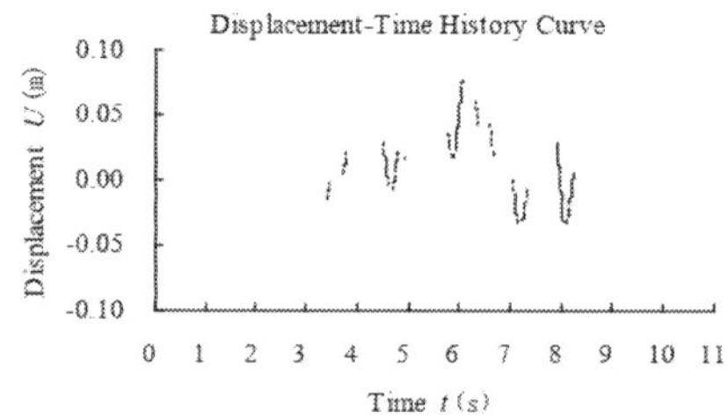

Figure 3. Displacement-Time History Curve of Incident Wave

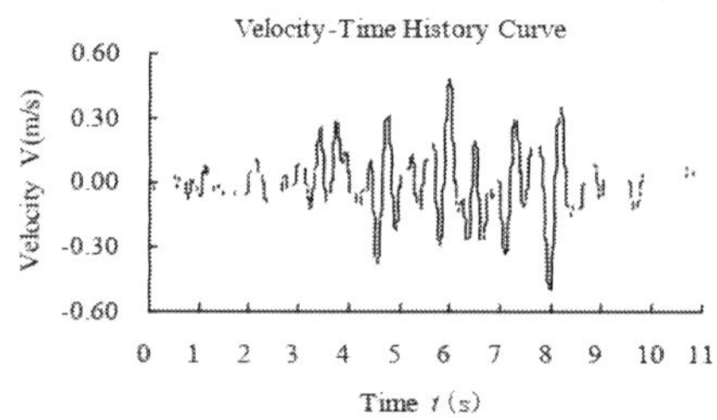

Figure 4. Velocity-Time History Curve of Incident Wave

4. Results and Analysis of Moment Envelope

Moment envelope is a very representative seismic response. The envelope of the tunnel lining moment in each work condition are shown respectively in Figure 5 for D1 rock mass group and Figure 6 for D3 rock mass group, and the moment of the arch foot is shown in Table 2. According to Figure 5 and Figure 6: the envelope of bending moments in D1 and D3 rock mass groups are very similar, and the extreme value of moments are all located at the arch foot, but the values are different. Further, according to Table 2: when the rock stratum is horizontal or vertical, the moment envelope is left-right anti-symmetric; and when the rock stratum is inclined, the moment envelope is non-anti-symmetric.

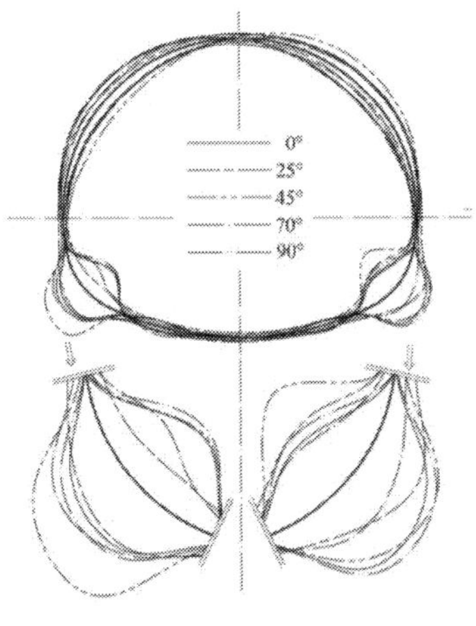

Figure 5. Moment Envelope of tunnel lining in D1 Rock Mass Group

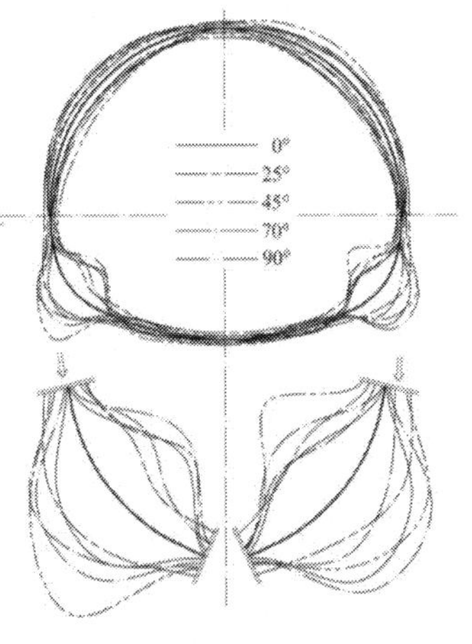

Figure 6. Moment Envelope of tunnel lining in D2 Rock Mass Group

Table 2. Peak Moment Comparison Table for qSV Waves

Cond-ition[a]	Left Foot		Right Foot		Cond-ition[a]	Left Foot		Right Foot	
	Moment (kN.m)	Time (s)	Moment (kN.m)	Time (s)		Moment (kN.m)	Time (s)	Moment (kN.m)	Time (s)
D1-0°	1420.09	6.71	-1420.44	6.71	D2-0°	1388.80	6.18	-1389.91	6.18
	-1961.00	8.21	1961.48	8.21		-1938.47	8.20	1937.30	8.20
D1-25°	1352.31	8.31	-1709.51	8.31	D2-25°	1345.22	8.19	-1524.94	8.34
	-854.32	5.97	1480.02	5.98		-1203.42	5.85	1269.87	6.00
D1-45°	2801.40	8.35	-2795.12	8.35	D2-45°	2992.65	8.36	-2907.72	8.36
	-2017.87	6.85	2024.10	6.85		-2136.09	6.86	2034.92	6.86
D1-70°	1773.55	8.27	-1365.46	8.27	D2-70°	1975.84	8.27	-1484.78	8.27
	-1472.44	6.77	1088.13	6.77		-1620.57	6.77	1214.69	6.77
D1-90°	1632.05	6.71	-1631.93	6.71	D2-90°	1865.78	6.70	-1864.01	6.70
	-2131.78	8.21	2132.02	8.21		-2268.35	8.20	2266.84	8.20

[a] Work condition number agreement as follows: Take D1-0° as an example, D1 is the rock mass group, 0° is the dip angle, and the rest is analogy.

To analyze the relationship between the non-anti-symmetry and dip angle, the parameter Moment Sum of Left and Right Foot χ^{M} is defined as:

$$\chi^{M} = \text{Moment of Left Foot} + \text{Moment of Right Foot} \quad (1)$$

The maximum and minimum values of arch moment envelope are substituted into equation (1) to calculate the sum of bending moments and draw Figure 7. According to Figure 7, the relationship between the non-anti-symmetry of the bending moment envelope and the dip angle is non-linear: when the dip angle θ^0 is about 45°, the absolute value of the bending moment on the left and right arches is almost equal, and when θ^0 is 0° or 90°, the absolute value of the bending moment on the left and right arches is strictly equal, i.e. the moment of seismic response is anti-symmetric strictly. Overall, the absolute value of left arch moment is larger than that of right arch when $\theta^0 \in (0°, 45°)$, and the non-anti-symmetry increases first, then decreases with the increase of dip angle, and the highest at about 25°; the absolute value of left arch moment is smaller than that of the right arch when $\theta^0 \in (45°, 90°)$, and the non-anti-symmetry first increases and then decreases with the increase of dip angle, and the highest at about 75°.

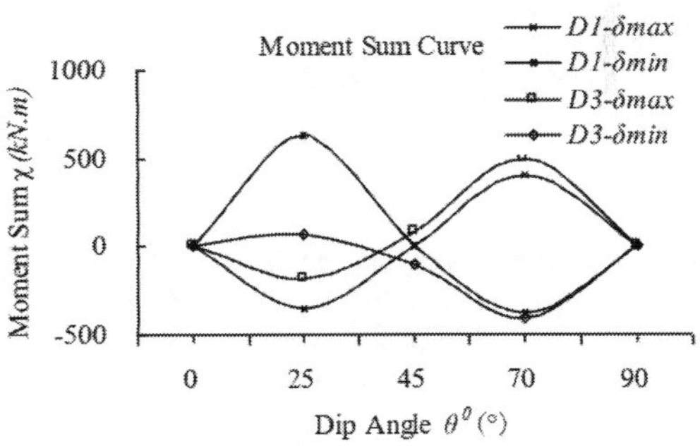

Figure 7. Moment Sum of Left and right foot (Envelope)

5. Results and Analysis of Relative Displacement

The layout of tunnel lining section displacement detection points A~H is shown in Figure 8. The relative displacement between the detection points in the section is shown in Table 3. In table 3: d_{BG} is the relative displacement between the detection points B and G at the same time, other analogies; And the relative distance between the points is a positive number when it is larger than the original distance, on the contrary, it is negative.

Table 3. Relative Displacements (unit: mm)

Cond-ition	Time/s	Diagonal relative displacement			Horizontal relative displacement		
		d_{BG}	d_{CF}	d_{AH}	d_{BC}	d_{DE}	d_{FG}
D1-0°	6.71	46.67	-46.66	0.12	0.04	0.04	0.03
	8.21	-63.98	63.98	0.21	0.07	0.07	0.06
D1-25°	8.31	38.60	-62.94	-19.12	-6.58	-7.70	-4.90
	5.97	-16.60	33.36	41.03	-8.03	-13.50	-9.48
D1-45°	8.35	90.00	-91.66	-14.87	6.79	10.63	6.26
	6.85	-64.92	66.33	12.14	-5.09	-7.92	-4.63
D1-70°	8.27	54.92	-42.62	17.70	0.17	-0.79	-1.44
	6.77	-37.33	31.70	-23.34	6.82	10.91	8.14
D1-90°	6.71	54.52	-54.52	0.17	0.04	0.04	0.04
	8.21	-11.83	11.84	0.02	0.01	0.01	0.01
D2-0°	6.18	45.25	-45.25	0.08	0.06	0.01	0.05
	8.2	-62.09	62.09	0.17	0.08	0.01	0.07
D2-25°	8.19	10.71	9.29	64.19	-16.45	-25.87	-17.67
	5.85	-18.68	14.34	-33.58	12.14	18.50	12.78
D2-45°	8.36	85.99	-89.06	-32.99	14.85	23.36	13.77
	6.86	-61.71	63.55	22.35	-10.10	-15.71	-9.07
D2-70°	8.27	50.95	-47.97	23.10	-7.91	-12.38	-9.65
	6.77	-39.06	39.13	-20.88	9.01	13.95	10.52
D2-90°	6.7	62.37	-62.36	0.21	0.06	0.06	0.05
	8.2	-75.65	75.65	0.30	0.09	0.09	0.08

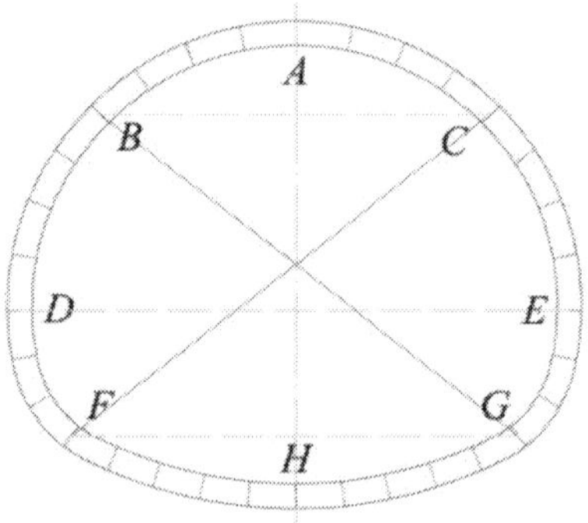

Figure 8. Lining Unit Number and Detection Points

According to table 3: when the rock stratum is horizontal or vertical, the relative displacements is left-right anti-symmetric; and when the rock stratum is inclined, the relative displacements is non-anti-symmetric.

To analyze the relationship between the non-anti-symmetry and dip angle, the parameter Diagonal Displacements Sum of Left and Right U^{M}, is defined as:

$$U^{M}=d_{BG}+d_{CF} \quad (2)$$

The relative displacements are substituted into equation (2) to calculate the sum of diagonal displacements and draw Figure 9. According to Figure 9, the relationship between the non-anti-symmetry of the bending moment envelope and the dip angle is non-linear: when the dip angle θ^0 is about 45°, the absolute value of the diagonal relative displacements on the left and right is almost equal, and when θ^0 is 0° or 90°, the absolute value of the diagonal relative displacements on the left and right is strictly equal, i.e. the diagonal relative displacements of seismic response is anti-symmetric strictly. Overall, the absolute value of left arch diagonal relative displacements is larger than that of right when $\theta^0 \in (0°, 45°)$, and the non-anti-symmetry increases first, then decreases with the increase of dip angle, and the highest at about 25°; the absolute value of left relative displacement is smaller than that of the right when $\theta^0 \in (45°, 90°)$, and the non-anti-symmetry first increases and then decreases with the increase of dip angle, and the highest at about 70°.

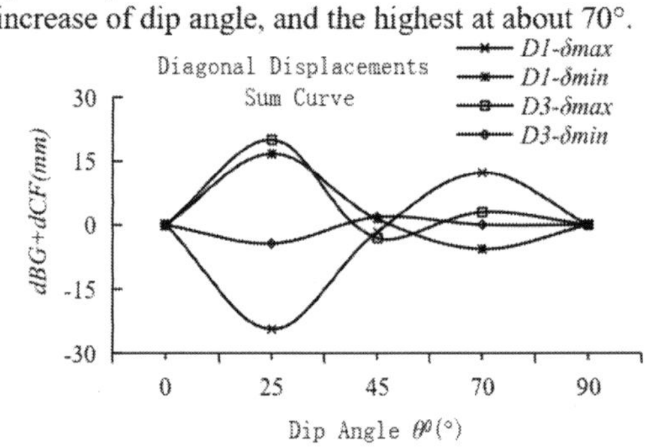

FiFigure 9. Diagonal Displacements Sum (Meanwhile)

6. Conclusion

The influence of dip angle on non-anti-symmetry response of tunnels in bedding rock mass is due to seismic qSV wave propagation:

(1) The non-anti-symmetry of the seismic response obeys the non-linear relationship with the dip, and with segmentation. And the moment of seismic response is anti-symmetric strictly when the rock layer is vertical or horizontal, and it is non-anti-symmetrical when the rock layer is inclined.

(2) In all work conditions of this paper, when the dip angle θ^0 is about 45°, the non-anti-symmetry behaves as a non-strict anti-symmetry, and when $\theta^0 \in (0°, 45°)$, the non-anti-symmetry first increases and then decreases with the increase of dip angle, and the strongest non-anti-symmetry is at about 25°; when $\theta^0 \in (45°, 90°)$, the non-anti-symmetry first increases and then decreases with the

increase of dip angle, and the strongest non-anti-symmetry is at about 70° or 75°.

(3) The influence of dip angle on the non-anti-symmetry of tunnel seismic response is significant and requires further study.

Acknowledgement

The authors appreciate the support from the National Natural Science Foundation of China (Project No. 51378436, 61703351), and the Fundamental Research Funds for the Central Universities(No. 2682017CX022).

Reference

1. Hu H, Zhou X and Wang J 2017. *CJRME*. **36** 1373-83
2. Hashash Y, Hook J, Schmidt B and Yao J 2001. *TUNN. UNDERGR. SP. TECH.* **16** 247-93.
3. Esmaeili M, Vahdani S and Noorzad A 2006. *TUNN. UNDERGR. SP. TECH.* **21** 511-9.
4. Park K, Tantayopin K, Tontavanich B and Owatsiriwong A 2009. *TUNN. UNDERGR. SP. TECH.* **24** 231-5.
5. Cilingir U and Madabhushi S 2011. *SOIL DYN. EARTHQ. ENG.* **31** 452-62.
6. Kouretzis G, Sloan S and Carter J 2013. *SOIL DYN. EARTHQ. ENG.* **46** 41-51
7. He C, Li L, Zhang J, Geng P and Yan Q 2014. *Chinese Journal of Geotechnical Engineering*. **36** 427-34. in Chinese
8. Han S, Zhang W and Zhang J 2017. *Geophys. J. Int.* **210** 1560–1569
9. Cheng J and Kang W 2016. *GEOPHYSICS*. **81** C39–C52
10. Brijendra Paswan B, Sahu S and Chattopadhyay A 2016. *Acta Mech.* **227** 3687–3701
11. Sharma J and Kaur R 2014. *LAT AM J SOLIDS STRU.* **11** 2141-2170
12. Ba Z, Lee V, Liang J and Yan Y 2017. *SOIL DYN EARTHQ ENG.* **98** 120-140
13. Ba Z and Liang J 2017. *ENG ANAL BOUND ELEM.* **76** 40-58
14. Modi C, Kumari P and Sharma V 2016. *APPL MATH MODEL.* **40** 8737–8749
15. Liu J, Du Y and Yan Q 2007. *Journal of Disaster Prevention and Mitigation Engineering* **27(Supplement)** 37-42. in Chinese

Characterization of the Undrained Shear Strength of Expansive Soils of High Water Content

Zheng Su[1, 2], Daokun Qi[3], Xinju Guo[3], Xiaojuan Xi[3] and Liang Zhang[3]

[1] Department of Geotechnical Engineering, Tongji University, Shanghai 200092, China;
[2] Key Laboratory of Geotechnical and Underground Engineering of Ministry of Education, Tongji University, Shanghai 200092, China;
[3] State Grid Henan Economic Research Institute, Zhengzhou 450052, China

Abstract. In recent years, engineering constructions increase rapidly in western and central areas of China, where expansive soil widely distributes. Since expansive soil is sensitive to water content, the characterization of its shear strength should be carefully conducted. For simplicity and ease of use, the Mohr-Coulomb criterion is often adopted to describe the shear strength of expansive soil. In this paper, the physical meaning of the cohesion and frictional strength of expansive soil are explained, and the variations of the strength parameters with water content are investigated. By fitting to the experimental results from direct shear test and triaxial tests, the changing characteristics of cohesion and friction angle with water content are obtained.

1. Introduction

Expansive soil widely distributes in the western and central China. Due to the season changes and temperature variation, the water content changes often. The mechanical behavior, i.e. shear strength and deformation characteristics, is highly affected by its water content. The shear strength increases with the shrinkage and decreases with the swelling effect. Since the increasing construction in expansive soil area, there is a great engineering demand on studying the shear strength of unsaturated expansive soil.

The influence of the water content on the shear strength of granular material has been studied experimentally and theoretically. By using X-ray microtomography, Scheel et al. [1] studied the influence of liquid content on the mechanical properties of glass beads; the findings provide a profound understanding on the wet granular material, including soil. Lu et al. [2] studied this problem in silty sand (50-200μm), fine sand (80-400μm), and medium sand (200-900μm), and found that the results are sensitive to particle size. Results of silty sand showed that the tensile strength increases dramatically to a peak and then decreases gradually with the increase in the saturation. Based on the direct shear tests and 3D molecular dynamic simulation of wet sand and glass beads with diameter of 0.1-1mm, Richefeu et al. [3] found that with the gradual increase in the water content, the cohesion of wet granular materials increases firstly and then tends to be unchanged, while the friction angle is irrespective of water content. Mani et al.[4] proposed a model to study the redistribution process of liquid in unsaturated granular material in plane shear; results showed that the shear rate profile would affect the liquid migration pattern. Tang et al. [5] investigated the influence of water content and dry density of compacted clay soil on its tensile strength, a peak value of the tensile strength characteristic curve (TSCC) was observed at the critical water content around 11.5%. Based on direct shear tests, Elkady and Abbas [6] studied the influence of two loading paths on the drained strength behavior of expansive soil of the northeast region of Saudi, Arabia; the loading path was shown to affect the strength parameters slightly. Wu and Wang [7] conducted direct shear tests and swelling tests on undisturbed expansive soil in Yanji Basien, and the influence of water content on strength parameters and swelling pressure was obtained; the strength parameters decrease with water content. H. Louati [8] studied the influence of particle size and saturation on shear behavior of partially wet granular material. It was found that shear strength increases for smaller particles and increases with liquid up to a critical value and then decrease.

Based on soil mechanics and theoretical analysis, this paper characterizes the shear strength of expansive soil. The first part is the description of Mohr-Coulomb criterion and the physical explanation of the strength parameter. The second part is the calibration of the criterion by experiments; the existing tests and consolidated undrained traxial tests were used. The varying characteristics of cohesion and friction angle with the water content were clear shown.

2. The shear strength of unsaturated soil

The Mohr-Coulomb criterion has been widely used in

describing the shear strength of soils. The stress used in the criterion can be effective stress or total stress; the effective stress can be calculated by the Terzaghi's principle in fully saturated soil. Since the expansive soil is often unsaturated and the pore water pressure cannot be easily obtained in engineering filed, the criterion is described by total stress for simplicity

$$\tau = c_{total} + \sigma * tan\varphi_{total} \tag{1}$$

where c_{total} and φ_{total} are the equivalent cohesion and friction angle.

The cohesion is caused by the inter-particle physicochemical forces, i.e. van der Waals attraction, electrical double layer repulsion or attraction, cementation due to solute precipitation, and capillary stress; the capillary stress can be induced by the negative pore pressure or the surface tension of liquid [2]. The first two components are important parts for fine grained soils especially for clay [9], the third component is important for some soil, i.e. loess, and the capillary stress are significant for unsaturated soil. The equivalent cohesion c_{total} can be described as

$$c_{total} = c_{va} + c_{ea} + c_{ce} + c_{cp} + c_{ct} \tag{2}$$

where c_{va} is van der Waals attraction, c_{ea} is electrical double layer repulsion or attraction, c_{ce} is cementation due to solute precipitation, c_{cp} is capillary stress due to the negative pore pressure, c_{ct} is capillary stress due to the surface tension of liquid.

Since the water contend affects the interaction between the water and particles, the schematic description of the particle-water-gas system in microscopic point of view is shown in Figure. 1. The capillary attraction is significant when the soil contains very low water content, and the reason is that the water concentrates among the small voids and forms capillary bridges under this condition (as shown in Figure.1-(b)). The macroscopic capillary attraction or tension strength is relatively small because of the interconnectivity of the voids in soil. With the increase in saturation, more small voids in soil are filled by water and connected with each other to be capillary clusters (as shown in Figure.1-(c)). As a result, the macroscopical capillary attraction is higher under this condition. Nevertheless, the macroscopic capillary attraction decreases when the saturation goes higher for the voids remained empty in soil under this condition is the macro-voids, which contributes to a relatively small capillary attraction. Finally, the capillary attraction goes to zero when the soil is saturated.

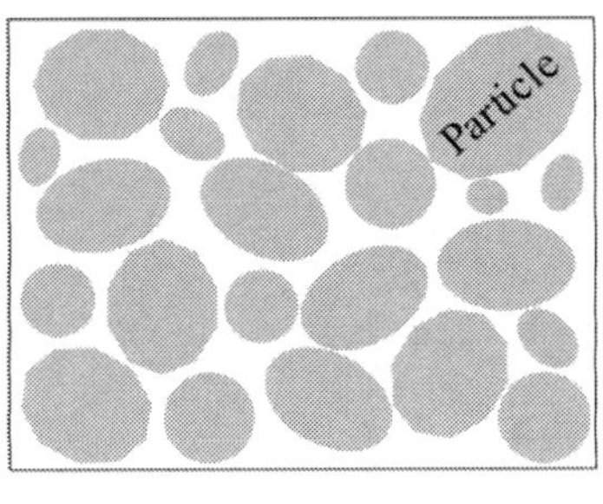

(a) dry

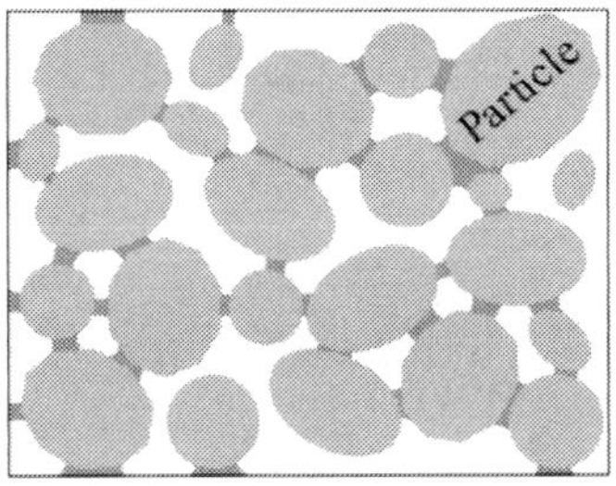

(b) low saturation

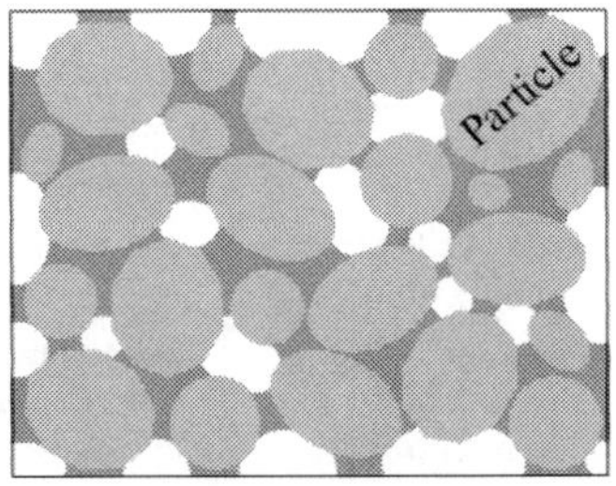

(c) high saturation

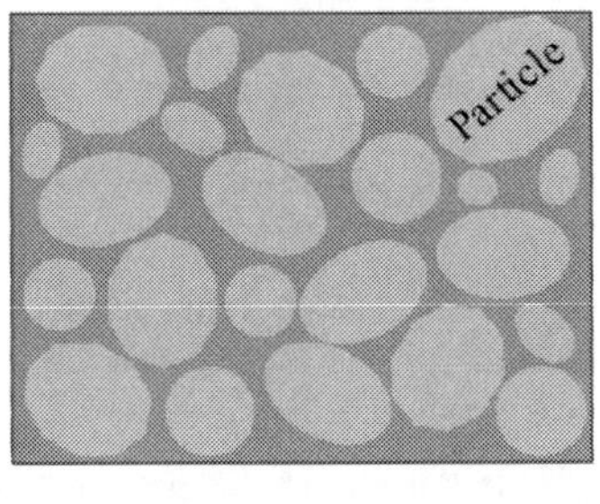

(d) fully saturated

Figure.1. Schematic description of unsaturated soil.

Since the friction angle φ_{total} comes from the friction between particles, it will be inevitably affected by water content. The existence of water between particles decreases the sliding friction for the liquid lubrication. The equivalent friction angle φ_{total} can be defined as

$$\varphi_{total} = \varphi_{fr} + \varphi_{oc} \tag{3}$$

where φ_{fr} and φ_{oc} are components induced by the sliding friction and interlocking between particles. It is apparent that φ_{fr} is sensitive to water content due to the lubrication effect.

Based on the theory mentioned above, the influence of water content on the strength parameters of soil is described in table 1 from micro perspective.

Table 1 Influence of water content on strength parameters

Water content	Cohesion	Friction angle
Low water content	The capillary stress would increase when water content increases from zero to a lower value. As	Both the sliding friction and interlocking friction are barely affected by variation of water

Water content	Cohesion	Friction angle
	a result, cohesion increases slightly in the meantime.	content due to the disconnectivity of the water in voids between soil particles.
Medial water content	The capillary stress decreases with water content increase gradually, and cohesion decreases.	The φ_{total} would decreases with water content increases, because the sliding friction is affected by water content obviously.
High water content	The capillary stress decreases to be zero when the water content goes to very high or the soil is approximate to saturated. Hence, the cohesion decreases further on this occasion.	The influence of water content on sliding friction tends to be ignorable in high water content. As a result, the friction angle φ_{total} tends to be stable on this occasion.

Based on the theory mentioned above, the parabolic and metatypic Pearl growth curve formulas were proposed to describe the relationship between water content and the equivalent strength parameters as following:

$$c_{total} = mw^3 + nw^2 + pw + q \quad (4)$$

$$\varphi_{total} = \frac{1}{a+b*e^{(w-d)}} + k \quad (5)$$

Where, m, n, p, q, a, b, d, k are fitting parameters, and $q>0$, $a>0$, $b>0$, $\omega_{sat}>d>0$, $k>0$. w is water content.

3. Calibration by experimental results

3.1 Existing experiments

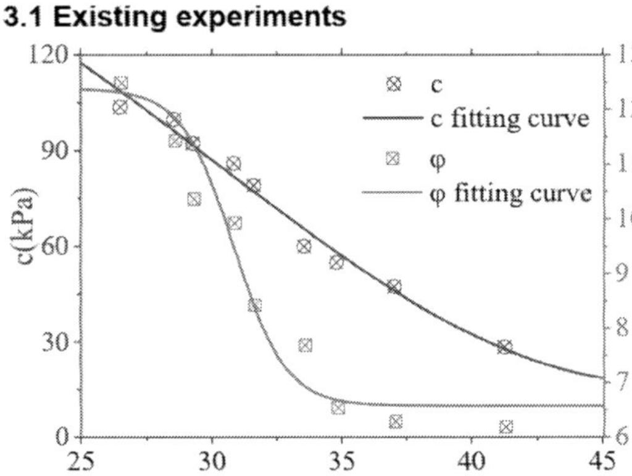

(a) Yichang expansive soil (after Kong et al. (2000)[10])

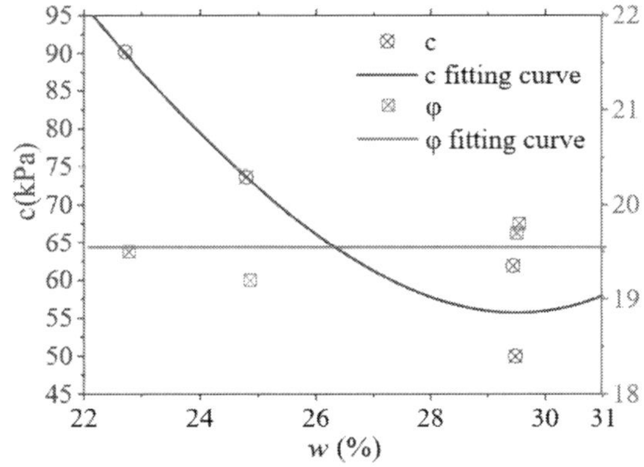

(b) Guangxi expansive soil (after Miao et al. (2007)[11])

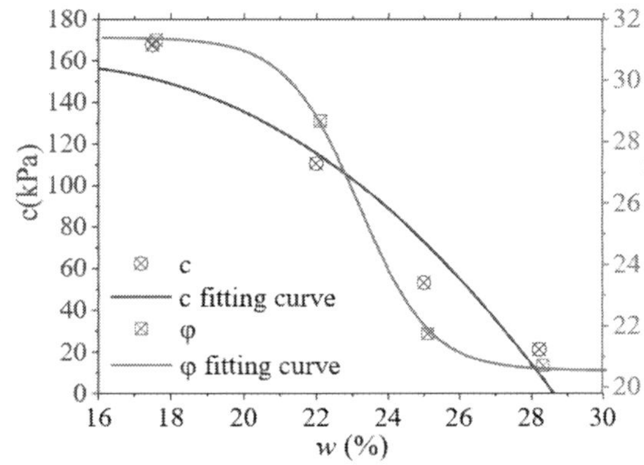

(c) Hunchun expansive soil (after Wu et al. (2017)[7])

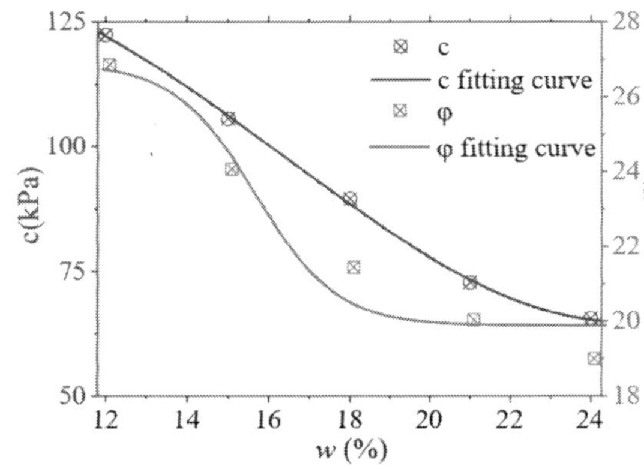

(d) Hanzhong expansive soil (after Jia et al. (2014)[12])

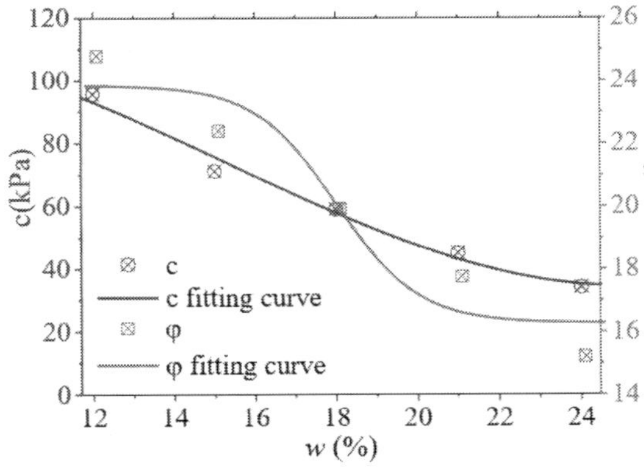

(e) Ankang expansive soil (after Jia et al. (2014)[12])

Figure.2. Strength parameters of expansive soil.

Table 2 Tests of expansive in different areas

Type of expansive soil	Sample type	Test method	Water content
Yichang expansive soil [10]	Remoulded sample	Direct shear test	Medial-high water content
Guangxi expansive soil [11]	Remoulded sample	Triaxial consolidated drained test	High water content
Hunchun expansive soil[7]	Undisturbed sample	Direct shear test	Medial-high water content
Hanzhong expansive soil[12]	Remoulded sample	Triaxial consolidated undrained test	Medial-high water content
Ankang expansive soil[12]	Remoulded sample	Triaxial consolidated undrained test	Medial-high water content

To validate the above formulas, three experimental results of the shear strength of partially saturated expansive soil were adopted. The first one is the direct shear test on the expansive soil in Yichang, Hubei [10]. The obtained shear strength is shown in Figure.2(a); with the increase in water content, φ decreases firstly and then tends to be a constant, and the cohesion monotonically decreases from 103.67kPa to 28.38kPa. The strength parameters decrease with water content similarly. The second one is the triaxial tests conducted on Guangxi expansive soil [11]. The obtained cohesion and friction angle under different water content conditions are plotted in Figure.2(b). The cohesion decreases approximate linearly with the saturation degree, however, the friction angle remains constant (around 19°). The third one is the direct shear test on the expansive soil in Hunchun, Jilin[7]. The fourth and fifth are Hanzhong and Ankang expansive soil in Sichuan respectively.Similar pattern can be found in Figure.2(a), (b), (c), (d) and (e),the experimental results match the theory when the water content of the expansive soil is high.

3.2 Consolidated undrained triaxial tests on Nanyang expansive soil

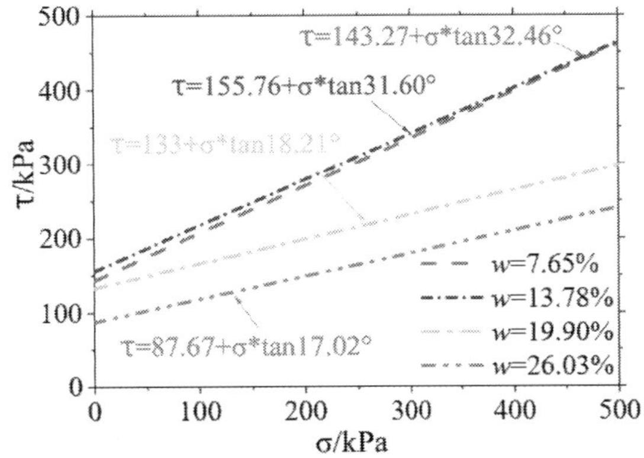

(a) Strength envelope lines

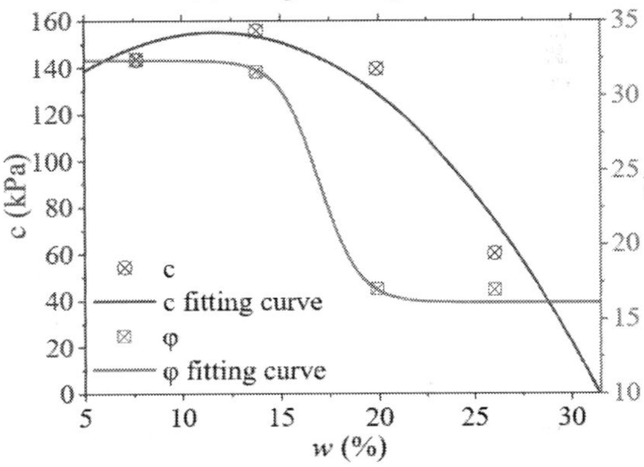

(b) Strength parameters

Figure.3. Failure envelope lines and strength parameters of Nanyang expansive soil.

To further validate the propose failure criterion, a series of consolidated drained triaxial tests were conducted. Soil samples were remodeled by using the expansive soil from Nanyang city, Henan province. The experiments were conducted on the GDS controlled stress path triaxial instrument at Tongji University. The air in the samples was allowed to be exhausted and the specimen was kept undrained. The confining pressures were 50, 100, and 200 kPa and the water content of the sample were ω =7.65%, 13.78%, 19.90%, 26.03% (with Sr=25%, 45%, 65%, 85% respectively). According to the stress-strain relationship from the test results, the peak shear stress under different confining stress and water content can be obtained. As shown in Figure. 3(a), the strength envelope lines were obtained by fitting the Eq. (1) against the experimental data. The cohesion and friction angle of the expansive soil with different initial water content is shown in Figure. 3(b). With the increase in the water content, the cohesion and friction angle show decreasing trends as a whole. Besides, fitting parameters of the six kinds of expansive soil are listed in table 3.

Table 3 fitting parameters of different expansive soil

	Nanyang	Yichang	Guangxi	Hunchun	Hanzhong	Ankang
m	-0.001	0.0068	0.031	-0.007	0.028	0.022
n	-0.342	-0.603	-1.807	-0.408	-1.346	-1.032
p	8.39	11.61	25.08	15.93	15.87	9.79
q	105.25	98.00	86.65	32.87	77.83	85.65
a	0.062	0.172	0.317	0.092	0.142	0.132
b	35.46	$2.52*10^{-5}$	0.95	0.039	0.197	0.096
d	23.33	21.95	0.034	22.2	15.91	17.61
k	16.13	6.576	19.55	20.54	19.88	16.26

4. Summary

The Mohr-Coulomb criterion was adapted to describe the undrained shear strength of unsaturated expansive soil, and the cohesion and friction angle in total stress state were obtained. The physical meaning of cohesion and frictional strength were interpreted microscopically. Existing experiments on unsaturated expansive soil were adopted to study variations of cohesion and friction angle with water content. Triaxial tests of unsaturated expansive soil in Nanyang with were conducted, and the obtained cohesion shows a peak value in the cohesion-saturation curve, while the friction angle doesn't. However, the shear strength characteristic of expansive soil at very low saturation and the application of the theory proposed in the paper for expansive soil in other places in the world are not yet studied in this paper, and it deserves further study.

Acknowledgement

The financial support by the State Grid Corporation of China (Grant No. 5217L0160001) and the Fundamental Research Funds for the Central Universities (0200219221) is gratefully acknowledged.

Reference

1. M. Scheel, R. Seemann, M. Brinkmann,et al. Morphological clues to wet granular pile stability. *Nature MaterialsNature Materials*,(2008), **7**(3):189-193.
2. Ning Lu, Bailin Wu, Chee P. Tan. Tensile Strength Characteristics of Unsaturated Sands. *Journal of Geotechnical and Geoenvironmental Engineering*,(2007), **133**(2).
3. V. Richefeu, M. Sa Amp, Iuml. Shear strength properties of wet granular materials. *Physical Review E*,(2006), **73**:51304.
4. R. Mani, D. Kadau, H. J. Herrmann. Liquid migration in sheared unsaturated granular media. *Granular Matter*,(2012), **15**(4):447-454.
5. Chao-Sheng Tang, Xiang-Jun Pei, De-Yin Wang,et al. Tensile Strength of Compacted Clayey Soil. *Journal of Geotechnical and Geoenvironmental Engineering*,(2015), **141**(4):4014122.
6. T. Y. Elkady, M. F. Abbas. Shear Strength Behavior of Highly Expansive Soil. GeoCongress 2012: State of the Art and Practice in Geotechnical Engineering,(2012):2532-2541.
7. Xiong Wu, Zhuo-ran Wang. The relationship between the swelling pressure and shear strength of unsaturated soil: the Yanji Basin as a case study. *Arabian Journal of GeosciencesArabian Journal of Geosciences*,(2017), **10**(15).
8. H. Louati, D. Oulahna, A. D. Ryck. Effect of the particle size and the liquid content on the shear behaviour of wet granular material. *Powder Technology*,(2017), **315**:398-409.
9. J. P. K. Seville, C. D. Willett, P. C. Knight. Interparticle forces in fluidisation: a review. *Powder Technology*,(2000), **113**(3):261-268.
10. Ling-wei Kong, Luo-rong Tan. Study on shear strength and swelling- shrinkage characteristic of compacted expansive soil. *Unsaturated Soils For Asia*,(2000):515-519.
11. Linchang Miao, Sandra L. Houston, Ying Cui,et al. Relationship between soil structure and mechanical behavior for an expansive unsaturated clay. *Canadian Geotechnical JournalCanadian Geotechnical Journal*,(2007), **44**(2):126-137.
12. Wencong, Jia. Experimentle Study on the Strength and Deformation Characteristics of Unsaturated expansive soil. Master's thesis, *Northwest Agriculture & Forestry University*. Yangling Shaanxi, China. (2014) (in Chinese)

Study of Excavation Stability and Treatment Measures of the Side Slope of Anti-Dip Bedding Highway

Qian Shen , Liu Liu , Jinxiu Yuan

Department of Civil Engineering, Hebei Jiaotong Vocational and Technical College, Shijiazhuang Hebei China 050091

Abstract: In order to improve the slope safety of the side slope of anti-dip bedding highway, this paper analyzes the stability of the excavation by using the finite difference numerical simulation technique. The development process of the disaster under the three excavation conditions and the whole process of the slip surface of the anti dip bedding rock slope are revealed; The method of determining the potential slip surface is introduced, that is to say, the slope is unstable and slip by reducing the material parameters of the slope, and then the position of the slip surface is judged by the difference of the displacement value. The research results have important significance for understanding the development mechanism of slope disaster and adopting corresponding engineering measures for different stages of slope.

1. Introduction

The slope is a dome with a certain slope and it is made to ensure the stability of the roadbed on both sides of the roadbed [1]. In recent years, with the increasing mileage of highway, the stability of slope excavation in mountainous area of highway is becoming more and more serious, which brings great losses to people's travel and life and property. Therefore, it is necessary to study the stability of anti -dip highway slope excavation.

At present, some experts have done some researches on the stability of slope excavation. The main research works are as follows: the deformation and failure mechanism of anti dip bedding slope during excavation and unloading is studied [2-3]. The stability of bedding rock slope during excavation is studied [4-5]. The application of strength reduction method in slope stability analysis is studied [6-8]. The above studies have been carried out on the stability of anti dip and bedding slopes and the determination methods of slip surfaces. However, the stability and slip surface determination method of the side slope of anti-dip bedding highway are not studied. Therefore, This paper mainly study the stability of the slope during the excavation of the side slope of anti-dip bedding highway by using the inite difference numerical simulation technique. The research results can provide reference for excavation stability study of similar projects.

2. Analytical Model

2.1 Geological conditions of slope

The strata in the slope area are mainly purple red mudstone of the lower three series. The slope area is inclined anticline structure, the axis of anticline is inclined to 45 degrees with the slope, and the rock stratum is inverted, and the changes of attitude is larger.

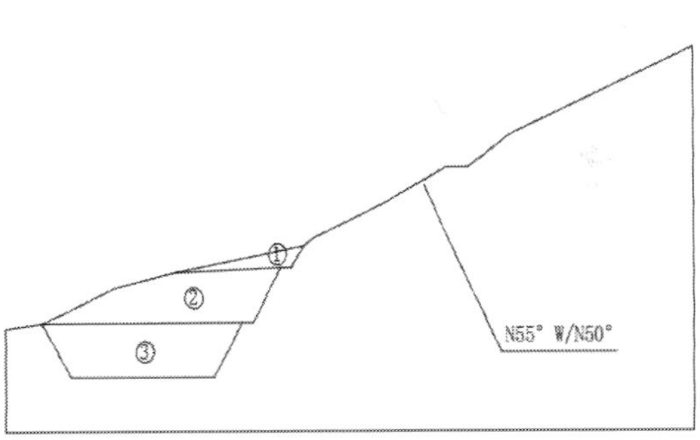

Fig. 1: Generalized geological model

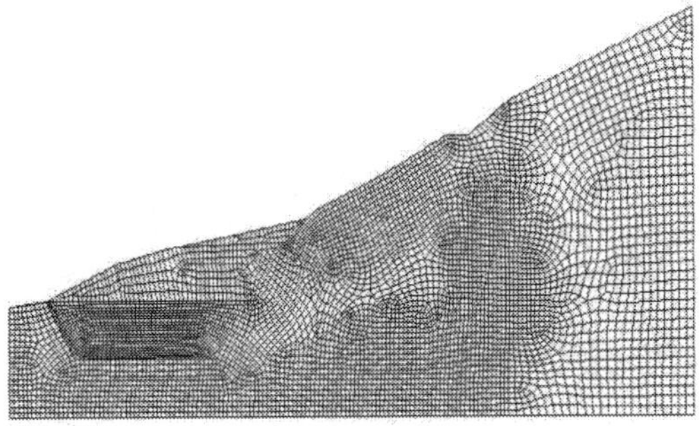

Fig. 2: Discrete calculation model

2.2 Computational model

The horizontal distance of the model is 91.71m, the vertical height is 45.92m, and the vertical width is 1m. The effect of the lateral pressure on the right boundary is triangular distribution, the bottom boundary is constrained by vertical displacement, the left boundary is constrained by horizontal displacement, and the lateral boundary is constrained by horizontal displacement in the direction of Y. The slope rock mass is layered rock mass, and the intraformational joint controls the deformation of landslide body. Ubiquitous-joint model is used in the calculation. The calculation model is shown in figure 2.

Table 1: Calculation steps

Step	Study object
1	Excavation （①）
2	Excavation （②）
3	The potential slip surface is determined after step 2
4	Excavation (③)

2.3 Calculation parameters

The calculation parameters of slope are shown in table 2.

3. Calculation results analysis

3.1 First excavation

The calculation results after the first excavation are shown in figure 3.

Since the slope is an inverted stratified rock slope, the stability of the slope is basically not affected after excavation, and there is no plastic unit in the slope. As shown in Figure 3, the maximum displacement after excavation is only 3.94mm, which is located in the middle of the excavation horizontal plane, the displacement contour is circular and downward gradually, and the displacement is caused by elastic rebound of rock mass unloading. The scope of the slope excavation is very limited, and it is not enough to cause too much impact on the stability of the slope, so no further analysis.

3.2 Second excavation

The calculation results after the second excavation are shown in figure 4.

As shown in Figure 4, the maximum value of the displacement occurs in the middle part of the new excavation slope, the maximum value is 0.015m, and the displacement isoline decreases gradually to the slope, and the displacement values are greater than 0.010m. The displacement isoline decreases gradually to the slope, and the displacement values are greater than 0.010m. The plastic zone is only distributed in the toe, and it was straight upward inclined to extend to the inside of the slope, the farthest extends about 12M, but the slope without plastic element, the plastic zone is not through the slope.

Table 2: The calculation parameters

Slope structure	E (MPa)	μ	γ (KN/m^3)	C (KPa)	φ (°)	T (KPa)
Rock	11000	0.28	22	300	35	300
Joint	11000	0.28	22	40	28	40

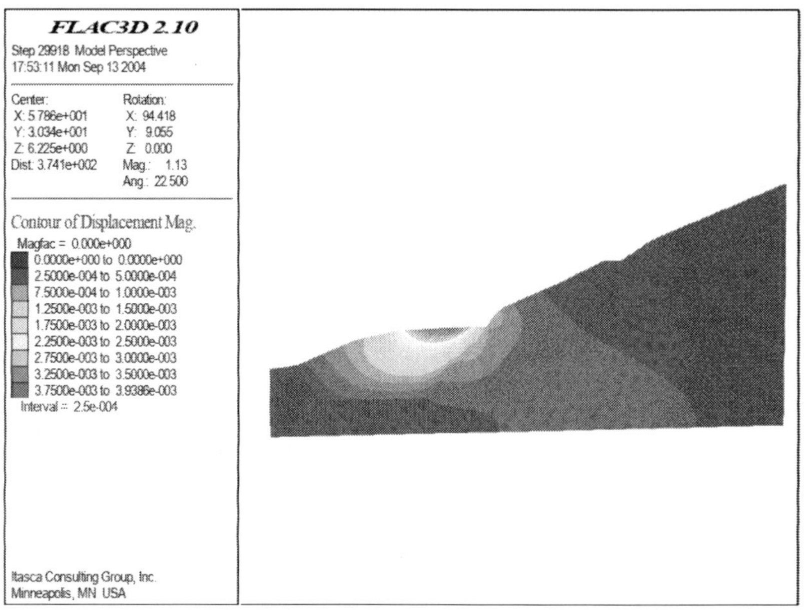

Fig. 3: Displacement (step 1)

 https://doi.org/10.1051/matecconf/201820601003

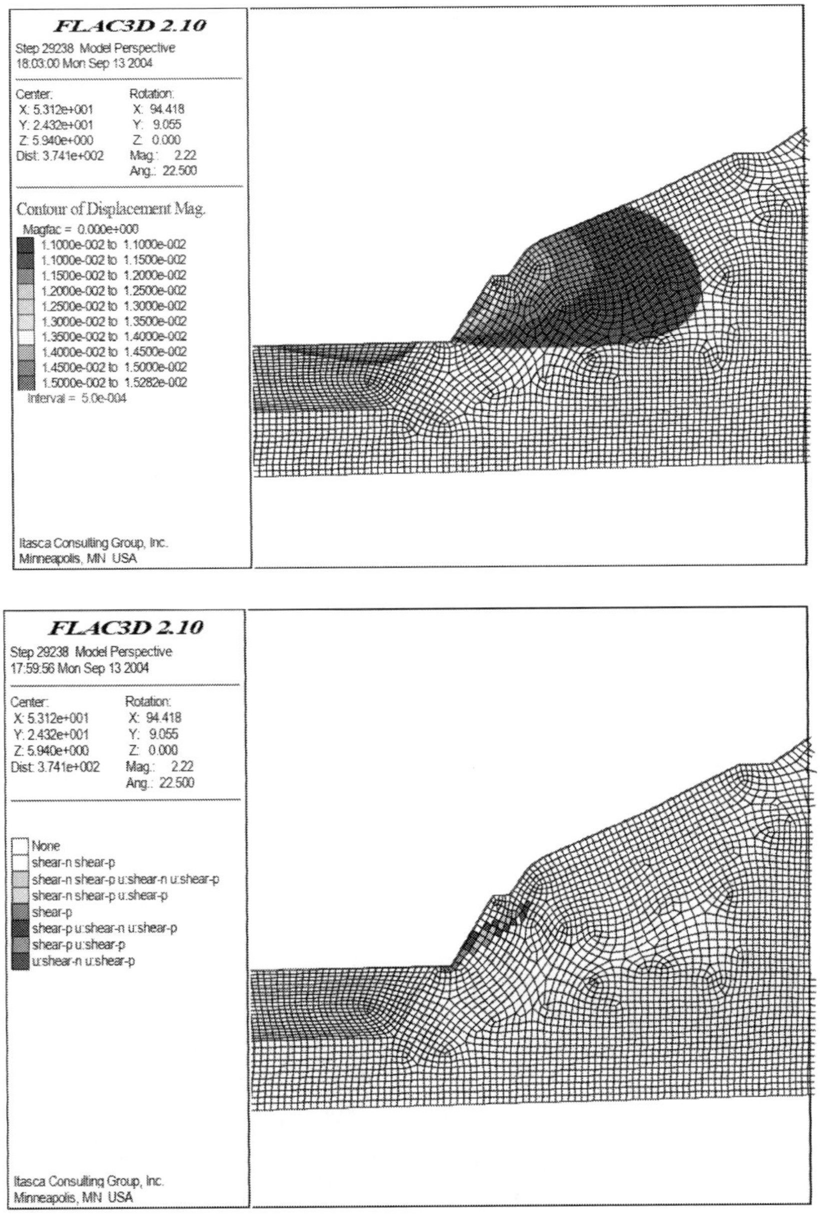

Fig. 4: The calculation results of step 2

To sum up, the plastic zone in the slope is limited to a limited range, and the deformation zone shown in Figure 4 is mainly due to the adjustment of the elastic stress caused by excavation unloading, so the displacement is not large. Because the slope is an anti dip rock stratum, and it is beneficial to the stability of slope. The artificial slope stability of the second excavation is better, only local plastic deformation, no overall slip.

3.3 Determination of potential slip surface after second excavation

In order to further analyze the stability of slope after the first excavation and the second excavation, and to find the potential slip surface of slope under the influence of various external factors. It reduces the rock and soil parameters of slope are reduced.

In the calculation, the smaller reduction factor is used to test the slope, and the slope slip surface does not form under a certain ninety percent off reduction factor. If the slope slip surface is not formed under a certain reduction factor, the calculation will be continued by increasing the reduction factor, and until a uniform complete slide surface appears.

3.4 Determination of potential slip surface

This paper mainly studied the process of plastic damage to the slip surface.When the reduction factor K is 2.5, a uniform slip surface is formed in the slope. The calculation results are shown in Figure 5.

 https://doi.org/10.1051/matecconf/201820601003

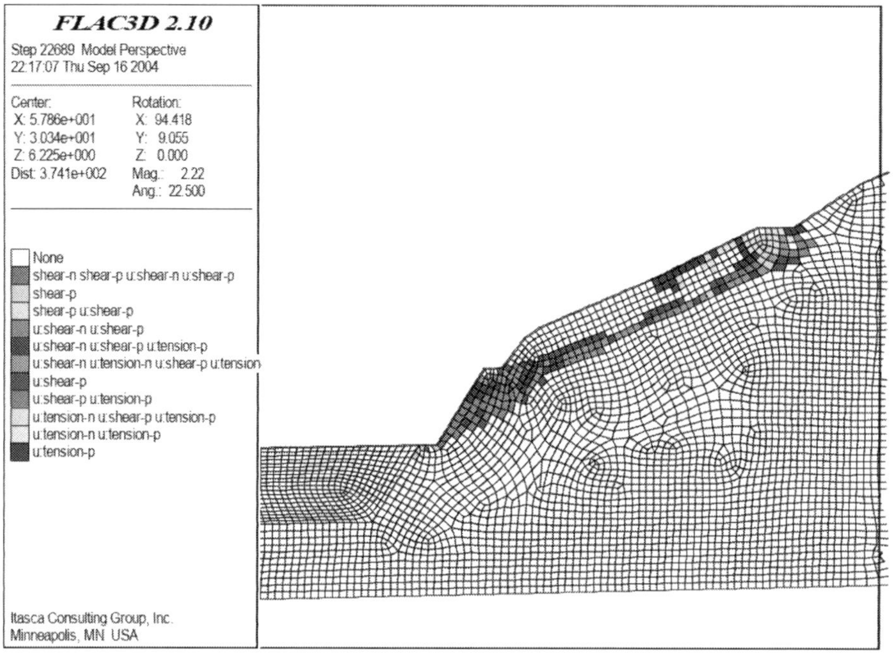

(a) The plastic zone at the beginning of the calculation

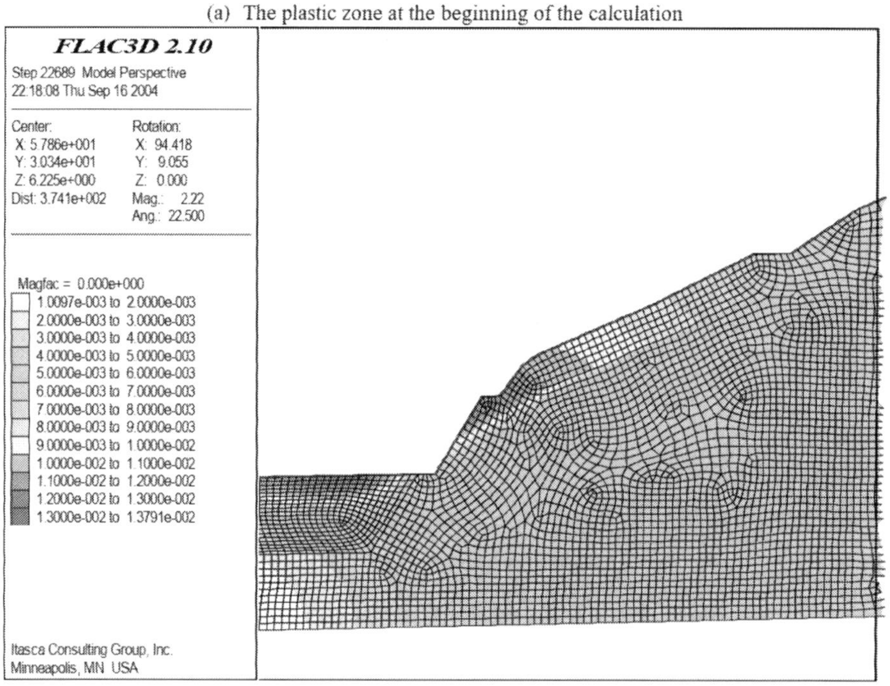

(b) The displacement at the beginning of calculation

https://doi.org/10.1051/matecconf/201820601003

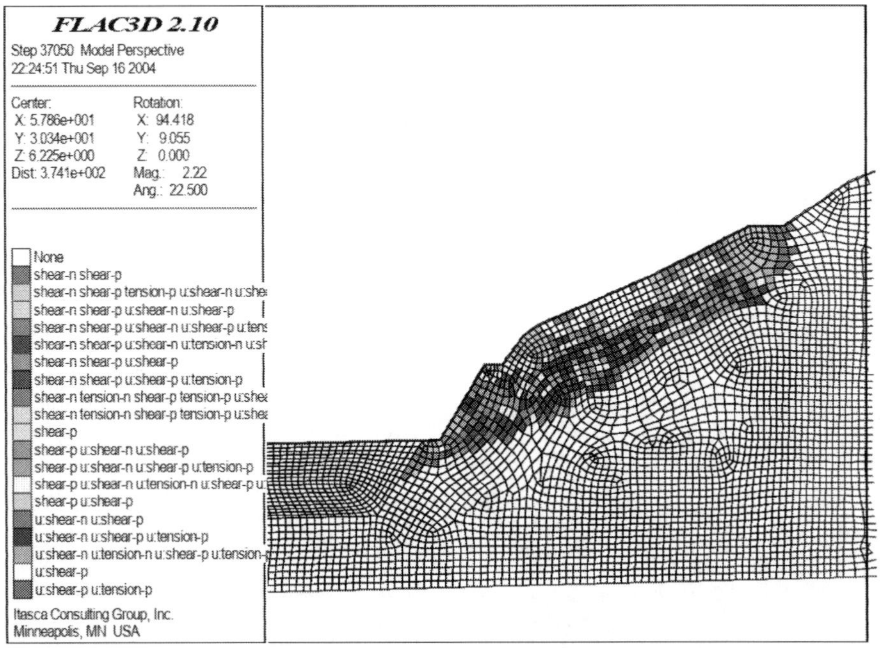

(c) The plastic zone at the end of the calculation

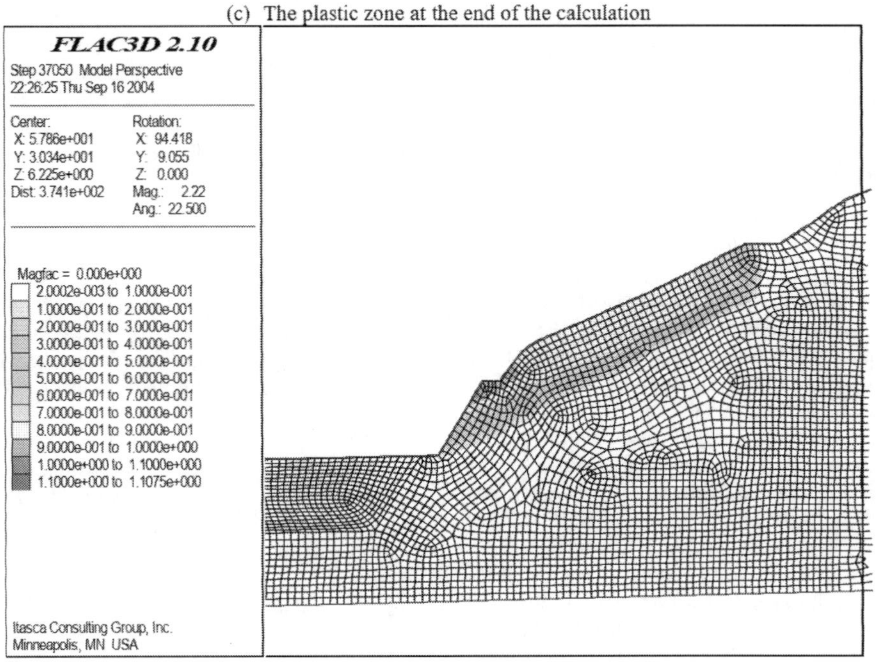

(d) The displacement zone at the end of the calculation

Fig. 5: The calculation results of step 3

As shown in Fig. 5, when the calculation starts, the plastic unit in the slope and the plastic unit on the slope have been integrated, and the plastic influence range of the slope deformation caused by the excavation scope has been preliminarily revealed. At this point, the displacement of the slip surface can not be determined from the displacement of the slope, and the displacement vector contour is shown in Figure 5 (b). The maximum displacement in the plastic zone of the slope is almost the same as the maximum displacement at the excavation horizontal plane, and the maximum value is 13.791mm. From the displacement in the plastic zone, the upper displacement on the leading edge of the slope is greater than the displacement at the bottom, and the position of the displacement interface have seen indistinctly, but the displacements interface between the front and back is not clear. By contrast (a) and (c), it can be seen that after the formation of the slip surface, the plastic zone in the slope has little change, and there is almost no expansion before and after the slip surface, but only the depth increases more.

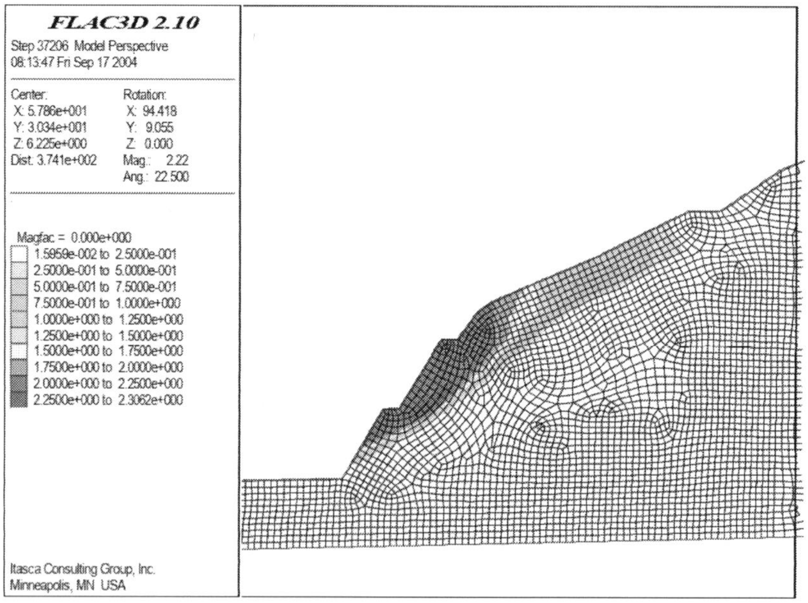

Fig. 6: The calculation results of step 4

At the end of the computation, slip surface has been formed, The slide surface begins to extend obliquely from the foot to the top of the slope. The slip surface of the trailing edge becomes basically vertical, and the thickness of the sliding body is almost the same.

3.5 Third excavation

The calculation results after the third excavation are shown in figure 6.

As shown in Fig. 6, after the calculation, the slip surface of the slope revealed by the displacement is very clear, and the displacement of each point in the sliding body is more than 0.25m, and the boundary of the slip surface is very clear. The result of stress adjustment of the slope after excavation is that a landslide is formed in the slope. According to the different displacement values in the landslide, the landslide body is divided into two grades, the former landslide displacement is larger, and the latter landslide displacement is small. The maximum displacement of the sliding body is 2.306m, and the overall displacement of the front sliding body is near the maximum, and the displacement of the back sliding body is more than 0.25m, and the displacement change is obvious.

4. Treatment measures

The stability of the slope is affected by geological conditions, rainfall, excavation and other factors. Therefore, under different influence factors, for a specific slope, choosing the reasonable and effective engineering and technical measures is the key to achieve the desired goal of slope treatment.

(1)At present, the main measures of slope treatment areas follows: Anti sliding retaining wall, anchor and prestressed anchor cable, anti slide pile, prestressed anchor pile, etc.

(2)According to the stability of the anti-dip bedding slope, the stability of slope decreases during excavation. The toe of the slope stability is the worst, and it is the most dangerous part of the slope. Therefore, combined with the failure characteristics of the slope, in practical engineering, the anti slide pile, anti slide pile and prestressed anchor rope can be used to reinforce the slope.

5. Conclusion

(1)The slope stability of the first excavation and the second excavation is better, it produces only local deformation at the foot of the slope, and no overall slip. After the third excavation, a plastic zone is formed in the slope, and the slope gradually loses its stability and forms a landslide.

(2)The development process of the disaster under the three excavation condition reveals the whole process of the whole process of the formation of the slip surface of the anti-dip bedding slope, and the specific position of slip surface is determined. After the slip surface is formed, the plastic zone in the slope does not continue to develop, but the displacement continues to increase.

(3)In practical engineering, it is suggested to use the method of anti slide pile, anti slide pile and prestressed anchor cable.

Acknowledgements

This work was financially supported by the Natural Science Foundation of HebeiProvince(No.E201619002), Hebei Talent Training Fund(A201500116), Key R & D Project of Hebei Province(172776471).

References

1. JIANG Deyi, WANG Guodong. *Engineering*

geological classification of highway engineering Slope, Journal of Chongqing University(Natural Science Edition), Vol.26, No.11, pp. 113-116, 2003.
2. YANG Xianwen. Stability analysis of excavation slope and study on deformation and failure mechanism, Railway Construction Technology, No.02, pp. 100-103+112, 2015.
3. LENG Xianlun, SHENG Qian, LIAO Hongjian, et al. *Study on deformation and failure mechanism of high slope excavation in anti-dumping*, Chinese Journal of Rock Mechanics and Engineering,Vol.23, No.S1, pp. 4468-4472, 2004.
4. FENG Jun, ZHOU Depei, Li Anhong. *Research on stability of rock bedded slopes*, Chinese Journal of Rock Mechanics and Engineering, Vol.24, No.9, pp. 1474-1478, 2005.
5. YU Fei, CHEN Shanxiong, YU Heping. Numerical simulation study on progressive destruction and failure mechanism of bedding rock slopes, Rock and Soil Mechanics, Vol.26, No.2, pp. 36-40, 2005.
6. LIU Xiaodong, ZHANG Zhongping, SUN Shuwei. Applicat-ion of strength reduction method in high-steep rock slope engineering, Railway Engineering, Vol.26, No.7, pp. 97-99, 2012.
7. CHENG Canyu, LUO FUrong, QI Chengzhi, et al. *Comparative analysis of slope stability by strength reduction method*, Rock and Soil Mechanics, Vol.33, No, 11, pp. 3472-3478, 2012.
8. WU Yingxiang, LIU Dongsheng, SONG Qianghui, et al. *Reliability analysis of slope dynamic stability based on strength reduction method*, Rock and Soil Mechanics, Vol.34, No.07, pp. 2084-2090, 2013.

A Research for Deviation Estimation Analysis for Driven Piles Foundation: A Case of Penang Second Marine Bridge

Kang Huang[1, 2]
1 China Southeast University, Sipailou 2#,Nanjing, Jiangsu Province,China
2 China highway Planning and Design Institute, No.85 De sheng men wai Street,Xi cheng District,Beijing,China

Abstract. The aim of this preliminary research is to study the accept tolerance of driven piles with position's deviation. More than five thousand driven piles were installed in the approach bridge, which adopted two classes, i.e.Φ1.6m steel pipe pile and Φ1.0m spun pile. The original position acceptance criteria was deviation tolerance less than 75mm. This was not reasonable enough to take into account the marine construction conditions. According to a detailed comparison of the existed different standards and simplified calculation with representativeness, The recommended acceptance criteria could be adjusted to 150mm. This study makes up for the lack of Malaysia pile foundation technical specification especially for large diameter driven piles and it will effectively promote the practical application of ones in Southeast Asia infrastructure works.

1 Introduction

The Penang Second marine bridge located at Penang island channel is approximately 22km in length with a total of 16.5km of bridgeworks of which around 15km is over water. It is the largest bridgework in the southeast Asia. The bridge is made up of multiple approach span viaducts from both sides with a signature structure featuring longer spans and a higher clearance envelope over the southern navigation channel. The general view of the bridge is shown in *Figure 1*.

The Malaysian Public Works Department (Malay: Jabatan Kerja Raya Malaysia), abbreviated JKR, is responsible for construction of public infrastructure in West Malaysia. The foundation of approach bridge mostly consisted of driven piles, i.e.Φ1.6m steel pipe pile and Φ1.0m spun pile. The number of 5168 spun piles and 396 steel pipe piles were installed in whole project. The proportion with deviation exceed original acceptance criteria, i.e. 75mm ruled by JKR specification, reached around 85 percent. Acceptance piling work faced specific challenges. In order to avoid delaying the time limit for the project, Appropriate revision of existing control standards would be essential.

2 Specifications used in different countries

2.1 Malaysia JKR standard specifications for roadworks

The Employer Requirement of this case was followed by Malaysia JKR Standard[1]. The Clause 10.2.2 of Section 10 Piling Works stipulate as follows:"For a pile cut off at or above ground level the maximum permitted deviation of the pile center points shown on the Drawings shall not exceed 75mm in any direction."This specification did not specify the difference of pile deviation between land piles and marine piles, therefore the tolerance was generally 75mm.

2.2 Hong Kong general specification

In the Hong Kong General Specification[2], both the installation tolerances for marine and land bored piles are given. The tolerances are 150mm and 75mm for marine and land piles respectively.

Figure1. General view of Penang Second Marine Bridge.

2.3 British standard BS 8400 and BS6349

BS8400[3]: Clause 7.4.1.3.1 General of Section 7 Pile Foundations stipulate as follows:"When driven, piles should not deviate more than 75 mm from their designed positions; a greater tolerance may be permitted for piles driven over water."

BS8400: Clause 7.4.2.5.4 Position and Alignment Tolerances of Section 7 Pile Foundations stipulate as follows:"Piles should not deviate more than 75 mm from their designed positions at the working level of the piling rig. Greater tolerance may be prescribed for piles driven over water and for raking piles." & "Any pile deviating beyond these limits and to such an extent that the resulting eccentricity cannot be taken care of by a redesign of the pile cap or pile ties should, at the discretion of the engineer, be replaced or supplemented by one or more additional piles."

BS6349[4]:Part 2: Clause 6.13.1 Pile Head "The centre of piles at the junction with the superstructure should be within 75 mm of the specified position for structures built on land or in sheltered waters. For piles driven through rubble slopes this tolerance may be increased to 100 mm. For access trestles and jetty heads a tolerance of 75 mm to 150 mm should be adopted, depending on the exposure of the site. The design of piles and deck members should take into account the specified tolerances."

Both standards specify the deviation of land piles is 75mm, and allow appropriate greater tolerance of deviation for marine piles. In BS6349, it is specified a tolerance 150mm for piles in jetty or access trestles.

2.4 Habour engineering standard of China

Clause 8.5.2.1 of the Specification of Pile Foundations for Port Engineering (JTJ-254-98) [5] is stipulated as follows in *table 1*:

This specification clearly specifies the different pile deviations at different regions, based on different operating conditions of marine piling works. Currently similar projects in China strictly follow the requirements of condition "Coastal Water without Cover" in *Table 1* to control the pile deviation. In addition, the condition "Coastal Water without Cover" is more in line with the sea conditions of this project and can be reference of this project.

3 Factors affecting pile deviations

Considering the actual situation of this project and summaries of similar projects[6], the major reasons for the deviations of driven pile using the current piling workmanship are listed in *Table 2*.

According to the servey of pile deviations' statistics, vertical piles have smaller deviations which basically can satisfy the requirement of 75mm. The deviations for raking pile are larger and are difficult to comply with the requirement of 75mm. The pile deviation for raking pile are within 150mm generally and individual pile exceeds 200mm.

Combining the pile arrangements of the bridgework and the data of completed piles, the following preliminary conclusions can be drawn for the influence to pile foundations due to pile deviations:

- As the spatial piles are adopted generally, it is possible to have deviations in the twist angle, rake and plan coordinate of single pile. In addition, the probability of having deviations in plan coordinates is the largest and the magnitude is the largest. It is the major factor that results in the force-taking of pile foundations different from the design value.
- Table 2 reveal that, the non human manageable factors that resulting plan coordinate deviations are more. It leads to large discreteness of deviated pile quantity, direction and deviation amount within a single pilecap. This shows that, by estimation to determine the maximum allowable tolerance for pile, it is difficult to cover different situations of actual piling works.
- With the provision of the discreteness of pile deviation, if a individual pilecap has more supporting pile and basically symmetrical pile arrangement, the actual pile group capacity after driving will have less difference against design value.
- From the pile material point of view, steel pile is obviously better than spun pile[7]. The former has better adaptability after the occurrence of pile deviation. However, the water for steel pipe pile zone is deeper and the steel pile body is longer. It makes the deviation control for steel pile is much difficult than that for spun pile. It is estimated that the deviation tolerance for steel pile can be larger than that of spun pile.

Table 1. Requirements for Pile Foundations Deviation Unit: mm

Pile Type	Spun Pile		Steel Pile	
	Vertical	Rake	Vertical	Rake
River and Coastal Water with Cover	150	200	100	150
Coastal Water without Cover	200	250	150	200
Offshore Water without Cover	250	300	250	300
Coastal water means distance to coast < 500m; offshore water means distance to coast ≧ 500m				

Table 2. Reasons for Deviations of Pile

Item	Reasons for Deviations
1	Accuracy of GPS measurement

2	Piling barges uses anchor and cable for positioning. Certain degree of waving and even shifting will be resulted under the effect of wind, wave and water flow.
3	Accuracy of setting-out before driving and errors

4 Analysis

4.1 Selection of Critical Pier Type

As this project has many type of pile arrangements, representative pier types with consideration of number of piles, scour depth and type of arrangement (see *Table 3*) are selected to perform estimation.

According to the comparison result in Table 3, Type 3 of steel pile and Type 4 and 6 of spun pile is selected as control pier type to carry out calculation.

4.2 Assumed Load Case

Considering the results from detailed design, distance that much possible appeared in pile deviation and the worst favorable deviation direction are selected to perform calculation. The determination of formation of pile deviation is listed in *Table 4*.

The influential factors considered in calculation and calculation procedure are as follows:

Step A. Select appropriate estimated deviation amount (Factor 1, Symbol A shown in *table 4*)

Step B. Select unfavorable pile deviation direction (Factor 2, Symbol B shown in *table 5*)

Step C. Select external control load case (Factor 3, shown in *table 6*) → *Perform Assessment to Analysis Result*

These 3 factors do not have relationship with each other, accurate calculation should take into account the envelop value from different combination of these 3 factors. However, with the limitation of huge amount of actual combinations and calculation results may not show any regular pattern, it can only consider the adverse conditions in 3 factors separately together with actual conditions in the analysis.

Table 3. Control Pier Type Comparison

Pile Type	Pier Type	Control Pier Type	Among same Pilecap type		
			Min. No. of Pile	Max. Scour Depth	Non-symmetrical Pile Arrangement
Steel Pile	Type 1/2/3	Type 3	√	√	√
Spun Pile	Type 4/5/6	Type 4	√	√	√
		Type 6	-	√	√

Table 4. Selection of Appropriate Estimated Deviation Amount (Symbol A) Unit: mm

Deviation Amount	A1 < 75mm	A2 < 150mm	A3 < 250mm	Remark
Relevant Standards	JKR-	Hong Kong General Spec. & JTJ-254-98	JTJ-254-98	-
Occurrence % in Actual Piling	10 %	80 %	10 %	Currently the quantity of completed steel pile is small, lack of actual deviation data
Appropriate Pile Type	Spun Pile	Spun Pile	Steel Pile	
Amount A selected in analysis	√	√	√	

Table 5. Selection of Unfavorable Pile Deviation Direction (Symbor B)

Deviation Direction	B1a(b) – All Longitudinal (Transverse) Deviation	B2 – All Radial Deviation

Diagram	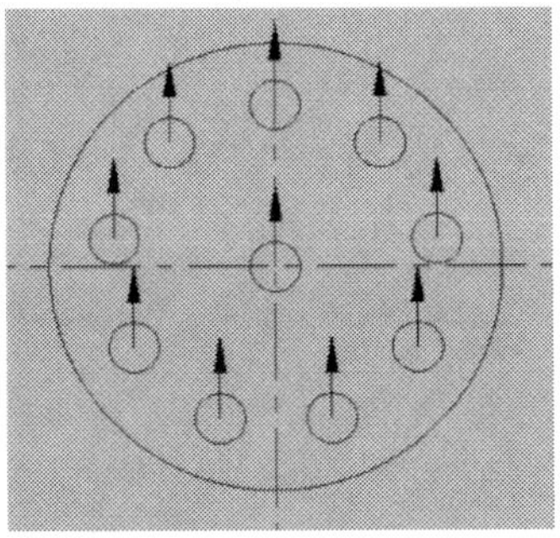	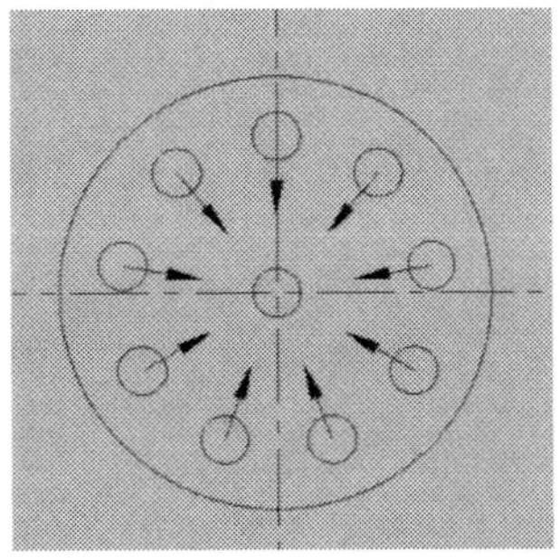
Description	Longitudinal and transverse deviation, separate analysis	All radial deviation (towards centre), individual analysis

Table 6. Selection of External Control Load Case

Pile Type	Spun Pile			Steel Pile	
Check Items	Single Pile Capacity	Stress Check under SLS	Strength Check under ULS	Single Pile Capacity	Pile Body Strength Check
Analysis Load Case	Operation Stage / Construction Stage	Operation Stage (wind + wave)	Operation Stage (seismic)	Operation Stage / Construction Stage	Operation Stage (seismic)

4.3 Summary of Analysis Result

Considering the deviation amount, deviation direction and different combinations of control load cases, the analysis combinations and corresponding results are tabulated as follows:

Considering spun pile, The above summary tables show that the deep water region like Type 5 and 6, due to longer pile free length, the control item after pile deviation is pile body stress under SLS conditions which is closed to the limit of pile. For most of pile in Type 4, the single pile working load is closed to the limit of 3800kN under certain load cases. Based on the presumed deviation, the working loads of pile in above combinations are closed to presumed design capacity and the pile group is still found to be structurally acceptable under the assumed pile deviation. Analysis Results of spun pile types are listed in *table7* and *table8*.

Considering steel pile, The above reason shows that the steel pile zone in deep water region, when large pile deviation happens, the single pile working load becomes large while the pile body strength still have certain amount of capacity. Based on 200mm assumed deviation for analysis, the working loads of pile in above combinations are comparatively large. However, if it is calculated based on 150mm deviation, the pile load will be well within the 7300kN capacity. Analysis Results of steel pile type is listed in *table9*.

Table 7. Analysis Results for Type 4 Pile Deviation Unit: kN,m MPa

Analysis Description	Single Pile Working Load	SLS Stress Check	ULS Strength Check		Conclusion
Analysis Combinations		Max. Tensile Stress	Relevant Axial Force	Moment	
A1&B1a	3527/4234	8.7	5354	1455	Single pile working load is closed to max. 3800kN in some combinations
A1&B1b	3547/4332	9.1	5337	1425	
A1&B2	3627/4361	8.4	5267	1410	
A2&B1a	3446/4134	9.2	5463	1498	
A2&B1b	3513/4375	8.6	5403	1454	
A2&B2	3712/4388	8.9	5274	1411	

Table 8. Analysis Results for Type 6 Pile Deviation Unit: kN,m MPa

Analysis Description	Single Pile Working Load	SLS Stress Check	ULS Strength Check		Conclusion
Analysis Combinations		Max. Tensile Stress	Relevant Axial Force	Moment	
A1&B1a	3425/3628	11.1	5421	1588	Tensile stress is

A1&B1b	3324/3533	8.4	5562	1605	close to max. 11.9MPa under in SLS conditions
A1&B2	3507/3713	10.9	5493	1551	
A2&B1a	3358/3566	11.2	5493	1625	
A2&B1b	3274/3575	8.6	5546	1648	
A2&B2	3523/3737	10.8	5494	1549	

Table 9. Analysis Results for Type 3 Pile Deviation Unit: kN,m MPa

Analysis Description / Analysis Combinations	Single Pile Working Load	ULS Strength Check		Conclusion
		Relevant Axial Force	Moment	
A3&B1a	7527	11074	3231	Single pile working load slightly exceeds 7300kN in some combinations, about 3%
A3&B1b	7488	11264	2709	
A3&B2	7139	11043	2834	

5 Conclusions

The conclusions of this research could be listed as followed:

- The analysis results are based on simplified assumed conditions. The deviations in actual piling process are randomized. The abovementioned extreme cases should not exist.
- The deviations for spun pile affect single pile axial force and pile body tensile stress. Based on the estimation from current combinations, *the maximum deviation for spun pile is limited within 150mm should be practical.*
- The main effect due to the deviations for steel pile is increase of pile body axial force. Based on 200mm assumed deviation for analysis, the basic results in some combinations slightly exceed 3% of limitation value. However, considering the deviations in actual piling process are randomized and the presumed deviation should not exist, *the maximum deviation for steel pile is limited within 150mm should be acceptable.*
- It is recommended for any pile group with piles deviated not more than 150mm, additional calculation will not be conducted for demonstration of acceptability. However, *for pile group with pile deviation greater than 150mm, Reviewing calculation shall be carried out to justify the structural acceptance.*
- As this project adopts precast R.C. shell to be the external formwork for pilecap construction, the extended length of the R.C. shell below the bottom of pilecap is comparatively long. *Over-deviated pile foundations will result to difficulty of R.C. shell installation. Therefore the Contractor shall strictly control the deviation direction and amount.* Every endeavour shall be tried to comply with 75mm of deviation requirements which is stipulated in JKR Standard Specifications for Road works.

This study is providing an approaching to solve the problem of driven pile foundation acceptance criteria and promoting the updating of local bridgework technical specification. The recommendation could be adopted for the similar huge marine bridgeworks which installed large diameter driven piles and will effectively promote the practical application of ones in Southeast Asia infrastructure works.

References

1. Malaysia Jabatan Kerja Raya 2005 *STANDARD SPECIFICATIONS FOR BUILDING WORKS 2005 Edition* (Kuala Lumpur:Malaysian Public Works Department) p 19
2. Geotechnical Engineering Office 2006 *FOUNDATION DESIGN AND CONSTRUCTION publication No. 1/2006* (Hong Kong: Geotechnical Engineering Office of Civil Engineering and Development Department, Hong Kong SAR Government)p 166
3. British Standard Institution committee 1986 *BS 8004: Code of Practice for Foundations* (London: British Standard Institution)p 122
4. British Standard Institution committee 2000 BS6349: Code of Practice for Maritime structures Part 2: Design of quay walls, jetties and dolphins(London: British Standard Institution) p 64
5. The third harbour engineering investigation and Design Institute of China Ministry of Communications 1999 *The Specification of Pile Foundations for Port Engineering (JTJ-254-98)* (BeiJing:China Communications Press) p 17
6. Budi, Gogot Setyo, Charles,Wijaya,Hendry 2015 *Deviation of Position of Piles Foundation from Its Original Designed Location.* (Jakarta Indoensia:Proceedings 10th Indonesian Geotechnical Engineering Conference and 19th Annual Scientific Meeting)p 469-474
7. K Paik,R Salgado,J Lee,B Kim 2003 *Behavior of Open- and Closed-Ended Piles Driven Into Sands 129(4)* (RESTON, USA:Journal of Geotechnical & Geoenvironmental Engineering) p 296-306
8. Li Zuojun 2010 *Preliminary Study on Pile Deviation Vol.40, Supplement* (China:Industrial Construction) p 1036-1041
9. Liu Jifu,Guo Shuyang,Xiao changjian 2015 Tubular Pile Vertical Deviation Detection and Its Influence on Vertical Bearing Capacity,Vol.5 (China: GuangDong Highway Transportation) p 39-42

Study on the Treatment Timing of Large Deformation of the Tunnel in Swelling Rock

Zhongmin Yang[1], Yongtao Gao[1], Ziqiao Cheng[2], Zijie Cong[1]

1. School of Civil and Resource Engineering, University of Science and Technology Beijing, Beijing 100083, China;
2. PowerChina Roadbridge Group Co., Ltd, Beijing 100048, China

Abstract. When large deformation of the tunnel occurred in the swelling rock, the large deformation treatment will not only greatly delay the construction time, but also lead to instability or even collapse of the tunnel. Selecting the reasonable timing for large deformation initial support replacement can solve this problem effectively. Based on the LiRang tunnel, the deformation and stress of the tunnel in anhydrite were analysed by FLAC software after excavation. Then replace the deformed first liner at different vault settlement value. The displacement and stress of surrounding rock were analysed when the model reaches equilibrium state again. The results showed that stress concentration at the arch wall and arch foot of the tunnel, and the stress release at the vault and invert is very large after excavation. If the large deformation was treatment when the crown settlement of the tunnel reaches 90% ~95%of the final predicted settlement value, the tunnel deformation and the final stress state could be in a small value. This study can provide a reference for the treatment of large deformation in the swelling rock tunnel.

1 Introduction

Swelling rock is a kind of rock that the water content and the volume increases with time after the physical and chemical reaction with water. In the tunnel engineering, the swelling rock produces large swelling force when it meets water, and causes continuous deformation of rock mass. After applied the first liner, the restrained swelling heave leads to the development of swelling pressure that damages the tunnel lining, even lead to the tunnel collapse. The hydraulic conductivities in the rock mass around the tunnel can increase greatly due to tunnelling; it can be increase by up to 6 orders of magnitude compare to the intact rock [1, 2]. In recent years, the tunnel engineering hazards caused by swelling rock increased sharply with the rapid development of highway and railway construction. Marco Barla[3] introduced a laboratory-based approach to predict the swelling behaviour around a tunnel through numerical simulation. Dehnavi R N etc.[4], Butscher C etc.[5], Pimentel E etc.[6] have studied the deformation mechanism of swelling rock, and have found that the probability of large deformation can be greatly reduced by the measurement of implement rock anchors, reinforced lining, mechanical resistance, reducing the swelling pressure by allowing deformation et. al[7-9]. However, due to the complex condition of the surrounding rock, large deformation events caused by swelling rock still often occur. Making repeated rehabilitation work necessary. Although scholars have also studied the treatment methods of large deformation tunnel, few studies have been done on the treatment timing. The usual method is to apply temporary support to reinforce the deformed first liner, then replace the deformed first liner when the deformation of the tunnel gradually decreases and tends to be stable [10].

However, this treatment method has the following disadvantages: (1) after applied the temporary support, in the deformation late stage of surrounding rock, the deformation rate decreased slowly. It needs a long period to the rock mass getting stable, for example, the deformation rate become less than 3.0mm/d. This will seriously delay the construction schedule. (2) As the swelling rock deformation, the strain energy become very large under the constraint of strong support, and the swelling potential become very high. If remove the support, the sudden stress release easily causes the tunnel instability. Therefore, it is necessary to study the timing of replacement of large deformation, so that the deformation of tunnel can be the smallest during the replacement process and the replacement, and the stress become the lowest when the support and the surrounding rock reaches equilibrium again.

2 Engineering background

The rock of LiRang tunnel is composed of Jurassic rhyolite and tuff of Jurassic Tao Bei Ying group (J3t). Along the axis of the tunnel, there are 150m gypsum rocks. The gypsum swelling rock mass scope is very large, so the swelling problem is a very serious problem in this tunnel. During the construction, the cracks get open and the enriched groundwater enters the gypsum rock, which is easy to cause the swelling deformation of the tunnel. In order to solve the large deformation problem in a fast and safe way, study the timing of large deformation treatment using the FLAC code. Through the gypsum rock softening freshwater immersion test [11], the mechanical properties of gypsum rock under the natural state are as follows: σ_c=31.157MPa, E=4.001MPa, μ=0.128, C=3.64MPa, internal friction angle is 45.

3 Numerical simulation

The numerical model was set up and calculated using the 2D finite difference method FLAC, as fig 1. The model size is 80m*80m, the tunnel span is 15.6m, the height is 9.7m, and the overburden thickness is 153m. The monitoring point is shown in fig 1. Due to the rock swelling in tunnel excavation, swelling potential and plastic deformation will seriously influence the rock stability and liner bearing capacity, it need to choice the elastic-plastic model which can consider the influence of swelling force. This paper adopts the Swell constitutive model based on the Mohr-Coulomb constitutive. Its mechanics and deformation mechanism based on non-associated plastic flow shear and tension associated plastic flow rule, and the swelling mechanism is the three-dimensional swelling constitutive under the control of the double semi logarithmic linear relationship.

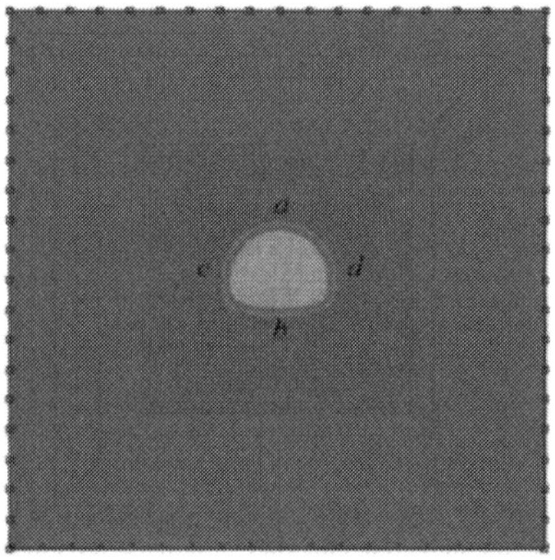

Figure 1 The geometric model of tunnel and location of monitoring points

The wetting-induced strain can be expressed by the following logarithmic function of the total compressive stress σ_{yy}. The mechanical parameters of the swelling rock mass are obtained by the test result and the field conditions, as shown in Table 1.

$$\varepsilon_{yy} = c_1 \log[a_1 - (-\sigma_{yy} / p_a)]$$

$$\varepsilon_{xx} = \varepsilon_{zz} = c_3 \log[a_3 - (-\sigma_{yy} / p_a)]$$

Where: a_1, a_3, c_1, c_3 -soil swelling properties;

p_a -the atmospheric pressure

4 Result and discussion

4.1 Stress analysis of tunnel in swelling rock

The strata direction of the swelling rock is horizontal in this model. When the swell rock expands with water, it is mainly swelling in the direction perpendicular to and parallel to the stratum. According to the parameters in Table 1, the swelling strain of the swelling rock in the vertical direction was larger than that in the horizontal direction. From the numerical results in figure 2, the stress release was occurred at the edge of the arch waist of the tunnel, while the stress increased greatly in the surrounding rock of the arch foot and the arch waist. The maximum principal stress reaches 6.00MPa, and the minimum principal stress reaches 2.50MPa. The stress concentration appears at the arch foot with large curvature. The stress release is large in the vault and arch bottom. Although there is no tensile stress, the stress value is much less than the original rock stress, which will cause larger deformation value of the tunnel. Therefore, special attention should be paid to the support of roof and floor, so as to prevent large arch crown settlement and arch bottom heave.

Table 1. Parameters of swelling rock mass.

Density (kg/m^3)	Elastic Modulus (GPa)	Poisson's ratio	Cohesion (MPa)	friction angle (°)	a1	c1	a3	c3
2270	1.2e9	0.13	1.10e6	39	1.533	-0.5e-2	0.5	-0.2e-2

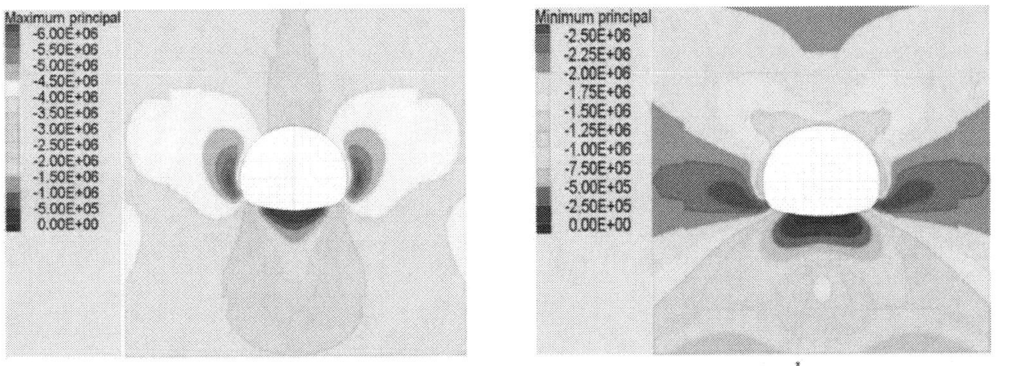

Figure 2 The maximum and minimum principal stress distribution after excavation

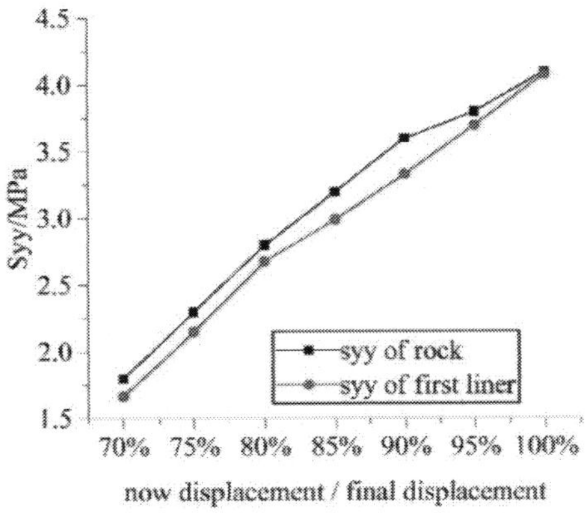

Figure 3 The displacement of tunnel when large displacement treatment at different time

4.2 Study on replacement timing of deformed lining

The monitoring stress of the surrounding rock and support in different deformation periods shown that the stress increased with the increase of deformation, as shown in figure 3. The sudden release of large stress and stored strain energy drop will seriously reduce the stability of the tunnel, and cause the rock block falling even tunnel collapse. In order to study the best treatment time of large deformation. Firstly, the predicted final displacement d_0 was calculated when the tunnel gets stable. Then the deformed first liner was removed when the displacement reached $70\%d_0$, $75\%d_0$, $80\%d_0$, $85\%d_0$, $90\%d_0$, $95\%d_0$, $100\%d_0$. Finally, the displacement and the stress were recorded when the model become balanced again after applied the strength first liner.

As shown in figure 4, if replace the deformed first liner when the displacement reach 70% and 75% of the final displacement, there still will be a large deformation after the tunnel get balance finally. The settlement of vault decreases gradually with the delay of deformation time. When the treatment timing reached 90% of the final displacement, the displacement after treatment become a low value, and the it is almost the same as replacing at 95% and 100% of the final displacement. From the view of the convergence, the value is small, and its impact on the tunnel stability is small.

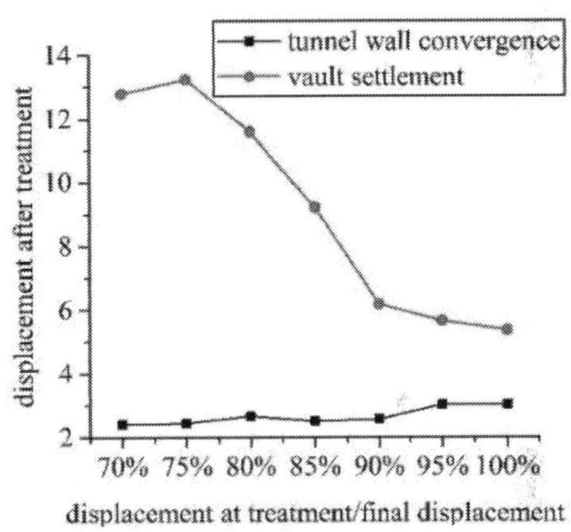

Figure 4 The tunnel settlement and convergence after large deformation treatment at different timing

Since the swelling of the rock in the vertical direction is greater than in the horizontal direction, only the stress in the vertical is calculated. Figure 5 represents the stress of the rock mass after treatment. The stress in the arch waist is greater than that at the vault and arch bottom. From the monitoring data in the bottom and the arch, as shown in Figure 5 (a), if manage the large deformation when the displacement get between 80% and 90% of the final deformation, the vertical stress has a great growth. As the deformation exceeds 90%, the stress decreased sharply. Because this stress will be directly applied on the first liner, it is directly affect the force of the liner and the stability of the tunnel. It needs special notice of the stress in these locations. From the figure 5 (b), the final stress has a large drop as the delay of the treatment time. The stress drops to the minimum when the deformation reaches 90% of the final deformation. But if the treatment time reaches 95% or 100% of the final deformation, the stress increased slightly. It indicated that it is beneficial for the final stress environment to replacement the deformed first liner at an earlier time. Considering the displacement and the final stress, it is better to replace the deformed first liner when the displacement exceeds 90% of the final predicted deformation. Next, we can study the stiffness and strength of the first liner by the swelling property of the surrounding rock and the timing of treatment.

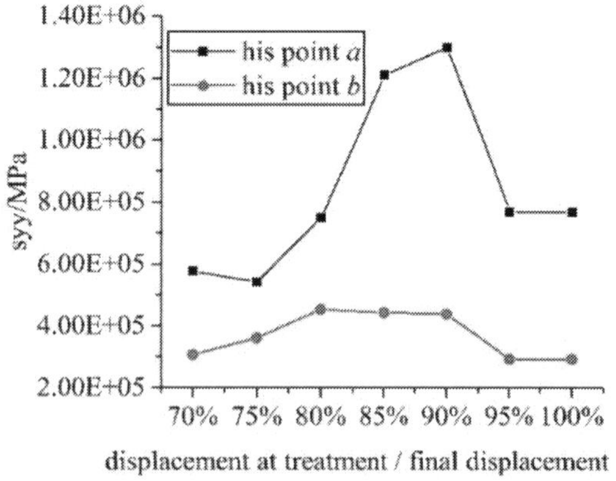

a

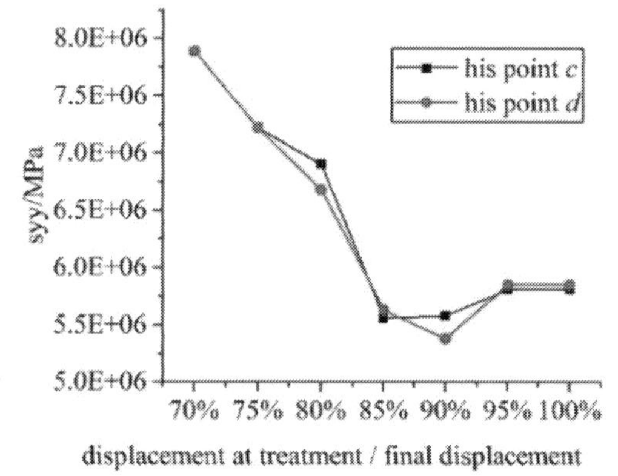

b

Figure 5 The stress of the rock mass at different location after treatment at different timing

5 Conclusion

In this paper, a typical tunnel in the swelling as the representative, through the numerical simulation, the following conclusions:

(1) In the swelling rock which strata is along the horizontal direction, tunnel arch wall and foot will appear stress concentration after the excavation. While the stress in the vault and the arch bottom decreases sharply. It is necessary to pay particular attention to settlement of vault and arch bottom bulge;

(2) With the deformation of the tunnel, the stress of the rock mass and support in the vault increases gradually. The strain energy stored in the surrounding rock. It may be more dangerous if remove the deformed first liner when it gets the final displacement.

(3) The displacement and stress was analysed after replace the deformed first liner in different deformation timing. The result shown that it is better to do the treatment when the vault settlement is between 90% and 95% of the final settlement. This makes both the deformation and stress smaller when the tunnel become balance again after treatment and the tunnel become more stable.

References

1. Butscher C, Scheidler S, Farhadian H, etc. Swelling potential of clay-sulfate rocks in tunneling in complex geological settings and impact of hydraulic measures assessed by 3D groundwater modeling [J]. Engineering Geology. 2017, 221 pp143-153.
2. Tsang CF, Bernier F, Davies C. Geohydromechanical processes in the Excavation Damaged Zone in crystalline rock, rock salt, and indurated and plastic clays - in the context of radioactive waste disposal [J]. International Journal of Rock Mechanics and Mining Sciences, 2005, 42(1) pp 109-125.
3. Barla M. Numerical simulation of the swelling behaviour around tunnels based on special triaxial tests [J]. Tunnelling and Underground Space Technology incorporating Trenchless Technology Research, 2008, 23(5) pp 508-521.
4. Dehnavi RN, Sadeghi M. Deterioration of weak rocks over time and its effect on designing tunnel support systems [J]. Bulletin of Engineering Geology and the Environment, 2017 pp 1-12..
5. Butscher C, Mutschler T, Blum P. Swelling of Clay-Sulfate Rocks: A Review of Processes and Controls [J]. Rock Mechanics & Rock Engineering, 2016, 49(4) pp 1533-1549.
6. Pimentel E, Anagnostou G. New Apparatus and Experimental Setup for Long-Term Swelling Tests on Sulphatic Claystones [J]. Rock Mechanics & Rock Engineering, 2013, 46(6) pp 1271-1285.
7. Vergara M R, Triantafyllidis T. Swelling behavior of volcanic rocks under cyclic wetting and drying [J]. International Journal of Rock Mechanics & Mining Sciences, 2015, 80 pp 231-240.
8. Butscher C, Huggenberger P, Zechner E, Einstein HH. Relation between hydrogeological setting and swelling potential of clay-sulfate rocks in tunneling [J]. Engineering Geology, 2011, 122(3) pp 204-214.
9. Butscher C, Huggenberger P, Zechner E. Impact of tunneling on regional groundwater flow and implications for swelling of clay–sulfate rocks [J]. Engineering Geology, 2011, 117(3) pp 198-206.
10. Pu Wenming, Chen Fan, Ren Song, etc. Research of Swelling Rock and Summarize of Tunnel Construction [J]. Chinese Journal of Underground Space and Engineering, 2016, 12(s1) pp 232-239.
11. Ren Song, Deng Gaoling, Wu Jianxun, etc. Immersion tests on gypsum rocks using fresh water [J]. Rock and Soil Mechanics, 2017, 38(4) pp 943-950.

Wind Directional Effect on a Single Storey House Using Educational Wind Tunnel

S S Zaini[1], N Rossli[1], T A Majid[1], S N C Deraman[1] and N A Razak[2]

[1] *Disaster Research Nexus, School of Civil Engineering, Universiti Sains Malaysia, Engineering Campus, 14300 NibongTebal, Penang, Malaysia*
[2]*School of Aerospace Engineering, Universiti Sains Malaysia, Engineering Campus, 14300 NibongTebal, Penang, Malaysia*

Abstract. Wind tunnel testing of single-storey isolated building with 1: 100 scale down model was carried out in an open circuit wind tunnel without roughness elements facilities. The gable roof building model with 30° roof pitch was studied for wind directions of 0°, 30°, 45°, 60° and 90°.Pressure measurements were performed on all the walls and the roof (Zone 1, 2, 3 4 and 5) of the building model with wind speed of 12 m/s. The results showed that the high suctions were generally induced by the 90° wind direction for Zone 1 and 60° and 90° wind directions for Zone 2. Mostly, high suction was also observed in case of 45° and 60° wind direction in Zone 3. In zone 4 and zone 5, high suction was generally induced by the 0° wind direction.

1. Introduction

Wind engineering covers the study on the effect of wind flow in natural environment. Moreover, this engineering discipline investigates the possible damage that may be caused by a strong wind event towards a structure, wind conform and the energy that can be extracted for mankind. High wind speed events can cause widespread destruction as reported by [1].

Most of the houses all over the world are constructed as low-rise building 1 where the mean roof height is less than the least horizontal dimension and also less than 18.3 m 2. In Malaysia, severe damage on low-rise building always occurred in the rural area especially in the northern region of Peninsular Malaysia due to windstorm events. Most of the damages occurred on the roofing system, caused by high local suctions and large pressure fluctuations around the roof structure 3.

Wind pressure may not necessarily exert to the surface of the building wall at 0° as it can vary at different angles and the pressures are dependent on the wind direction 4. When the approaching wind is normal to the ridge, the windward wall and the leading edge of the sidewalls experience high pressure and high suctions, respectively. Holmes 5 studied the wind flow on a single-storey house with 10° pitch roofs using wind tunnel facilities. The wind directions were set at 0°, 60° and 90°. The results showed that the 60° wind direction produced the critical mean suction on the gable roof due to conical vortex along the edges and corner of the roof.

Jing and Li 6 investigated the wind pressure distribution on a 1:40 scale low-rise house model using wind tunnel facilities. They reported that for 0° wind direction, the largest absolute values of peak negative wind pressure occurred on the edges of windward wall, the leading edges of windward roof and the leeward roof ridge. In the case of 45° wind direction, the largest absolute values of peak negative wind pressure occur on the corner and edge of the windward wall, the windward roof corner, and the leeward roof ridge due to the formation of conical vortex and separation of airflow.

The number of houses being damaged during windstorm events increased annually in Peninsular Malaysia This phenomenon becomes worse where the surrounding economic activities are mainly agriculture. Such locations are large area of flat land for paddy plantation. This type of open terrain provides no obstacle to shield the wind flow.

The low-rise buildings in rural area are more susceptible during monsoon seasons which compared to that in the urban area where wind flow is commonly obstructed or shielded by other building structures. In an open area, wind can impinged on a building from any direction and as such, there is a need to study the effect of wind flow from several angles.

This study aims to investigate the distribution of the pressure coefficients along the profile of a down scale isolated low rise building model using wind tunnel test facilities. All tests pertaining to the effect of wind direction conducted by previous researchers used boundary layer wind tunnel test facilities. Where the model is a scaled single storey house model, there is a possibility to use educational wind tunnel test without using roughness element. This approach reduces the experimental time and cost.

2. Experimental setup and procedure

The tests were conducted using an open circuit wind tunnel or Eiffel wind tunnel. The wind tunnel has a working section of 300 mm wide, 350 mm high and 550 mm long as shown in figure 1. This type of tunnel has a very low turbulence level, approximately 0.15%. All tests were performed at wind speed of 12 m/s.

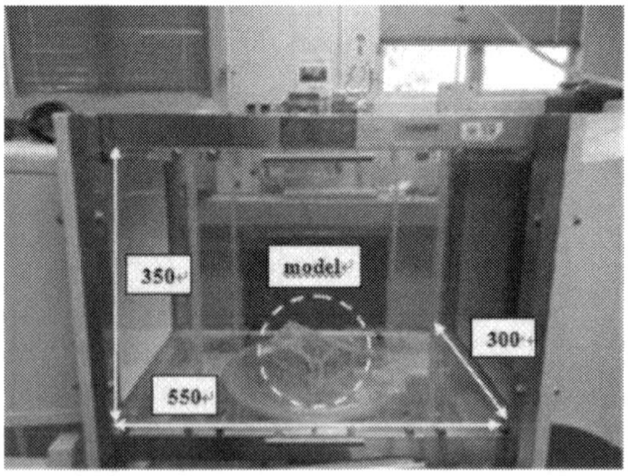

Figure 1. Test section (all dimensions in mm).

The building model was fabricated using 6 mm thick Perspex. The dimension for model is 80 mm width, 120 mm length and 40 mm height following 7. The locations of the pressure taps placed on the walls and roofs of the models are shown in figure 2. The blockage ratio for the models in the wind tunnel ranged from 2.62 to 7.21%. 100 cm long PVC tube with 0.1 cm internal diameter and 0.2 cm external diameter were inserted into the holes drilled into the Perspex. The model was placed on top of the turntable located at the middle of the test section. Each pressure tap on the building model was connected to differential pressure sensor. The channel box allowed simultaneous monitoring of 8 pressure tap readings. The measured values of static pressure were converted to pressure coefficient, Cp, which is defined as

$$C_P = \frac{P - P_\infty}{0.5\,\rho U_\infty^2} \qquad (1)$$

P is the pressure at a given location of model, P_∞ is the free stream static pressure, U_∞ is the free stream velocity and ρ is the air density. Since the wind tunnel is able to generate wind in 0° only, the assignment of 30°, 45°, 60° and 90° directions were done by turning the model to the desired angle.

3. Results and discussion

Figure 3 compares the results between Ozmen et al. 7 and present study. It can be seen that the use of educational wind tunnel facilities is able to produce similar trend along the profile of roofs and walls. The differences in the results occurred especially at the end of wall A and leading edge of roof B. However, the disparity is due to the location and quantity of the pressure taps. The present study needs to maintain several pressure tap location as required by the following test. The ability of the educational wind tunnel facilities to produce reliable results was extended to the investigation on the effect of wind direction.

Figure 4 shows the pressure coefficient along the walls and roofs for wind directions 0°, 30°, 45°, 60° and 90° in Zone 1. In the case of 0° wind direction, high pressure was developed on the windward wall (region A) where the maximum positive pressure coefficient was recorded to be 0.96. However, the pressure coefficient reduced as it approaches the leading edge of the wall and converted to suction in region B with maximum suction of -0.89. The change from pressure in region A to suction in region B is because the air flow has already separated at the leading edge of the windward roof forming a separation zone 6. In region C and D, the suction remains almost constant with an average reading of -1.44.

Figure 4 shows the pressure coefficient along the walls and roofs for wind directions 0°, 30°, 45°, 60° and 90° in Zone 1. In the case of 0° wind direction, high pressure was developed on the windward wall (region A) where the maximum positive pressure coefficient was recorded to be 0.96. However, the pressure coefficient reduced as it approaches the leading edge of the wall and converted to suction in region B with maximum suction of -0.89. The change from pressure in region A to suction in region B is because the air flow has already separated at the leading edge of the windward roof forming a separation zone 6. In region C and D, the suction remains almost constant with an average reading of -1.44.

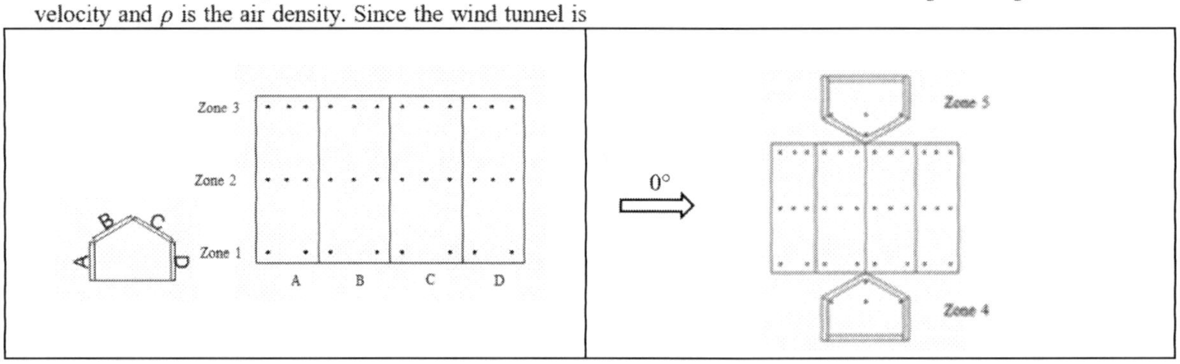

Figure 2. The building model showing the location of pressure taps for Zone 1 to Zone 5

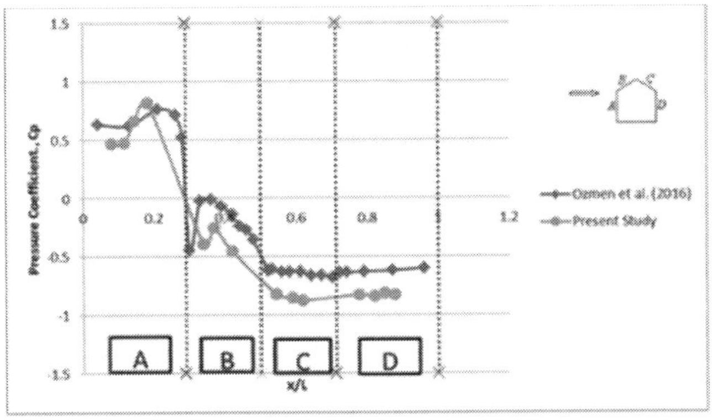

Figure 3. Comparison between Ozmen et al. [8] with present study.

The same trend was also observed for 30° and 45° wind directions. However, the positive pressure coefficient at the lower end of the wall in region A was higher (approximately 0.91) compared to the 0° wind direction (0.30). On the contrary, the pattern of the pressure coefficient in region A for 60° wind direction immediately reduced as it approaches the leading edge of the roof and converted to suction with a maximum value of -2.18 in region B. The suction decreased in region C and maintained almost the same magnitude in region D. In the case of 45° wind direction, the wind impinged the model at a skewed angle. As a result, conical vortex was formed at the leading edge and created suction in region B, C and D. As for the 30° and 60° wind direction, the conical vortex was developed at the roof corner.

Finally, in the case of 90° wind direction, the distribution of the pressure coefficient changed drastically where all the surfaces only exhibited negative pressure coefficients 6. The highest suction showed by the 90° wind direction at the leading edge of roof in region B can be associated to the formation of separation point at that particular area leading to the development of separation zone 8. Similarly, the reduction in the negative pressure coefficient observed in region C and D (-2.07) can be related to the reattachment process of the separation zone occurred as the wind approached the leeward surfaces as reported by 9.

Figure 5 shows the pressure coefficient for wind speed at 12 m/s with the variation of wind directions 0°, 30°, 45°, 60° and 90° in Zone 2. In the case of 0° wind direction high pressure was occurred on the windward wall (region A) where the maximum positive pressure coefficient was recorded to be 0.76. However, the pressure coefficient decreased as it approaches the leading edge of the wall and converted to suction in region B. It can be seen that the magnitude of the suction slightly fluctuated. The maximum suction in region B showed to be approximately (-0.84). In region C the suction remains almost constant with an average reading of (-1.43). This occurrence continues in region D. In general, the same trend was observed as in the case of 0° wind direction in Zone 1.

The same trend was also observed for 30° and 45° wind directions. However, the positive pressure coefficient at the lower end of the wall in region A was higher (approximately 0.53) compared to the 0° wind direction (0.44). The suction increased slightly in region C and remained almost the same magnitude in region D. The higher suction showed in region C for 60° wind direction can be related to the formation of larger separation zone closer to region C.

Finally, in the case of 90° wind direction, the distribution of the pressure coefficient changed drastically where all the surfaces only exhibited negative pressure coefficients (suction). The highest suction showed by the 90° wind direction is located at the ridge of roof in region C (-1.80).

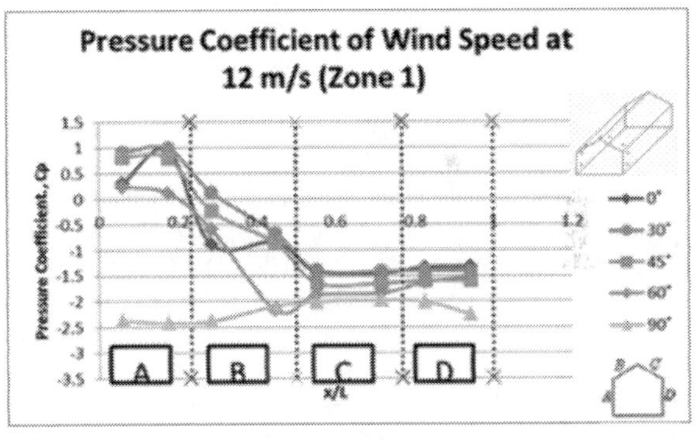

Figure 4. Pressure coefficient of wind speed at 12 m/s (Zone 1).

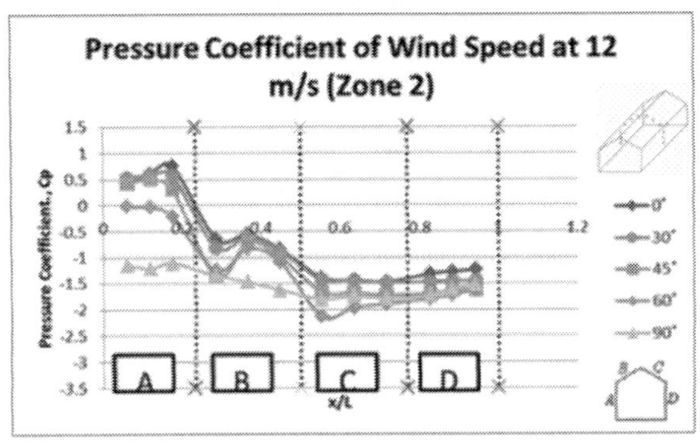

Figure 5. Pressure coefficient of wind speed at 12 m/s (Zone 2).

Figure 6 shows the pressure coefficient for wind speed at 12 m/s with the variation of wind directions 0°, 30°, 45°, 60° and 90° in Zone 3. In the case of 0° wind direction, high pressure was developed on the windward wall (region A) where the maximum positive pressure coefficient was recorded to be 1.01. However, the pressure coefficient reduced as it approaches the leading edge of the wall and converted to suction in region B (maximum suction -1.02). In region C and D, the suction remained almost constant with an average reading of -1.45. The same trend was also observed for 30° and 45° wind directions. However, the positive pressure coefficient in region A was lower (approximately 0.50) compared to the 0° wind direction (0.99). The 90° wind direction showed similar trend with Zone 1 and Zone 2. The high suction was located at the middle of roof in region C and remained constant with an average reading of -0.81 in region D.

Figure 7 shows the pressure coefficient for wind speed at 12 m/s with the variation of wind directions 0°, 30°, 45°, 60° and 90° for Zone 4 and Zone 5. In Zone 4, the positive pressure coefficients occurred in 45°, 60° and 90° wind directions. In the case of 60°, high pressure was developed on the leading edge of wall (tap 2) as reported where the maximum pressure coefficient was recorded to be 0.69. Similar trend was observed as in the case of 45° wind direction. In the case of 90° wind direction, the high pressure occurred on the middle of wall with maximum pressure coefficient was recorded to be 0.77.

Hence, in the case of 0° and 30° wind directions, the distribution of the pressure coefficient changed drastically where all the surfaces only exhibited suction. The highest suction showed by 0° wind direction at the leading edge of wall (tap 2) with a maximum value of -1.13. This can be associated to the fact that airflow already separated at the leading edge of the windward wall forming a separation zone. In the case of 30° wind direction, the high suction occurred at the middle of wall (tap 3).

Finally, in Zone 5, suction occurred in the case of 0°, 30°, 45°, 60° and 90° wind directions. The highest suction showed by the 0° wind direction at the leading edge of windward wall (-1.78). Similar trend was also observed as in Zone 4. In general, high suction was developed at the edge of wall. This is particularly true in all cases except for 0° and 90°wind directions.

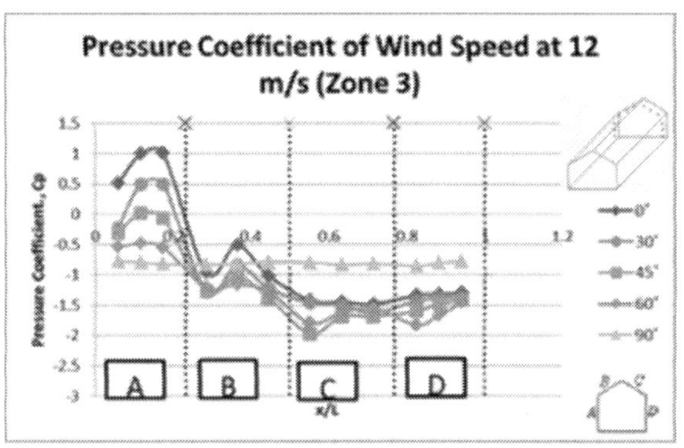

Figure 6. Pressure coefficient of wind speed at 12 m/s (Zone 3).

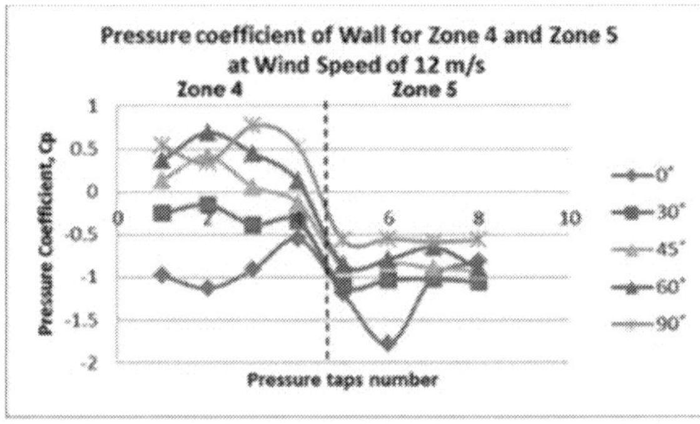

Figure 7. Pressure coefficient of wind speed at 12 m/s (Zone 4 and Zone 5).

4. Conclusion

The positive pressure coefficients were recorded in region A for all wind directions except for 90° wind direction in Zone 1. Suction in region B, C and D was developed for all wind directions. The suction decreased in the case of 30° wind direction and increased in the case of 45°, 60° and 90° wind direction for all regions. In Zone 2, most of the suctions were lower compared to Zone 1 in region B, C and D for 0°, 45° and 90° wind direction. However, the overall distribution of the pressure coefficient was found to be relatively similar to Zone 1. On the contrary, for Zone 3, the positive pressure coefficients were recorded in region A for 0° and 30° wind directions only. The suction increased in the case of 30° wind direction and increased in the case of 45° and then decreased again, in the case of 60° and 90° wind directions. In Zone 4, the positive pressure coefficients were recorded for 45°, 60° and 90° wind directions. Suction was developed for 0° and 30° wind directions. The high suction occurred at one of the edge of wall for 0° wind direction and at the middle of wall for 30° wind direction. High suction occurred at the edge of wall for Zone 5 in the all cases except for 30° and 90° wind directions.

Acknowledgments

The authors express their gratitude for the financial support of the Fundamental Research Grant Scheme (FRGS) from the Ministry of Higher Education Grant (203.PAWAM.6071317), Universiti Sains Malaysia and MyBrain15 Scholarship from the Ministry of Education Malaysia and School of Aerospace Engineering, USM for the wind tunnel laboratory.

References

1. Prasad D, Uliate T and Ahmed M R 2009 Wind loads on low-rise building models with different roof configurations Int. J. Fluid Mech. Res **36**pp 231-243
2. American Society of Civil Engineers (ASCE) Standard 2002Minimum design loads for buildings and other structures*New York, USA*

3. Ramli N I, Majid T A, Ali M I, Saad M S H, Hashim M and Zakaria I 2010Wind related disaster risk reduction activities in Malaysia
4. Meecham D, Surry D and Davenport A G 1991The magnitude and distribution of wind-induced pressures on hip and gable roofs*J. Wind Eng. Ind. Aerodyn.***38**pp 257-272
5. Holmes J D 1994 Wind pressures on tropical housing *J. Wind Eng. Ind. Aerodyn.***53**pp 105–123
6. Jing X and Li Y 2013Wind tunnel tests for wind pressure distribution on gable roof buildings *Sci. World J.* 2013
7. Ozmen Y, Baydar E and van Beeck J P A J 2016 Wind flow over the low-rise building models with gabled roofs having different pitch angles*Build. Environ***95**pp 63–74
8. Liu H 1991Wind Engineering A Handbook for Structural Engineers New Jersey: Prentice-Hall, Inc,
9. Holmes J D 2001*Wind Loading of Structures*, Second Edi London and New York: Taylor & Francis

Finite Difference Method in Stress Analysis of Anchorage Zone of Highway Extradosed Cable Stayed Bridge

Tang Yingying[1,2], Li Yingshuai[1], Niu Yanwei[1], Chen Hong[1], Huang Pingming[1]

[1]*School of Highway, Chang'an University, 710064 Xi'an, China*
[2]*School of Science, Chang'an University, 710064 Xi'an, China*

Abstract. With the development and wide application of extradosed cable stayed bridge, an effective method is needed for simulation of the stress distribution in anchorage zone, which plays a vital role in transferring force from the cable to the pylon and diverting the cable direction. Based on the Finite Difference Method (FDM), an efficient and practical method of stress filed simulation was presented in the paper. First, a plane finite element model was established using ABAQUS, to determine the boundary condition for FDM. Based on this, FDM was used to simulate the stress distribution of the concrete in the anchorage zone. Finally, on the simulation result was calculated and in comparison with finite element model result. It has been found that: The vertical compressive stress of concrete in the anchorage zone gradually reduces from the middle to the two sides, and the stress assumes double peak type in transverse. Compared with the finite element solutions, the approximate solution simulated by FDM improve the computational efficiency with certain accuracy keeping, except the loading area. In vertical orientation, the concrete in the place of H (the width of the pylon) from cable force acting position is in a state of uniform stress.

1 Introduction

Finite difference method (FDM) is one of the simplest and oldest numerical simulation methods. The advent of finite difference techniques in numerical applications began in the early 1950s, and then it was widely used with constant development. The principle of FDM is close to numerical schemes used to solve ordinary differential equations. FDM usually included three steps: (1) the solution domain is divided into grid, and (2) the differential equation is approximated by the difference equation at the node of the grid, (3) the direct method or iterative method is used to solve the difference equations to get the approximate solution of the differential equation [1].

As a well-established numerical analysis method, FDM is widely used in the fields of civil engineering, material forming and other fields. Liao D M et al simulates the stress field, fluid flow field, temperature field and stress field simulation all based on FDM [2]; Chuang K P et al analyse cylindrical shell roofs with FDM [3]; Shen J B simulates the thermal behaviour of classical or composite Trombe solar walls with FDM[4]; Appelö D proposed a stable finite difference method for the elastic wave equation on complex geometries with free surfaces[5]; Griffin D S et al solved some plane elasticity problems [6].

2 Problem of stress distribution in anchorage zone of extradosed cable stayed bridge

As shown in Fig.1, Compared with the conventional Cable-Stayed Bridge, cables in extradosed cable stayed bridge could be anchored in the pylon directly, or through the pylon by sub-steel-pipe and double casing saddle and then anchored in the girder on the other side [7]. The anchorage zone in the pylon is a key part of the bridge design especially for saddle anchorage form, and its stress is so complex that researchers need to pay more attention to it.

Most recent results of the stress distribution researches were based on the general finite element software or measured in the model test [8-11]. Although the simulation results obtained possesses the advantages of high accuracy and stability, there are some defects. With the first approach, it needs proficient with the application of the software whose principle is very complex. For the second approach, researcher will spend more money and time to get the rule of the stress distribution. Therefore, a simplified the stress simulation method is needed to improve the efficiency of calculation, which will provide guidance for the design and construction of extradosed cable stayed bridge.

In this study, a new simulation method based on FDM is applied in the anchorage zone stress distribution simulation. This method could improve the efficiency of calculation with a few grids, and the results show good agreement with it finite element model results.

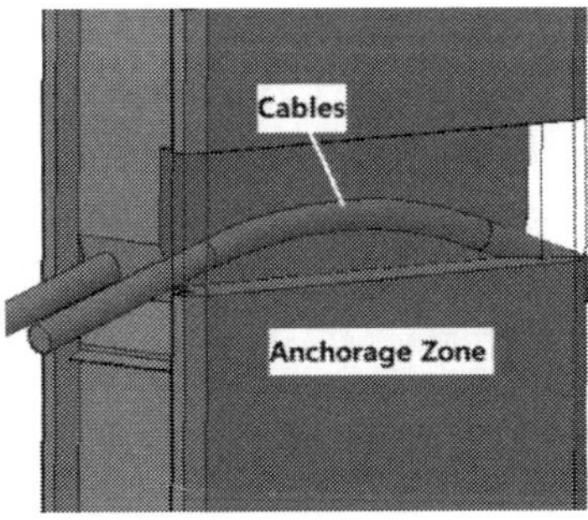

Figure 1. Anchorage zone

3 FDM simulation

According to the research results of Liu Z et al [12], the stress distribution in transverse direction could be simplified as a plane problem. Also, the FDM for simulating the stress distribution in this study was based on a plane model.

3.1 Stress boundary condition

In the analysis process, boundary condition has significant influence on result. Therefore, it is very important to determine it. The boundary condition includes the load boundary condition and the stress boundary condition. As the load boundary condition, according to Saint-Venant's Principle (the difference between the effects of two different but statically equivalent loads becomes very small at sufficiently large distances from load), the uniformly stress distribution of concrete is assumed in a position far away from the saddle, which means the boundary condition determination problem is transformed into distance determination problem. Although the stress distribution becomes more uniformity with the increase of distance, it could improve the simulation efficiency with a small domain.

In order to determine the nearest position with uniform pressure, a plane finite element model is established using ABAQUS as show in Figure 2. Besides, a new parameter called deviation rate was defined as follow:

$$DevR = \frac{(\sigma_{max} - \sigma_{min})}{\sigma_{AVG}} \times 100\% \qquad (1)$$

Where DevR= deviation rate, the smaller of the DevR means the distribution of the stress is uniformity, the DevR of uniform load is 0; = maximum stress, minimum stress and average stress, respectively.

Figure 2 shows the relationship between DevR and distance. The stress of the concrete in line A, B, C, D and E were obtained by the FEM. It can be seen that the DevR is getting smaller as the distance increases. When the distance is increased to H, at line C, DevR equals 4.01%, which could believe that the stress at line C is uniform. So it is suggested that the depth of the anchorage zone using for analyzing equals the anchorage zone's width. In other words, the analysis domain is recommended as a square. Besides, the effect of the saddle on the concrete was assumed as concentrated force according to Saint-Venant's Principle. The grid spacing equals H/6 for the calculation of example, and the coordinate axis is shown in Figure 3.

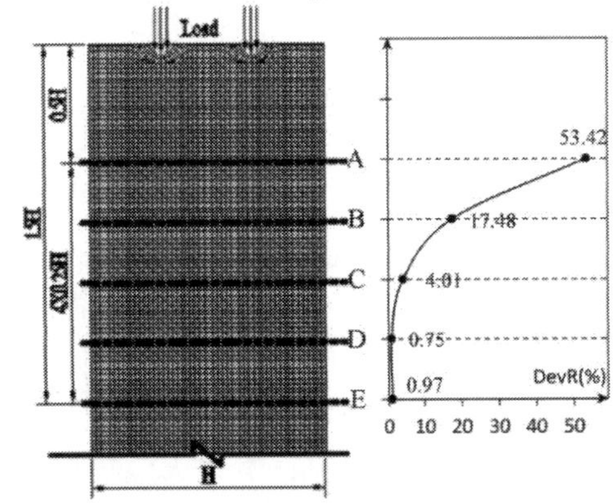

Figure 2. DevR

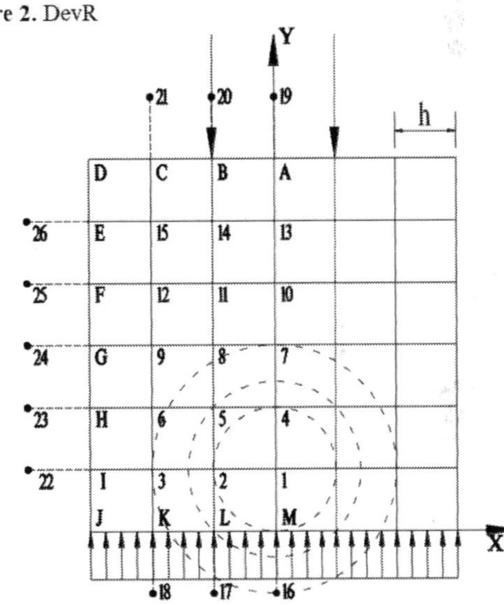

Figure 3. The grid

After the load boundary condition determined, the stress boundary condition is needed to be determined before establish the finite difference model. Only half domain is required to be calculated because of the symmetry. Take node A as the reference point, and let $\Phi_A = \left(\frac{\partial \Phi}{\partial x}\right)_A = \left(\frac{\partial \Phi}{\partial y}\right)_A = 0$.

In addition, some virtual nodes were added outside the boundary for calculation and the value of the values of those virtual nodes were calculated by the values of the nodes in the boundary and the nodes next to the boundary.

1) On the upper and lower sides, there is no surface force in X direction, which means $\frac{\partial \Phi}{\partial y} = 0$. So,

$$\Phi_{16}=\Phi_1,\ \Phi_{17}=\Phi_2,\ \Phi_{18}=\Phi_3,\ \Phi_{19}=\Phi_4,\ \Phi_{20}=\Phi_{14},\ \Phi_{21}=\Phi_{15} \quad (2)$$

2) On the left side, $\left(\frac{\partial\Phi}{\partial \mathbf{x}}\right)_{I,H,G,F,E}=3qh$,

$$\Phi_{3,6,9,12,15}=\Phi_{22,23,24,25,26}+2h\left(\frac{\partial\Phi}{\partial x}\right)_{I,H,G,F,E}.\ \text{So,}$$

$$\Phi_{22,23,24,25,26}=\Phi_{3,6,9,12,15}-6qh^2 \quad (3)$$

Where $\frac{\partial\Phi}{\partial x}$, $\frac{\partial\Phi}{\partial y}$ = the sum of the surface forces in the X, Y direction, respectably.

Table 1. Stress boundary condition

Node	$\frac{\partial\Phi}{\partial x}$	$\frac{\partial\Phi}{\partial x}$	Φ
A	0	0	0
B	$3qh$	0	0
C	$3qh$	0	$-3qh^2$
D	$3qh$	0	$-6qh^2$
E,F,G,H,I	$3qh$	0	$-6qh^2$
J	$3qh$	0	$-6qh^2$
K	$2qh$	0	$-3.5qh^2$
L	qh	0	$-2qh^2$
M	0	0	$-1.5qh^2$

3.2 Difference formulas

By applying Taylor series, Difference formulas could be derived 5. Take node 1 as an example:

$$\left(\frac{\partial^4\Phi}{\partial x^4}\right)_1=\frac{1}{h^4}\left[6f_1-4(f_2+f_P)+(f_3+f_R)\right] \quad (4)$$

$$\left(\frac{\partial^4\Phi}{\partial x^2\partial y^2}\right)_1=\frac{1}{h^4}[4f_1-2(f_4+f_M+f_2+f_P)+(f_5+f_L+f_O+f_Q)] \quad (5)$$

$$\left(\frac{\partial^4\Phi}{\partial y^4}\right)_1=\frac{1}{h^4}\left[6f_1-4(f_4+f_M)+(f_7+f_{16})\right] \quad (6)$$

3.3 Difference equations

The stress differential equation is:

$$\frac{\partial^4\Phi}{\partial x^4}+2\frac{\partial^4\Phi}{\partial x^2\partial y^2}+\frac{\partial^4\Phi}{\partial y^4}=0 \quad (7)$$

From equation (4) - (7), the difference equations could be obtained at node 1 with the symmetry considering as,

$$20\Phi_1-8(2\Phi_2+\Phi_4+\Phi_M)+2(2\Phi_5+\Phi_L)+(2\Phi_3+\Phi_7+\Phi_{16})=0 \quad (8)$$

Substituting the known Φ, equation (8) is reduced to:

$$21\Phi_1-16\Phi_2+2\Phi_3-8\Phi_4+4\Phi_5+\Phi_7+4qh^2=0 \quad (9)$$

The same procedure may be adapted to obtain the difference equations of all nodes. Finally, 15 difference equations with 15 unknowns were got at node 1to 15. And Φ of the node1 to 15 was obtained by direct solution (unit: qh^2).

$\Phi_1=-1.44$, $\Phi_2=-1.95$, $\Phi_3=-3.49$, $\Phi_4=-1.27$, $\Phi_5=-1.82$, $\Phi_6=-3.44$, $\Phi_7=-0.99$, $\Phi_8=-1.58$, $\Phi_9=-3.34$, $\Phi_{10}=-0.58$, $\Phi_{11}=-1.19$, $\Phi_{12}=-3.19$, $\Phi_{13}=-0.13$, $\Phi_{14}=-0.62$, $\Phi_{15}=-3.03$.

In addition, for the nodes at boundary and virtual nodes were calculated by equation (2)-(3):

$\Phi_{16}=-1.44$, $\Phi_{17}=-1.95$, $\Phi_{18}=-3.49$, $\Phi_{19}=-0.13$, $\Phi_{20}=-0.62$, $\Phi_{21}=-3.03$, $\Phi_{22}=-9.49$, $\Phi_{23}=-9.44$, $\Phi_{24}=-9.34$, $\Phi_{25}=-9.19$, $\Phi_{26}=-9.03$.

The vertical stress is much greater than the transverse stress, so the vertical stress is the key point what were concerned in this study. The vertical stress can be calculated:

$$(\sigma_y)_L=\frac{1}{h^2}\left[(\Phi_K+\Phi_M)-2\Phi_L\right]=(-3.5-1.5+2\times2)q=-1.0q.$$

$(\sigma_y)_1=-1.02q$, $(\sigma_y)_2=-1.02q$, $(\sigma_y)_3=-0.99q$, $(\sigma_y)_4=-1.10q$, $(\sigma_y)_5=-1.07q$, $(\sigma_y)_6=-0.94q$, $(\sigma_y)_7=-1.18q$, $(\sigma_y)_8=-1.17q$, $(\sigma_y)_9=-0.90q$, $(\sigma_y)_{10}=-1.22q$, $(\sigma_y)_{11}=-1.39q$, $(\sigma_y)_{12}=-0.81q$, $(\sigma_y)_{13}=-0.98q$, $(\sigma_y)_{14}=-1.92q$, $(\sigma_y)_{15}=-0.56q$. $(\sigma_y)_A=0$, $(\sigma_y)_B=-3.0q$, $(\sigma_y)_C=0$, $(\sigma_y)_D=0$, $(\sigma_y)_E=-0.06q$, $(\sigma_y)_F=-0.38q$, $(\sigma_y)_G=-0.68q$, $(\sigma_y)_H=-0.88q$, $(\sigma_y)_I=-0.96q$, $(\sigma_y)_J=-0.5q$, $(\sigma_y)_K=-0.56q$, $(\sigma_y)_M=-0.98q$.

4 Verification

In order to verify the accuracy of the finite difference method, the solution was compared with the solution of the finite element model and in-place field test data.

In the finite element model, q takes as 100N/m for the convenience of comparison. The vertical stress distribution in anchorage zone is show in Figure 4 and Figure 5. It can be seen that the simulation results of stress distribution by FDM and Finite element model is very similar. The vertical compressive stress of concrete in the anchorage zone gradually reduces from the middle to the two sides. The stress assumes double peak type in transverse, and the peak is under load position. In the meanwhile, the stress distribution rules got in the finite element model is in good agreements with it.

Figure 6 shows the error between FDM results and finite element model results in node 1-15. Most of the errors are less than 5%, and the maximum error is 10%. That is to say, the FDM could simulate the internal stress field distribution with high efficiency, while maintaining good accuracy. But for the loading node, the stress error of B point is 70%.

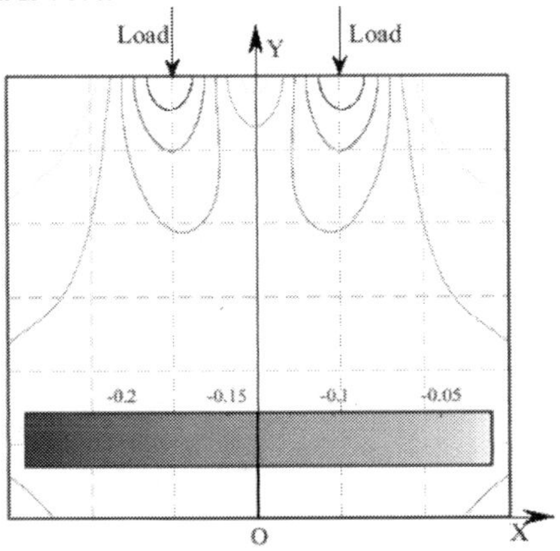

Figure 4. Vertical stress distribution simulated by FDM (unit: MPa)

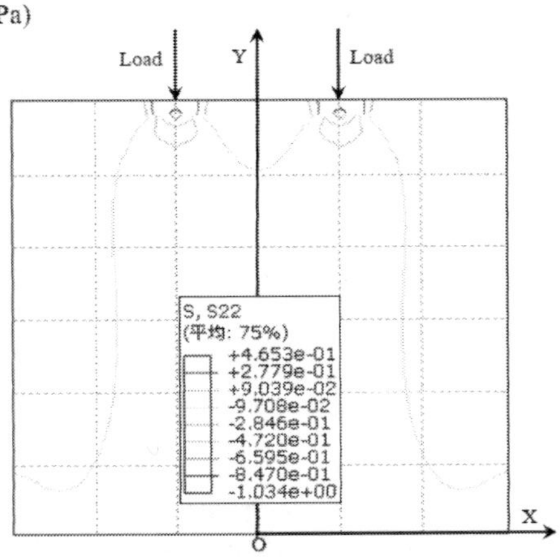

Figure 5. Vertical stress distribution simulated by simulated by Finite element model (unit: MPa)

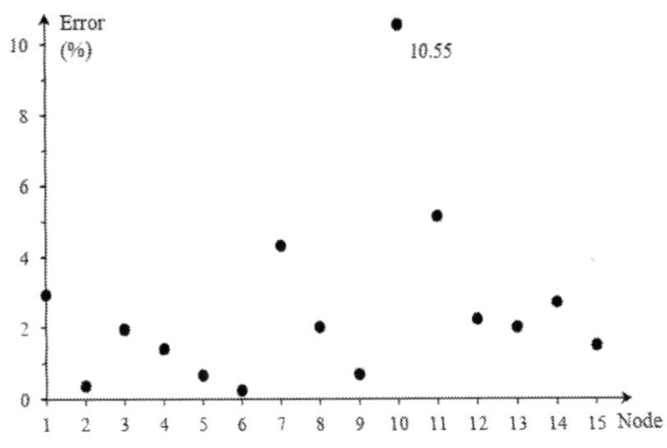

Figure 6. The stress error

The difference simulations of different grid spacing (H/4, H/8, H/10 and H/12) were established respectively to improve the simulation accuracy and analyze the effect of the grid density used for FDM. Figure 7 shows the relationship between the simulation error and the grid spacing. The decrease of internal nodes' stress error was not obvious because of the good accuracy at the beginning. Although the loading point is reduced quickly, it is still larger. So, it needs to further reduce the grids spacing to obtain a more accurate stress in the load position. But it will really affect the work efficiency to add enough grids. Therefore, it is suggested that when the difference method is applied to the simulation, the stress of the loading position should be properly enlarged before interpolating.

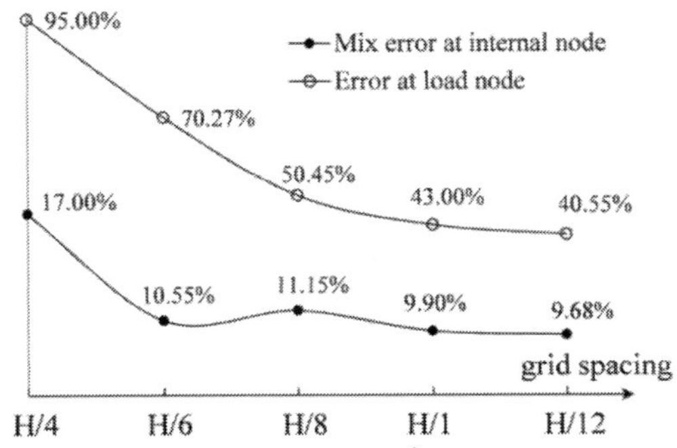

Figure 7. Relationship the simulation error and the grid spacing

A field test was done based on Nanpanjiang Bridge (108+180+108m) to validate the feasibility and effectiveness of the proposed method. Limited by the article space, the simulation results were only validated by correlation with in-place field test at the middle section of the anchorage zone, the comparison result was shown in Figure 8.

The stress distribution simulated by the FDM show a good agreement with the field test data. It is also noted that the simulated stress (Horizontal and Vertical stress) less than the test data. It was analyzed that the changes on stress attenuate quickly with distances and the grid was too little to keep good precision. Increasing the number of the grid could improve the precision at the cost of the equations number increasing, which will .reducing the efficiency of manual calculation. The future study will focus on this problem, and make the difference equation automatically generate with the help of computer. Then it will not only improve the precision by using more grids, but also improve the calculate efficiency.

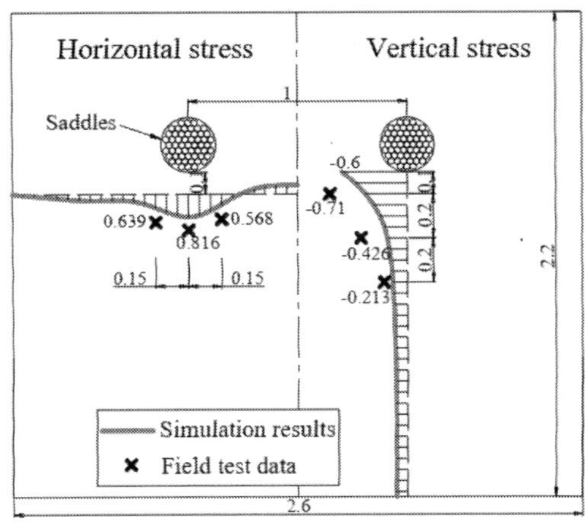

Figure 8. Simulation results and field test data (unit: m; MPa)

5 Conclusion

In this study, FDM was adopted to simulate the concrete in anchorage zone stress field. Based on the comparison of the FEA results, the following conclusions can be drawn: (1) The stress distribution in anchorage zone of extradosed cable stayed bridge simulated by finite difference method is in good agreements with the one simulated by the finite element model, the vertical compressive stress of concrete in the anchorage zone gradually reduces from the middle to the two sides.; (2) The difference method could increase the efficiency with keeping the accuracy of the simulation at most nodes except the load node; (3) The stress distribution becomes uniformity with a distance which equals the anchorage zone's width.

Acknowledgments

This study is supported by the National Science Foundation of PR China (51208056), and the Fundamental Research Funds for the Central Universities (310821161013, 300102218213).

Reference

1. H. Levy, F. Lessman. Finite Difference Equations. J. Physics Today, **14** (4):64-64 (2005)
2. D.M. Liao, B. Zhang, J.X. Zhou, R.X. Liu and L.L. Chen. Using finite difference method to simulate casting thermal stress. J. China Foundry, **08** (2):177-81 (2011)
3. K.P. Chuang and A.S. Veletsos. A Study of Two Approximate Methods of Analyzing Cylindrical Shell Roofs R. University of Illinois Engineering Experiment Station. (1962)
4. J. Shen, S. Lassue and L. Zalewski, et al. Numerical study on thermal behavior of classical or composite Trombe solar walls J. Energy & Buildings, **39** (8):962-974 (2007)
5. D. Appelö and N.A. Petersson. A stable finite difference method for the elastic wave equation on complex geometries with free surfaces J. Communications in Computational Physics, **5** (1): 84-107 (2009)
6. D.S. Griffin and R.S. Varga. Numerical Solution of Plane Elasticity Problems J. Journal of the Society for Industrial & Applied Mathematics, **11** (4):1046-1062. (1963)
7. F.K. Liu, P.Z. Lin and C. Quan, et al. Investigation of Characteristic Parameters of Cable-stayed Bridge with Low Towers. J. Engineering Mechanics, **21** (2):199-203 (2004)
8. S.Q. Tang, W.S. Cai, J.Z. Wang and S.Y. Wang. Section Model Test Research in Saddles of Zhangzhou War Preparation Bridge. J. Bridge Construction, (03): 9-13 (2002)
9. R.R. Guan, J.P. Zhang, A.R. Liu and M.Y. Zhou. Model Test on Main Tower Saddle Section of Extradosed Cable-Stayed Bridge J. Journal of Guangzhou University (Social Science Edition), **4** (5):449-453 (2005)
10. H.W. Zhang and Y.D. Li. Local Stress Analysis of Concrete Around Cable Saddle in Short-Pylon Cable-Stayed Br1dge. J. Railway Standard Design, (01): 42-44. (2009)
11. Z.W. Xiao. Spatial Stress Test and Analysis on the Cable Saddle of Extradosed Cable-stayed Bridge. D. Northeast Forestry University, (2013)
12. Z. Liu, S.P. Meng and H. Zang, et al. Model test and design investigation on saddle deviator zoneof extra-dosed bridge J. Journal of Southeast University (Natural Science), **37** (2):291-5 (2007)

The Framework of Combining Artificial Intelligence and Construction 3D Printing in Civil Engineering

Kang Tan

Dalian University of Technology, College of Civil Engineering, No.2 Linggong Road, Dalian City, P.R.China

Abstract. Construction 3D printing is becoming a significant force driving the transformation of civil engineering industry, but facing many problems for its immaturity. At the same time, the development of artificial intelligence and other advanced technology, especially BIM and virtual reality technology, provides new methods for solving these problems. Based on the defects of construction 3D printing and the advantages of these technology, this paper proposes a framework that combines artificial intelligence with construction 3D printing in five aspects: materials for 3D printing, automation design for 3D printing, digital construction for 3D printing, robots for 3D printing and BIM platform system for 3D printing. BIM platform system associates with the other aspects to work together efficiently during the life cycle of construction.

1 Introduction

In recent years, construction 3D printing is one of effective methods to build in a digital way, which brings great changes to the civil engineering industry. Construction 3D printing adopting the principle of layered materials, based on building information model(BIM), decomposes the model into layers with a certain thickness in one direction, then generates the corresponding algorithm and finally completes the automatic construction by the numerical control system in a specified path. Obviously, construction 3D printing is a multidisciplinary area, containing a variety of technologies such as numerical control, material molding and construction management.

Artificial intelligence is a technology that develops theories, methods, techniques, and applications to extend people's intelligence. Artificial intelligence is quite extensive that consists of various parts such as machine learning and computer vision. In general, the goal of it is to make the machine be capable of doing complex work which requires human intelligence in the past.

BIM is not only a kind of technical tools, but also a kind of management tools. It is designed to integrate all the concerned parts into a system in order to manage efficiently. For many projects, artificial intelligence combined with BIM will provide useful reference for decision making. Nowadays, BIM is mainly applied in design and construction, but we do not obtain information about structural damage or safety evaluation. For this reason, the integration between BIM and artificial intelligence is significantly important. It not only provides a visual platform for construction, but also provides vital reference for the management during life cycle of the infrastructure.

With the continuous development of artificial intelligence and BIM, at the same time, construction 3D printing, data mining, virtual reality(VR) and augmented reality(AR) are also flourishing. How to integrate all these advanced technology into a system in order to drive the transformation of traditional construction and improve efficiency of operation has become a hot research topic. Currently, construction 3D printing is immature, there are still many problems in theory, methods and applications. Artificial intelligence provides a new solution for this problem by applying a number of technologies, such as machine learning and computer vision, to construction 3D printing. This article proposes the framework in five aspects: materials for 3D printing, automation design for 3D printing, digital construction for 3D printing, robots for 3D printing and BIM platform system for 3D printing.

2 The former research of construction 3D printing and artificial intelligence

Construction 3D printing, proposed by Joseph Pegna, was a method for the construction of free-form components through cementitious material layer accumulation and selective solidification. The method derived from emerging field of additive manufacturing processes is investigated for its potential effectiveness in construction automation [1]. Behrokh Khoshnevis, a professor at the University of Southern California, presenting 'Contour Crafting' that enabled the layered printing of concrete by large three-dimensional extrusion devices. Contour crafting (CC) is a layered fabrication technology that has a great potential in automated construction of whole structures as well as subcomponents [2]. Monolite, a company in UK, put forward a new construction 3D printing method called 'D-shape' that selectively hardened and accumulated gravel powder layer by layer. This method divided structure model with horizontal parallel planes in predetermined path, prearranging an apparatus that deposits in alternation a layer of granular material and a liquid binder only in coincidence with the solid portions [3]. Richard Buswell at Loughborough University proposed 'Concrete Printing', which had higher three-dimensional freedom and smaller stacking resolution. In addition, he explores the issues surrounding design, data and process in terms of the impact of rapid manufacturing on conventional manufacturing [4]. Rouhana studied the effectiveness of this technique in reducing construction duration using computer simulation and comparing the conventional construction method and Contour Crafting. The findings suggested that contour crafting can reduce project duration due to automate the construction process [5]. Mechtcherine adapted a concrete pump to deliver material to specific positions autonomously and attached to the boom accurately with the help of custom-developed printing head [6]. Comparing the value stream maps revealed a 60% reduction of production time. Further analysis unveiled reduction in construction cost, especially labour cost, minimization of different types of waste as well as improvement of quality [7].

Artificial intelligence is applied to 3D printing in an intelligent, efficient, high quality and service-oriented production process and the printability of 3D objects can be determined by using machine learning. The prefabrication is accelerated through parallel slicing algorithms and the path planning is optimized intelligently. In the aspect of service and security, intelligent demand matching and resource allocation algorithms enable a cloud service platform and evaluation model to provide clients with an on-demand service and access to the collection of shared resources [8]. Yang presented a rapid design method of product modeling and structure condition selecting based on interactive genetic algorithm and 3D printing technology, which gave a design pattern to meet the function needs [9]. Makagonov proposed a solution to the problem of detecting collisions in the additive process of 3D printing through two methods, which are used in computer vision for recognition: it is the machine learning and tracking based on singular points [10]. Wu detected malicious infill defects in 3D printing process and the data extracted from images was input to two machine learning algorithms, Naive Bayes Classifier and J48 Decision Trees [11].

3 Materials for 3D printing

3.1 Performance requirements of materials

Compared with traditional properties of building materials, 3D printing materials should have controllable setting time, appropriate workability and superior mechanical properties. Application of P.O and R-SAC mortar for 3D printing in construction was investigated by Lin [12]. Different key powder parameters such as particle size distribution, powder bed surface quality, powder true/bulk densities, powder bed porosity, and binder droplet penetration behavior were used to quantitatively evaluate the printability of prepared geopolymer-based material [13]. Lediga presented an investigation into the criteria for an optimum concrete mix design to be used in a 3D concrete printing machine [14]. Shakor demonstrates the properties of different concrete mixture designs, showing their performance when used filament in 3D printing and representing a comparison of the results [15].

According to the past research, concrete and gravel powder are two materials used in construction 3D printing, they will be built on principle of layered stacking. The former is self-solidification accumulation, the latter is bonded with gravel powder. The defect of these materials is that their tensile properties are poor. Once tensile stress exceeds the tensile strength, they tend to crack, which may lead to serious damage to structure. Therefore, materials for 3D printing should have good anti-cracking ability, which can restrict the cracks caused by load. As the printing materials need to be transported through the pipe, they should be of good workability. During the process of printing, the upper layer of materials accumulation will impose pressure on the lower layer. If the lower layer does not form enough strength in a short time, it will deform under the pressure. To conclude, the printing materials should have faster initial setting time and higher initial setting strength.

3.2 Application of artificial intelligence to design 3D printing materials

The use of artificial intelligence to design new materials may lead to the revolution of materials, which is also suitable for 3D printing materials. To meet above requirements, new materials should be developed with these features: high compression and tensile properties, strong crack resistance and toughness, faster initial setting time and higher initial setting strength. Specifically, artificial intelligence is used to design new materials in following steps. Firstly, a large database is set up to record all the current materials information, including materials components, preparation methods and performance characteristics. Secondly, machine learning algorithms are used to select the materials with ideal performance according to the given requirements. Then the materials in database are compared and analyzed, figuring out the best materials after pattern recognition and extraction. The advantage of machine learning algorithms is that the failed experimental data can also be used as input for next, and then the algorithms are refined.

4 Automation design for 3D printing

In the early stage of project, artificial intelligence deals with a large number of relevant case data so as to determine whether the current project has good economic benefits. At the same time, the combination of artificial intelligence and BIM can also help owners to set project target costs through materials classification in order to provide precise budget for target cost control. In this way, materials can be finely managed, the quantity of materials required according to the schedule can be rationally guaranteed, and the amount of material used in each area can be reasonably controlled to reduce the waste caused by second handling.

Artificial intelligence can help architects complete site analysis, planning and architectural design, greatly reducing the workload of architectural design. It not only can be applied to early stages of design, including geographical location, climate, geometry, spatial relationships and conceptual design, but also further design and construction plan. After training a large number of data through related machine learning algorithms, artificial intelligence have ability to design. As long as people accurately describe their expectations, artificial intelligence can generate a variety of options for them to choose. In addition, with the help of rapid data analysis and computer vision, it will be much more efficient than humans.

The current method of structural design is mainly based on traditional buildings, but 3D printing building is a new type of structure. Therefore, many new problems should be taken into consideration in its design. In this case, artificial intelligence can learn the principle and method of structural design through a certain algorithm, and then complete the parametric modeling. After analyzing and comparing a variety of design schemes automatically, reasonable scheme will be selected. For example, taking the design of beam as an example, a number of different schemes can be generated to meet the given requirements such as strength, stiffness and stability.

The mechanical and electrical equipment should not only be designed for long-term operation, but as green as possible. It will be achieved by using artificial intelligence algorithms to operate the equipment both efficiently after learning a lot of data. In this way it can not only realize low energy consumption and high efficiency, but also intelligent controlling and recovery. The realization of these functions requires the application of artificial intelligence algorithms for their ability to control complex events and adjust system parameters.

5 Digital construction for 3D printing

During the construction phase, first of all, artificial intelligence assists people in making schedule plan, integrating data with BIM model, and dynamically displaying schedule plan in BIM model, so as to realize collaborative management. Secondly, BIM can simulate and optimize the construction plan. With features of visualization, each step in the construction process is simulated. For example, the arrangement can simulate the construction process of masonry and obtain the least amount and maximum efficiency. Third, BIM guides workers during construction. It can reduce the time delay and improve the quality of the project by using the BIM with various professional applications. For example, when installing pipelines, workers can use manual roaming or path roaming to acquire properties of components or local areas, so that collision of the actual construction can be found in advance, and the construction progress can be accelerated.

Digital construction for 3D printing is the most challenging and difficult process, involving wide range of aspects and complex conditions. The application of artificial intelligence to the construction process accomplishes the digital construction by adjusting to actual situation. First, artificial intelligence learns a large number of actual construction cases, and then revise the corresponding construction scheme based on the characteristics of the 3D printing. In addition, the robot should be able to automatically identify the characteristics of the environment by using advanced algorithm, then select the best plan in short time. After the completion of construction, the robot should be able to detect the quality of the construction and provide timely feedback.

Specifically, artificial intelligence is mainly used in the material identification and the optimization in construction. Firstly, it will judge the quality and quantity of materials, inspect the surface and internal quality of materials by computer vision algorithm. Secondly, artificial intelligence will be conducted to the dynamic optimization and the adjustment of construction process. If there are contradictions or conflicts in the process of construction, they will be solved by analyzing the different order of construction. Moreover, artificial intelligence can help decision-makers establish decision making model, provide various alternatives, compare and judge various schemes to achieve the maximum benefits. In addition, artificial intelligence can operate multiple elements properly such as time, investment, materials, equipment and space due to it's powerful data computing capabilities, which improves the efficiency of management significantly.

In the completion stage, firstly, we can accurately audit quantities of projects based on changes in construction with the help of BIM and artificial intelligence, knowing the way to increase profits and reduce costs. Secondly, the completion information should be added without loss and omission into BIM model for retrieval and analysis. The intelligent terminal can be used to collect data, and problems in the construction such as quality defects and risks which will be visualized in the BIM model. Thirdly, the BIM model should be maintained and updated regularly. It contains a large amount of data concerned with various specialized sections needed for operation and maintenance to ensure that the information in the model is consistent with the actual situation.

6 Robots for 3D printing

6.1 The defects of robots for 3D printing

The problem of robots for 3D printing exists in the cooperative work of numerical control between software and hardware. The software export control depends on the model, in many cases, it is difficult to print model accurately by CNC software-driven mechanical equipment. The second problem is concerned with printing process. If not based on the analysis of printed objects, the process of 3D printing will be cumbersome and inefficient, it's necessary to consider its characteristics and then optimize the process. In addition, artificial intelligence is needed to learn failed cases to deal with unexpected problems during the printing process. Taking this problem for example, it is not uncommon to see that inadequate layer adhesion results in deficiency of strength and stiffness.

6.2 Improvement and optimization for 3D printing robots

In order to properly plan a complex tool path, one needs to determine the dynamic mechanical response of the extruded filament material as the robot arm is moving in free-space. Numerical simulations are undertaken to illustrate the basic model and machine learning algorithm is developed to optimize the robotic system to deliver a prespecified structure [16]. A research model has been developed by Wolfs Rob, which allows 3D concrete printing to be studied in an efficient way. The model is set up in a parametric environment, combining a structural analysis with an optimization module based on the simulated annealing algorithm [17]. Gardiner employed a 6-axis gantry robot with interchangeable end effector tooling to coarsely 3D print and subsequently mill finish complex moulds for concrete at a construction scale [18].

To solve the problem, artificial intelligence algorithm is applied for model slicing and path planning to generate multiple schemes, then choose the best one with high efficiency and quality. After that, 3D model is imported into the numerical control system. The system deals with the model in two steps: a series of parallel 2D models are used to fit the 3D model called hierarchical slice processing, the sliced profile is converted into a printing path called path planning. After generating mechanical movement instructions, the model will be stacked layer by layer. Therefore, under the control of the planned path, robots ultimately complete the construction. In the process of printing optimization, BIM can be transformed into several processes. By using artificial intelligence algorithms and learning a large number of different process plans, the best solution will be selected after analyzing and comparing the advantages and disadvantages of various plans. This will be efficient to optimize the material consumption and reduce construction time. In addition, with the help of compute vision algorithms, robots can analyze the structure they are printing, once found wrong, they can immediately fix the problem. For instance, if there is a rough surface of the structure which can be automatically observed, robots will take measures to ensure smoothness soon.

7 BIM platform system for 3D printing

Integration of BIM method with 3D printing modeling will be effective for energy efficiency, better design, cost reduction and isolation of structure [19]. Ding explored a new BIM-based automated construction system (BIMAC) including composition of BIMAC, detail of execution setup, data issues, filling layer algorithm and shows printed example results, which are the highly customized building components [20]. Davtalab proposed a software platform for data retrieval and analysis from BIM models and utilizing it efficiently during various stages of the process [21]. Correa studied the impact of BIM in bridging the gap between design and construction and proposed a computer-aided system that translates a generic architectural project in a set of pieces to be fabricated with 3D printers [22].

System integration should be based on BIM platform, usually using the Revit API for secondary development, hardware integration should consider the matching and the convenience of regular system expansion. Network technology should be used in system integration, considering requirements of network topology, the system should adopt TCP/IP reference model.

It shows that the interaction is based on the BIM model, and the information in database is concerned with the components and objects in the model. Specifically, the four parts of main information exported into the BIM model are: design information, construction information, operation and maintenance information. Design information includes: architectural design, structural design and electrical equipment design. Construction information includes: construction process, construction quality and construction monitoring. Operation and maintenance information includes: structure and component status warning and assessment and electrical equipment status, structural testing and repair records.

The BIM model is based on cloud and can be connected with the BIM platform through the Internet and mobile Internet. All data obtained after being processed is stored in the BIM cloud, which can be inserted, extracted and modified in BIM in time to update the BIM model automatically. User interface interaction can adopt Client/Server mode or Brower/Server mode and should be used for network security measures such as encryption and hierarchical authorization. The system can grant remote access monitoring data and reports through the internet. Monitoring information management includes importing of monitoring information, exporting data in the form of graphics or files, the query of historical monitoring information and the visualization of monitoring information.

Monitoring and warning information are displayed in BIM, switching from different perspectives, transforming coordinate range, rotation, cutting and synchronization updating. The warning status information is classified into good, general and poor, by using three colors to represent the status: green, yellow and red. BIM cloud platform should be able to work with other software and have nice human-computer interaction. When one of the objects is operated in BIM, it can show the object's original data, monitoring data, changes in the curve. When loading the BIM model, it can realize parametric loading according to the user's concerns and highlight important details. At the same time, the conclusion and warning information of BIM monitoring system, combined with virtual data, are expressed by virtual reality technology (VR) and augmented reality technology (AR).

8 Summary

From above, artificial intelligence and BIM certainly provides some useful methods for the problems existing in construction 3D printing. It is foreseeable that the construction 3D printing combined with BIM and artificial intelligence will be the trend in future. Admittedly, this paper only proposes the application framework of the combination, there are still many aspects which is not discussed in depth. If they can be successfully applied to the construction 3D printing, it is not far to realize automatic construction.

References

1. J. Pegna, Automation in Construction, **5**, 427 (1997)
2. K. Behrokh, Automation in Construction, **13**, 5 (2004)
3. E. Dini, R. Nannini, M. Chiarugi, WO, EP1868793(2007)
4. R. A. Buswell, A. Thorpe, R. C. Soar, A.G.F. Gibb, Automation in Construction, **17**, 923(2008)
5. C. M. Rouhana, M. S. Aoun, F. S. Faek, M. S. Eljazzar, F. R. Hamzeh, *22nd Annual Conference of the International Group for Lean Construction* (IGLC, Oslo, 2014)
6. V. Mechtcherine, V. N. Nerella, BFT Magazine, **82**, 150(2016)
7. F. E. Sakka, F. Hamzeh, *Conference of the International Group for Lean Construction* (IGLC, Heraklion, 2017).
8. J. Yang, Y. Chen, W. Huang, Y. Li, *International Conference on Automation and Computing* (IEEE, Huddersfield, 2017)
9. Z. Yang, J. Shen, *Seventh International Symposium on Computational Intelligence and Design* (IEEE, Hangzhou, 2015)
10. N. G. Makagonov, E. M. Blinova, I. I. Bezukladnikov, *Young Researchers in Electrical and Electronic Engineering* (IEEE, Moscow, 2017)
11. M. Wu, V. V. Phoha, Y. B. Moon, A. K. Belman, *International Mechanical Engineering Congress and Exposition* (ASME, Phoenix, 2016)
12. J. C. Lin, X. Wu, W. Yang, R. X. Zhao, L. G. Qiao, *2nd International Conference on New Material and Chemical Industry* (IOP, Sanya, 2017)
13. M. Xia, J. Sanjayan, Materials & Design, **110**, 382(2016)
14. R. Lediga, D. Kruger, Solid State Phenomena, **263**, 24(2017)
15. P. Shakor, P. Shakor, *International Symposium on Automation and Robotics in Construction* (IAARC, Taipei, 2017)
16. T. I. Zohdi, Computer Methods in Applied Mechanics & Engineering, **331**, 343(2018)
17. R. Wolfs, T. Salet, *22nd Workshop of the European Group of Intelligent Computing in Engineering* (EG-ICE, Eindhoven, 2015)
18. J. B. Gardiner, S. Janssen, N. Kirchner, *International Symposium on Automation and Robotics in Construction* (IAARC, Auburn, 2016).
19. M. Sakin, Y. C. Kiroglu, Energy Procedia, **134**, 702(2017)
20. L. Ding, R. Wei, H. Che, Procedia Engineering, **85**, 123(2014)
21. O. Davtalab, A. Kazemian, B. Khoshnevis, Automation in Construction, **89**, 13(2018)
22. F. Correa, F. Correa, *International Symposium on Automation and Robotics in Construction* (IAARC, Auburn, 2016)

Simplified Design Equations for a Class of Rhombic Auxetic Plates

Teik-Cheng Lim

School of Science and Technology, Singapore University of Social Sciences, Singapore

Abstract. Equations for solving the deflection and bending moments of rhombic plates by exact method are known to be highly tedious. A set of simplified equations is developed for design purposes of such simply supported plates under uniform load. Curve-fitting from exact data allows the deflection and its second derivatives, evaluated at the plate centre, to be expressed in greatly simplified and yet sufficiently accurate empirical models for thin rhombic plates. Using the simplified model, it is shown that the maximum bending moment can be reduced by using auxetic materials. By including the effects of shear deformation for thick rhombic plates, it is demonstrated that the ratio of shear-to-bending deformation decreases as the rhombic plate approaches a square shape and as the plate's Poisson's ratio becomes more negative.

1 Introduction

The Poisson's ratio of materials is normally assumed to be positive, with a value of v = 0.3 being typically assigned for isotropic materials. Although the Poisson's ratio of anisotropic solids has no bounds [1], it has been shown that the Poisson's ratio in the case of isotropic materials are bounded within $-1 \le v \le 0.5$ [2]; solids with negative Poisson's ratio have been studied [3]. A number of auxetic materials applications have been suggested; interested readers are referred to related reviews [4-6] and a monograph [7] for further details of such materials. Investigations into the performance of plates with the use of auxetic materials have been carried out [8-11]. Although studies on auxetic plates have been done on rectangular [12-14], circular [15-17], triangular [18-20], elliptical [21], sectorial [22], hexagonal [23] and regular polygonal [24] plates–no work has been done on auxetic rhombic plates. This paper aims to provide a set of design equations that is simple to execute and sufficiently accurate for a class of rhombic plates, which are simply supported and uniformly loaded, with special emphasis on the use of auxetic materials.

2 Preamble to thin rhombic plates

With reference to Fig.1 for a simply-supported rhombic plate of thickness h, sides a and complementary angles of 2α and $(\pi - 2\alpha)$ under uniformly distributed load of q_0, the deflection w^K of such plate within the context of Kirchhoff plate theory

$$w^K = w_0 + w_1 \tag{1}$$

at any point (r, θ) is given by the summation of a particular solution w_0

$$w_0 = \frac{q_0}{D}\frac{r^4}{64}\left(1 - \frac{4\cos 2\theta}{3\cos 2\alpha} + \frac{\cos 4\theta}{3\cos 4\alpha}\right) \tag{2}$$

and a complementary function w_1

$$w_1 = \frac{q_0}{D}\sum_{m=1}^{\infty}(a_m + b_m r^2) r^{\lambda_m + 1}\cos(\lambda_m + 1)\theta \tag{3a}$$

$$w_1 = \frac{q_0}{D}\sum_{n=1}^{\infty}(a_n + b_n r^2) r^{\lambda_n}\cos\lambda_n\theta \tag{3b}$$

according to Morley's [25] or Warren's [26] nomenclature, respectively, where

$$\lambda_m = \frac{(2m-1)\pi}{2\alpha} - 1 \tag{4a}$$

$$\lambda_n = \frac{(2n-1)\pi}{2\alpha} \tag{4b}$$

while the plate's flexural rigidity is

$$D = \frac{Eh^3}{12(1-v^2)} \tag{5}$$

in which E is the Young's modulus and v is the Poisson's ratio of the plate material. One may note that the plate's Poisson's ratio in Eqs.(1) and (2) are manifested only through the flexural rigidity.

The methods for evaluating the coefficients b_m and a_m by Morley [25] and b_n and a_n by Warren [26] are tedious, and are summarized in the Appendices A1 and A2, respectively. Warren [26] appreciated the fact that the coefficients b_m and a_m are chosen by Morley [25] so as to minimize the integral square error over AB. This leads to numerous integrals which are tedious to evaluate, and then finally to the solution of M by M simultaneous linear equations for determining these coefficients. Warren [26] proposed the method of "point matching" that requires satisfying Eqs.(A11) and (A12) exactly at M discrete points along AB, which leads directly to two sets of M by M simultaneous linear equations for determining the said coefficients. Both ways are nevertheless tedious to the designer, and a simplified model would be of great practical importance. To provide simplified design equations for thin rhombic plates, results from Morley [25] are used for generating semi-empirical models that are very convenient and yet sufficiently accurate. Thereafter, the simplified design equations for thin plates form the basis for evaluating the shear deformation in thick rhombic plates.

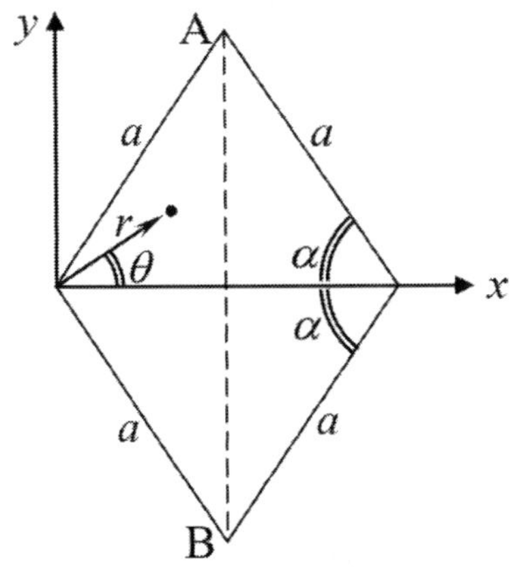

Figure 1. Schematic diagram of a rhombic plate.

3 Semi-empirical modelling for thin rhombic plates

Perusal to Eqs.(1) to (4), as well as Appendices A1 and A2, shows that the plate deflection is a function of load q_0, location (r,θ), half angle of obtuse corner α as well as the Young's modulus, plate thickness and Poisson's ratio. If one were to express the plate deflection in the form of $wD/(q_0 a^4)$, then this dimensionless plate deflection is independent from the material properties (E,ν) of the plate. Based on Morley's [47] numerical results for the dimensionless deflection as well as the bending moments at the center of the plate for $\nu = 0.3$ and $45° \le \alpha \le 75°$, a curve-fitting exercise yields the following semi-empirical models for the plate center:

$$\frac{w_{\max}^K D}{q_0 a^4} = \{1.907 \quad 3.578 \quad 2.069 \quad 3.397\} \begin{Bmatrix} +10^{-7}\alpha^3 \\ -10^{-5}\alpha^2 \\ +10^{-3}\alpha^1 \\ -10^{-2}\alpha^0 \end{Bmatrix} \tag{6}$$

$$\frac{D}{q_0 a^2} \begin{Bmatrix} \dfrac{\partial^2 w_{\max}^K}{\partial x^2} \\ \dfrac{\partial^2 w_{\max}^K}{\partial y^2} \end{Bmatrix} = \begin{bmatrix} 3.768 & 3.861 & 6.039 \\ 0 & -1.071 & -8.604 \end{bmatrix} \begin{Bmatrix} -10^{-5}\alpha^2 \\ +10^{-3}\alpha^1 \\ -10^{-2}\alpha^0 \end{Bmatrix} \tag{7}$$

where α in the semi-empirical models are expressed in degrees. Since Eqs.(1) to (3) are independent from the plate's Poisson's ratio when non-dimensionalized, it follows that this independence holds for Eqs.(6) and (7). Nevertheless the Poisson's ratio plays its role in the bending moment, as described in Eq.(8).

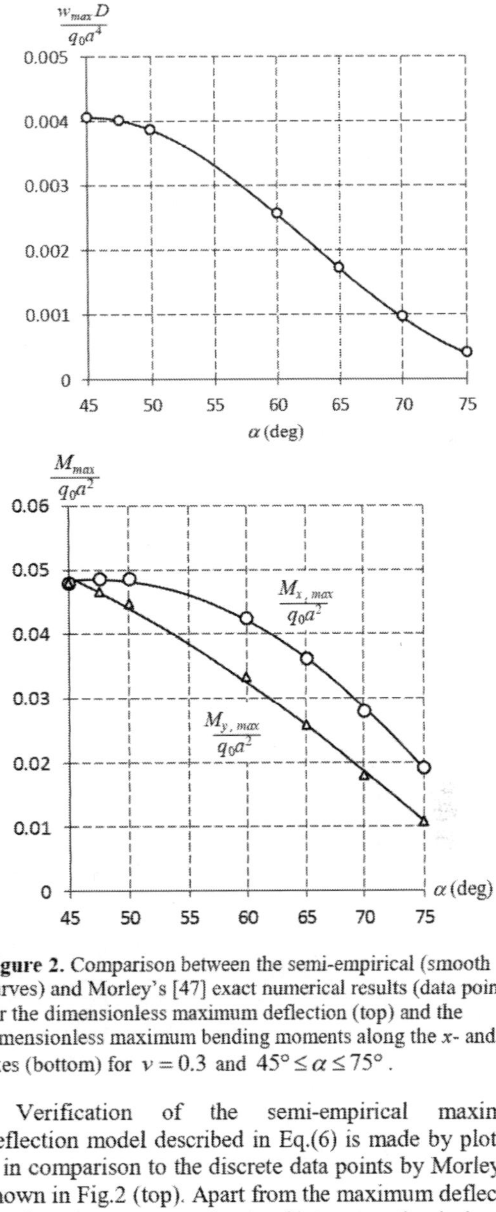

Figure 2. Comparison between the semi-empirical (smooth curves) and Morley's [47] exact numerical results (data points) for the dimensionless maximum deflection (top) and the dimensionless maximum bending moments along the x- and y-axes (bottom) for $\nu = 0.3$ and $45° \le \alpha \le 75°$.

Verification of the semi-empirical maximum deflection model described in Eq.(6) is made by plotting it in comparison to the discrete data points by Morley, as shown in Fig.2 (top). Apart from the maximum deflection, another plate response that is of interest to the designer is the maximum stresses. Since the bending stress is related to the bending moment, i.e. $\sigma_{\max} = 6M_{\max}/h^2$, it suffices to evaluate the maximum bending moments. Substitution of Eq.(7) and $\nu = 0.3$ into

$$\begin{Bmatrix} M_{x,\max} \\ M_{y,\max} \end{Bmatrix} = -D \begin{bmatrix} 1 & \nu \\ \nu & 1 \end{bmatrix} \begin{Bmatrix} \dfrac{\partial^2 w_{\max}^K}{\partial x^2} \\ \dfrac{\partial^2 w_{\max}^K}{\partial y^2} \end{Bmatrix} \tag{8}$$

allows the semi-empirical bending moments to be compared with the exact results, as plotted in Fig.2 (bottom), in which $M_{\max} = M_{x,\max} > M_{y,\max}$ for $\alpha > 45°$. Having established their validity, the semi-empirical models are used for plotting the loci of dimensionless maximum bending moments in the conventional $0 \le \nu \le 0.5$, moderately auxetic $-0.5 \le \nu < 0$ and highly auxetic $-1 \le \nu < -0.5$ regions. Reference to Fig.3 shows that, with all other parameters fixed, the maximum bending moment–and hence the maximum bending stress–is reduced by the extent of material auxeticity.

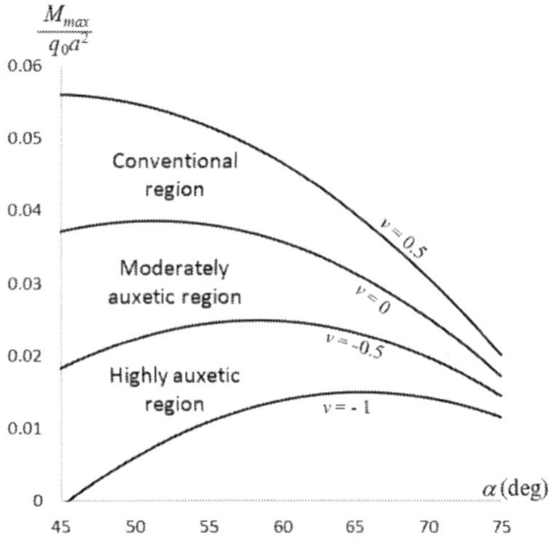

Figure 3. Loci of dimensionless maximum bending stress for conventional, moderately auxetic and highly auxetic regions.

4 First-order shear model for thick rhombic plates

The use of Kirchhoff plate theory, which accounts for the plate deflection on the basis of plate bending, in the preceding section is valid for thin plates. Thick plates experience pronounced transverse shear deformation. This section attempts to evaluate the shear deformation in thick rhombic plates using first-order shear deformation theory (FSDT). The FSDT by Mindlin [27] assumes constant out-of-plane shear strain, thereby incurring constant out-of-plane shear stress, across the plate's thickness. For this reason, a shear correction factor κ is required to correct the shear modulus G; the Mindlin plate deflection w^M has been shown to be related to the Kirchhoff plate deflection w^M as [28]

$$w^M = w^K + \frac{\mathrm{M}}{\kappa G h} \tag{9}$$

where the Marcus moment of the Kirchhoff plate is

$$\mathrm{M} = -D\nabla^2 w^K . \tag{10}$$

The deflection relationship between the Kirchhoff and Mindlin plate theories described in Eq.(9) can be expressed in a more tractable form as (e.g. [23,24])

$$w^M = w^K - \frac{h^2}{6\kappa(1-\nu)}\nabla^2 w^K . \tag{11}$$

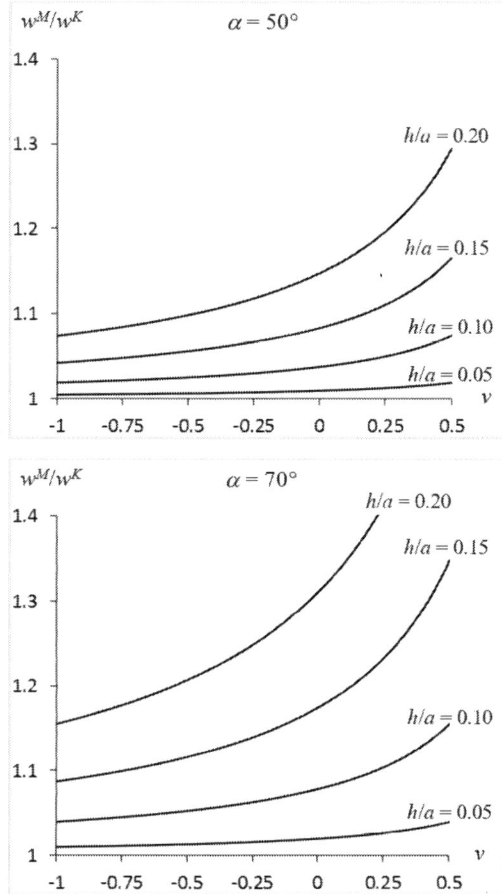

Figure 4. Plots of Mindlin-to-Kirchhoff plate deflection ratio versus Poisson's ratio at various relative plate thickness for $\alpha = 50°$ (top) and $\alpha = 70°$ (bottom).

From the semi-empirical models of Eqs.(6) and (7), as well as the shear correction factor $\kappa = 5/6$, the ratio of Mindlin-to-Kirchhoff plate deflection ratio at the plate center can be obtained as

$$\frac{w_{\max}^M}{w_{\max}^K} = 1 - \frac{h^2/a^2}{5(1-\nu)}\frac{f(\alpha)}{g(\alpha)} \tag{12a}$$

where

$$f(\alpha) = 3.768\times10^{-5}\alpha^2 - 2.79\times10^{-3}\alpha - 0.02565 \tag{12b}$$

and

$$g(\alpha) = 1.907\times10^{-7}\alpha^3 - 3.578\times10^{-5}\alpha^2 + 2.069\times10^{-3}\alpha - 0.03397 \tag{12c}$$

Figures 4 and 5 furnish plots of Mindlin-to-Kirchhoff plate deflection ratio at the plate center for moderately sharp ($\alpha = 50°$) and significantly sharp ($\alpha = 70°$) rhombic plates for various relative thickness h/a and Poisson's ratio ν. A full range of w^M/w^K plotted with respect to the rhombic plate's half angle α for various Poisson's ratio at $h/a = 0.2$ is given in Fig.6.

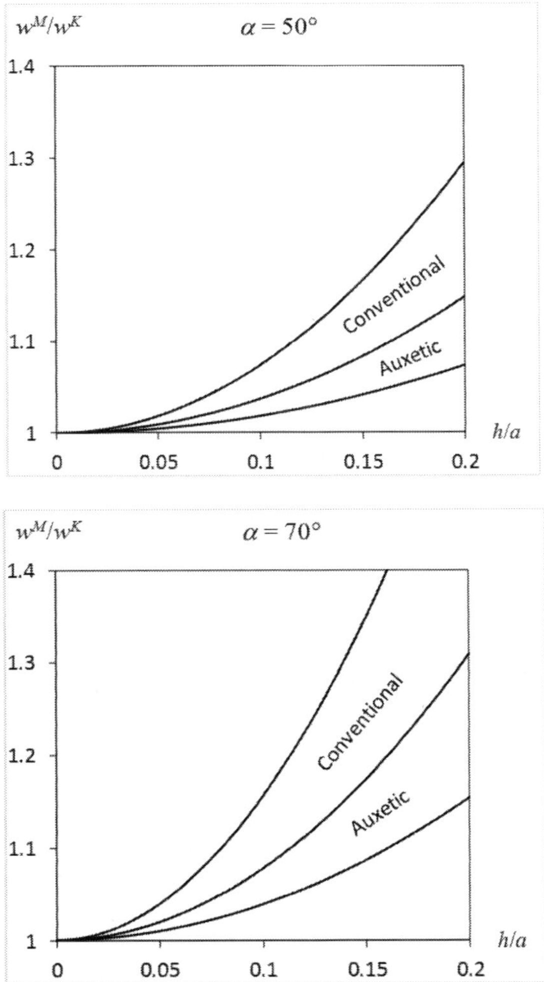

Figure 5. Demarcation of auxetic region from conventional region at $\alpha = 50°$ (top) and $\alpha = 70°$ (bottom).

As expected, the extent of shear deformation as a ratio of bending deformation diminishes with decreasing plate thickness; more interestingly, the shear-to-bending ratio decreases (a) as the plate Poisson's ratio becomes more negative, and (b) as the plate approaches a square, i.e. $\alpha \to 45°$.

5 Conclusions and recommendations

A set of simplified design equations for describing the deflection and second derivatives at the centre of a simply-supported uniformly-loaded thin rhombic plates has been developed herein by back calculation from numerical results based on exact model in dimensionless forms, which are independent from the Poisson's ratio.

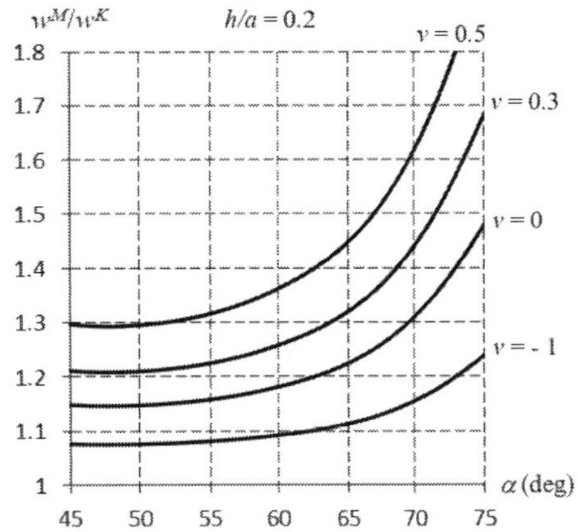

Figure 6. A family of w^M / w^K curves plotted against the half angles α of rhombic plates at dimensionless thickness of $h/a = 0.2$.

Results from the thin plate theory reveals that auxeticity decreases the dimensionless maximum bending moments, and hence the maximum bending stress, of thin rhombic plates. Extending this thin plate model to thick rhombic plates, it was found that the ratio of shear deformation vis-à-vis bending deformation increases with the dimensionless plate thickness *h/a*, half angle α, and Poisson's ratio. These observations suggest that the analysis of thick rhombic plates can be further simplified to thin plate theory if the thick rhombic plates are more regular and/or made from auxetic materials.

References

1. T.C.T. Ting, T. Chen, Quart. J. Mech. Appl. Math. **58**, 73 (2005)
2. L.D. Landau, E.M. Lifshitz, *Theory of Elasticity*, Pergamon Press, Oxford (1970)
3. K.W. Wojciechowski, Mol. Phys. **61**, 1247 (1987)
4. K.K. Saxena, R. Das, E.P. Calius, Adv. Eng. Mater. **18**, 1847 (2016)
5. T.C. Lim, Phys. Status Solidi RRL **11**, 1600440 (2017)
6. R.S. Lakes, Ann. Rev. Mater. Res. **47**, 63 (2017)
7. T.C. Lim, *Auxetic Materials and Structures*, Springer, Singapore (2015)
8. T.C. Lim J. Eng. Mech. **136**, 1176 (2010)
9. T.C. Lim, Smart Mater. Struct **25**, 054008 (2016)
10. P. Sobieszczyk, M. Majka, D. Kuźma, T.C. Lim, P. Zieliński, Phys. Status Solidi B **252**, 1615 (2015)
11. L. Boldrin, S. Hummel, F. Scarpa, D. Di Maio, C. Lira, M. Ruzzene, C.D.L. Remillat, T.C. Lim, R. Rajasekaran, S. Patsias, Compos. Struct. **149**, 114 (2016)
12. T.C. Lim, J. Mater.: Des. Appl. **227**, 111 (2013)
13. T.C. Lim, J. Eng. Mater. Technol. **136**, 031007 (2014)
14. T.C. Lim, Int. J. Mech. Mater. Eng. **11**, 13 (2016)
15. T.C. Lim, J. Mech. **29**, 121 (2013)
16. T.C. Lim, 2014 J. Eng. Mater. Technol. **136**, 021007 (2014)

17. T.C. Lim, J. Eng. Mater. Technol. **138**, 041011 (2016)
18. T.C. Lim, Mech. Based Des. Struct. Mach. **44**, 514 (2016)
19. T.C. Lim, J. Eng. Mater. Technol. **138**, 014501 (2016)
20. T.C. Lim, Smart Mater. Struct. **25**, 054001 (2016)
21. T.C. Lim, J. Mech. **32**, 413 (2016)
22. T.C. Lim, Phys. Status Solidi B **254**, 1600784 (2017)
23. T.C. Lim, Phys. Status Solidi B **254**, 1700014 (2017)
24. T.C. Lim, Proc. Inst. Civil Eng. - Eng. Comput. Mech. **170**, 167 (2017)
25. L.S.D. Morley, Quart. J. Mech. Appl. Math. **15**, 413 (1962)
26. W.E. Warren, AIAA J. **2**, 166 (1964)
27. R.D. Mindlin, J. Appl. Mech. **18**, 31 (1951)
28. C.M. Wang, W.A.M. Alwis, J. Eng. Mech. **121**, 1383 (1995).

Appendix A1: Evaluation of coefficients b_n and a_n by Morley (1962)

The calculations leading to the solutions of coefficients b_n requires

$$\delta \int_{-a\sin\alpha}^{a\sin\alpha} \left[\frac{\partial}{\partial x} \nabla^2 w^K \right]^2_{x=a\cos\alpha} dy = 0 \tag{A1}$$

that leads to a system of M equations

$$\begin{bmatrix} A_{11} & A_{12} & \cdots & A_{1M} \\ A_{21} & A_{22} & & \\ \vdots & & \ddots & \\ A_{M1} & & & A_{MM} \end{bmatrix} \begin{Bmatrix} b_1 \\ b_2 \\ \vdots \\ b_M \end{Bmatrix} = \begin{Bmatrix} A_1 \\ A_2 \\ \vdots \\ A_M \end{Bmatrix} \tag{A2}$$

where

$$A_{mn} = A_{nm} = 16(\lambda_m + 2)(\lambda_m + 1)(\lambda_n + 2)(\lambda_n + 1) \times \int_{-a\sin\alpha}^{a\sin\alpha} [r^{\lambda_m + \lambda_n} \cos\lambda_m\theta \cos\lambda_n\theta]_{x=a\cos\alpha} dy \tag{A3}$$

and

$$\frac{A_m}{4(\lambda_m + 2)(\lambda_m + 1)} = \frac{1}{2}\left(\frac{1}{\cos 2\alpha} - 1 \right) \int_{-a\sin\alpha}^{a\sin\alpha} [r^{\lambda_m + 1} \cos\lambda_m\theta \cos\theta]_{x=a\cos\alpha} dy , \tag{A4}$$
$$= \frac{(-1)^{m+1}(1-\cos 2\alpha)a^{\lambda_m + 2}}{\lambda_m + 1}$$

while calculations leading to the solutions of coefficients a_n requires

$$\delta \int_{-a\sin\alpha}^{a\sin\alpha} \left[\frac{\partial w^K}{\partial x} \right]^2_{x=a\cos\alpha} dy = 0 \tag{A5}$$

that leads to another system of M equations

$$\begin{bmatrix} A_{11}' & A_{12}' & \cdots & A_{1M}' \\ A_{21}' & A_{22}' & & \\ \vdots & & \ddots & \\ A_{M1}' & & & A_{MM}' \end{bmatrix} \begin{Bmatrix} a_1 \\ a_2 \\ \vdots \\ a_M \end{Bmatrix} = \begin{Bmatrix} A_1' \\ A_2' \\ \vdots \\ A_M' \end{Bmatrix} \tag{A6}$$

where

$$A_{mn}' = A_{nm}' = \frac{A_{mn}}{16(\lambda_m + 2)(\lambda_n + 2)} \tag{A7}$$

and

$$A_m' = (\lambda_m + 1) \int_{-a\sin\alpha}^{a\sin\alpha} [\chi r^{\lambda_m} \cos\lambda_m\theta]_{x=a\cos\alpha} dy \tag{A8}$$

in which A_{mn} is given by Eqs.(A3) while

$$\chi = \frac{r^3}{16}\left[\left(1 - \frac{1}{\cos 2\alpha} \right)\cos\theta + \frac{1}{3}\left(\frac{1}{\cos 4\alpha} - \frac{1}{\cos 2\alpha} \right)\cos 3\theta \right] + \sum_{m=1}^{M} b_m \{ (\lambda_m + 2)\cos\lambda_m\theta + \cos(\lambda_m + 2)\theta \} r^{\lambda_m + 2} \tag{A9}$$

Appendix A2: Evaluation of coefficients b_n and a_n by Warren (1964)

The coefficients a_n and b_n are chosen ideally so as to satisfy symmetry boundary conditions along the diagonal AB (See Fig.1). These conditions require

$$\left. \frac{\partial}{\partial x} \nabla^2 w^K \right|_{AB} = 0 \quad , \quad \left. \frac{\partial w^K}{\partial x} \right|_{AB} = 0 \tag{A10}$$

which give

$$\left\{ a\cos\alpha\left(1 - \frac{1}{\cos 2\alpha} \right) + 8\sum_{n=1}^{\infty} \lambda_n (\lambda_n + 1) b_n r^{\lambda_n - 1} \cos(\lambda_n - 1)\theta \right\}_{AB} = 0 \tag{A11}$$

and

$$\left\{ \frac{r^3}{16}\left[\left(1 - \frac{1}{\cos 2\alpha} \right)\cos\theta + \frac{1}{3}\left(\frac{1}{\cos 4\alpha} - \frac{1}{\cos 2\alpha} \right)\cos 3\theta \right] + \sum_{n=1}^{\infty} \lambda_n a_n r^{\lambda_n - 1} \cos(\lambda_n - 1)\theta + \sum_{n=1}^{\infty} b_n r^{\lambda_n + 1} \left[(\lambda_n + 1)\cos(\lambda_n - 1)\theta + \cos(\lambda_n + 1)\theta \right] \right\}_{AB} = 0 \tag{A12}$$

Landscape Design of River Channels in Small Towns based on Regional Culture-Case Study of Landscape Design of Duohua River

Man Lu[1,2], Chuyuan Wen[1] and Chuanyue Wang[1]

[1]School of Civil and Transportation, Hohai University, Xikang Road, Nanjing, China.
[2]School of Architecture, Southeast University, Sipailou, Nanjing, China.

Abstract. Water is the vital source of life while river channels constitute the arteries of the earth. As a key element of eco-structure of the Landscaping system, river plays a significant role in social life. Current river channels in small towns are managed only to achieve the traditional functions of flood prevention and irrigation while river channel landscape construction is confronted with problems such as replicating landscapes and degradation of ecological environment. River channel landscape should receive increasing attention after taking treatment measures including sewage interception, dredging, base soil treatment and revetment repair. This study offers ideas and strategies for landscaping designs of rivers in small towns that realize the value of regional culture in contemporary times on the basis of taking root in the concrete settings of the locations and the form of the design subjects in light of landscaping practices of Duohua River in Longli County which comb through regional elements and selectively keep some of them in the principle of respecting local history, natural conditions and culture in an effort to build a river landscape that represents regional culture and meets the requirements for sustainable development so that both the function and landscaping beauty can be better integrated.

1 Introduction

Current river channels in small towns are managed only to achieve the traditional functions of flood prevention and irrigation while river channel landscape construction is confronted with problems such as replicating landscapes and degradation of ecological environment. River channel landscape should receive increasing attention after taking treatment measures including sewage interception, dredging, base soil treatment and revetment repair. Compared with other countries, our country attaches importance to irrigation function of river channels but paid little attention to ecological environment construction. Recently with the increasing awareness of the importance of ecological balance and environment protection, the concept of ecological treatment of river channel gradually receives more attention. The treatment of river channel environment expands from mere treating water pollution to recovering ecological environment, wetland and ecological construction. However, the landscape design that integrates river channel treatment and landscape design remains at the experimenting stage. The article seeks to explore the application of urban river channel landscape design in real projects through real engineering examples.

2 Regional culture and river channel landscape

2.1. The concept of regional culture

The meanings of "region" are quite broad and can be classified in terms of national borders, weather patterns, ethnic cultures and etc. The boundary of region is determined by a particular shared characteristic. Therefore, the size of region varies. Culture, in broad sense, is the sum of material and spiritual achievement. Regional culture refers to the integration of natural landscape and history including its climatic conditions, topography, hydrogeology, flora and fauna, history, cultural, human activities, behaviours and etc. [1]. Regional culture is an organic whole that bases itself on the local traditional culture but actively absorbs foreign cultures.

2.2 The scope of river channel landscape

River channel, defined as the passageway of the river, refers to the entirety composed of river body, sandbars, beaches and the water across river shorelines. As the core of the river, it consists of three parts including river bed, mud flat and river terraces. Landscape refers to the combination of natural elements and cultural phenomena, the total of sights that from specific locations [2]. therefore, the landscape of river channel cannot be constituted by river alone but also the land area that interacts with the river as the core, generally referred to as the landscape within the radius of the river. The river channel landscape that is studied in this paper includes the river channel, buffer zone and the river bank slope [3]. Therefore, it is the regional complex composed of all natural elements and culture phenomena that are seen, heard and felt in the whole riparian zone.

2.3 The importance of integrating regional culture with river channel landscape design

The river in the city nourished humans living along its banks, pushing the progress of civilization of the society. The landscape elements of the banks contain abundant regional culture characteristics. However, in the context of globalization and cultural convergence, local culture and regional culture are bound to suffer different degrees of setback in their development. Current designers in China neglect the use of historical and cultural legacy and failed to settle down their minds and study landscape designs that feature local and regional features. Therefore, protecting and developing regional culture becomes ever more important in river channel landscaping practices. Amid current studies on river channel landscape design both at home and abroad, culture receives increasing attention with ideas of "rivers full of dynamics and culture", "river that reflects culture and new customs", "river featuring the balance of vegetation, water and culture" and etc. [4]. The formation of each river is accompanied by deep culture development. For the river channel landscape design to embody regional features, designers are required to materialize what is abstract and embody the regional culture as a whole so that such common ills as discontinuance of historical and culture ties and landscape convergence can be corrected and river landscape space that is full of vitality can be constructed.

3 Location analysis of Duohua River of Longli of Guizhou

3.1 Geographic background of Duohua River of Longli of Guizhou

Longli County, administered under Qiannan Buyi and Miao Autonomous prefecture of Guizhou Province, is located in the middle stretch of Miaoling Mountain range in the center part of Guizhou and also in the northwest of Qiannan Buyi and Miao Autonomous prefecture. Within the county, hills, low and average height mountains and river valleys are distributed in intervals from south to north. With neighbors of Guiding County, Fuquan County in the east, Huishui County in the south and Guiyang City in the west and north, it stands 1,080 meters above the sea level. It has a subtroFigureal monsoon humid climate with annual average temperature of 14.8 °C. By the end of 2012, the total population of the county is 224,266. It has multiple ethnic groups including Han, Buyi, Miao with ethnic minority of 91,754. The most famous tourist destinations in the county are Longjiashan Forest Park and Guan Mountain.

3.2 Analysis of the planning of Duohua River of Longli of Guizhou

In line with the principle of "treating the eco sensitive area with extra care", this paper proposes the idea of coexistence and thriving of urban development and regional culture- "sing praise to the landscape of Longli and restore the dynamics of the river".

Make the urban river system of Longli the bond that links fine tourist resource in Longli, the window through which Figureturesque scenery of small towns is displayed and the soil on which diverse ethnic group cultures thrive. Duohua River flows through the core section of the future new city and connects with national forest park. Fine river resources are valuable non-renewable resources. To protect the water quality and ecologic security of rivers in the city and realize its value, the old model of transforming rivers through engineering alone must be abandoned in planning.

By developing LID urban pipe network system, splitting rain from sewage, expanding the green area on both banks of the river and strictly managing the skyline and controlling on all other details, the future Duohua River becomes a "green waving zone" that circles the new Longli town where birds sing and flowers blossom. By connecting art gallery in the north, entrance to Longjia Mountain Forest Park, SenlinXipan waterfront town featuring local culture and Longjiashan Garden & reservoir, Duohua River landscape corridor system featuring regional culture can be constructed.

3.3 Analysis of status quo of Longli Duohua River of Guizhou

The surrounding area, in general, has beautiful natural scenery with tableland and mountains that are rich in topographic changes, minor interference from human activity, dense vegetation, dams and alluvial flat that are atheistically pleasant to the eye. But in some areas, the vegetation is in disarray or the revetment is damaged with poor landscaping effects. After meticulous analysis, the main problems are (1) embankment: currently in a poor state with some damaged; (2) bridge: lack of exchange across the river while some bridges hamper flood discharge and lack in locally suitable forms;(3) water bodies: small size of water body in some areas with severe stony desertification; (4) water surface: monotonous water surface morphology and lack of wavy changes; (5) Shoreline: aesthetically unsatisfying skyline and scenery line; (6) buildings: lack of cultural ties between newly built buildings and existing ones and no cultural adherence.

4 Regional strategies in landscape planning and design of Duohua River of Longli of Guizhou

The regional strategy is formed on the basis of field research and surveying of Duohua river basin with landscape elements at all levels extracted in the principle of respecting the geographic textures of the field.

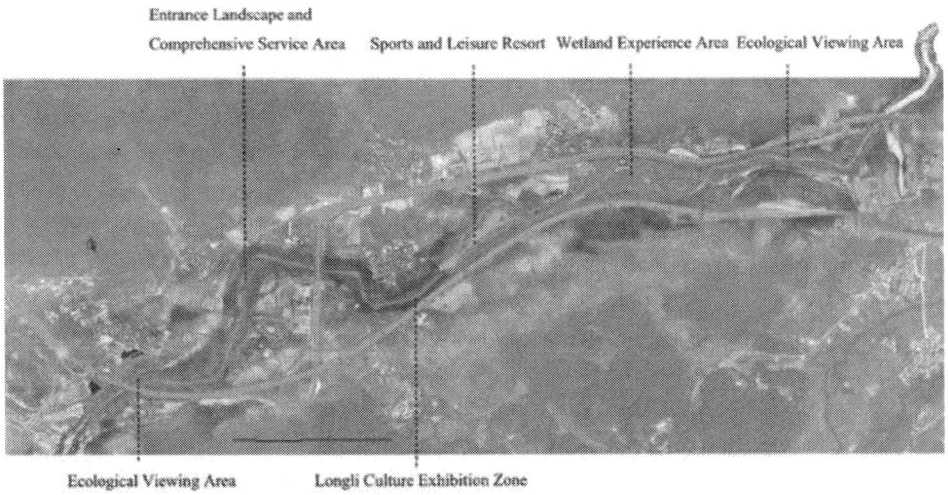

Figure 1. Landscape function zone chart

4.1 The establishment of the theme of the landscape of Duohua river channel

The theme of landscape design is the essence of the landscape of Duohua River. Not only the visual effects of the landscape need to be stressed, but also, more importantly, genius loci of the river channel must be brought out to reflect local culture. In the principle of keeping layout continuity and integrity, the themes of the sub zones should be determined by landscape structure, loci function, shoreline design and human behavior (Figure 1). Just as each symphony has a theme which is both the core and purpose of music, so does landscape design. To establish distinctive themes for each river section can help form culture axis with profound connotations in combination with local culture, so that people could have sense of pleasure and excitement as people move along the river and encounter different themes of landscape.

4.2 Eco-friendly revetment design that respects local features

Revetment is the basis of the landscape that surrounds the water body and also a critical component of the water conservancy and landscaping project. Firstly, the transformation of eco-friendly revetment is to meet higher standards of flood prevention. Local materials are used to treat the existing banks so that the embankment built earlier can be restored to a status that is nature friendly. Secondly, expand the water surface on both sides of river channel where conditions apply to overcome the shortcomings of hampering flood discharge and monotonous landscape so that the landscape space for both flood storage and sighting can be created. Thirdly, construct near-shore wetland space to smooth the shoreline where water space borders land area to achieve the dual functions of flood storage and adding landscape layers.

In light of the general requirements on the landscape of Duohua River, different revetment designs are adopted according to the cross section of the river channel. (1) Original type revetment: in sections where gradient is originally high or center area is large, consider keeping it in original status and consolidate the banks with the roots of aqua plants, as shown in Figure 2. (2) Natural type revetment: For sections where slope is steep and erosion is severe, not only grow plants, but also use local stones or wood for river bottom protection and enhance the ability of banks to fend off flood, as shown in Figure 3. (3) Oblique stone wall type revetment: in section where requirement on flood resistance is high and the center area small, gravity retaining wall must be built in a layer by layer approach. Then use materials such as reinforced concrete to make certain of its flood resistance ability on the basis of natural type revetment, as shown in Figure 4. (4) Steps stretching into water-type revetment: while ensuring flood resistance, water interaction space of the revetment varies as water level changes, as shown in Figure 5. (5) The viewing platform revetment: sightseeing platform combines with natural revetment, which ensures stability without causing major damage to its original look, as shown in Figure 6.

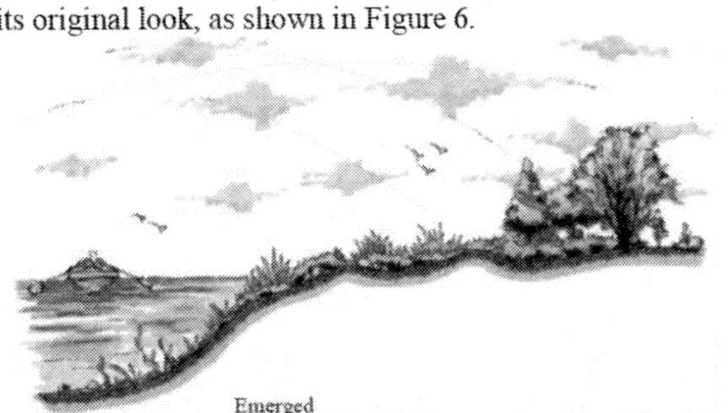

Figure 2. Original type revetment

Figure 3. Natural type revetment

Figure 4. Oblique stone wall type revetment

Figure 5. Sightseeing platform type revetment

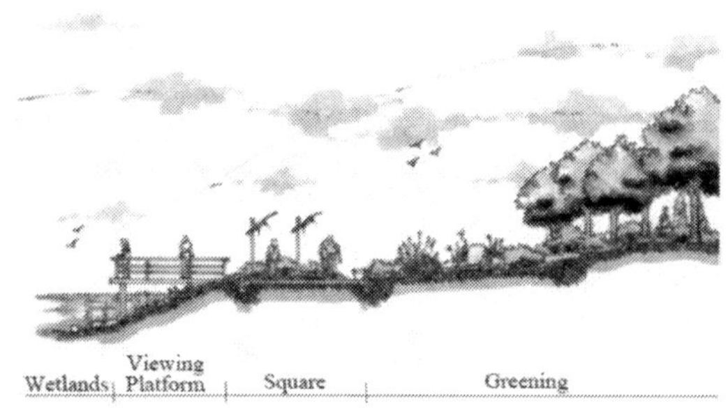

Figure 6. The viewing platform revetment

4.3 Regional transformation of buildings

The plane layout of buildings along the banks of Duohua river can be divided into three types: (1) Disperse type: with natural scenery as the dominant, buildings are distributed in clusters according to their functions so that they can integrate with the greater natural view as a whole; (2) Linear type: building complex is distributed along the shoreline. Also consider the combination and alternation of buildings with different function with special attention paid to sight line and skyline management. (3) Concentration type: in key areas of river channel, building complex is constructed with open space as the center to form building groups characteristic of the region so as to invoke a strong sense of belonging. To achieve that, manage the morphology of building along the banks from building skyline, sightline and morphology: (i) architecture skyline and sightline: as the margin of man-made landscape, the spatial form of skyline comes from the overlapping of the outlines of multiple clusters of buildings with those of the vegetation on mountains so that it becomes local architectural landscape; (ii) architecture morphology management: As a link in green infrastructure, the density and volume of buildings along the banks must be strictly limited to ensure small number of buildings and population and protect the ecological environment with due consideration on leaving the bottom of the building empty to save enough sight space for visitors to appreciate the river channel space together.

Furthermore, building in nearby villages of Duohua River should maintain the original architectural style that features the use of local materials such as wood and stones, as shown in Figure 7 Longli Mingju Residential Architecture. Given the functions and status quo of the existing buildings, different management practices are adopted, such as demolishing old buildings to restore the original look, transforming structure of the building where conditions apply and adopting the traditional facade design, cleaning the exterior facade and apply full scale of vegetation. Create local style buildings that both meet the needs of modern life and are rich in culture connotations such as buildings shown in Figure 8.

Figure 7. Longli Mingju Residential Architecture

Figure 8. Longli Mingju Residential Architecture

4.4 Regional expression of landscape features

Landscape features should be both entertaining and harmonious with the surrounding environment. Historic heritage must be protected. Local history, folk customs, art, religion, architectural style and even ethnic culture should be unearthed, understood and presented to materialize the abstract local culture, integrate culture into landscape and use landscape features to embody regional culture. Embody the artistic conception and theme through landscape features by certain artistic methods so as to make the landscape full of vitality. For example, the characteristic Tusi (system of appointing minority hereditary headmen) Culture of Longli (Figure 9 Tusi relic architecture sculpture) is presented through sculpture and other features displayed in cultural exhibiting zone in artistic methods including refinement and abstraction. Another example, Longli Wushan cave painting cluster is located near the living area of Wushan men of Guyuan Village of Gujiao Township of Longli County. It has in total more than 400 paintings stretching hundreds of meters with the size, number, themes, content, the rocks it covers boasting greatest in Guizhou. On the basis of protection of the authenticity of Wushan cave painting, the images on the cave paintings are adopted in designs of landscape features and barricades with the use of stones that are in abundance there. In addition, lotus roots are commonly seen in Duohua River which can also be conceptualized into a symbol and adopted in the design of interesting but live landscape features (Figure 10 Lotus root shaped landscape feature).

Figure 9. The characteristic Tusi Culture of Longli

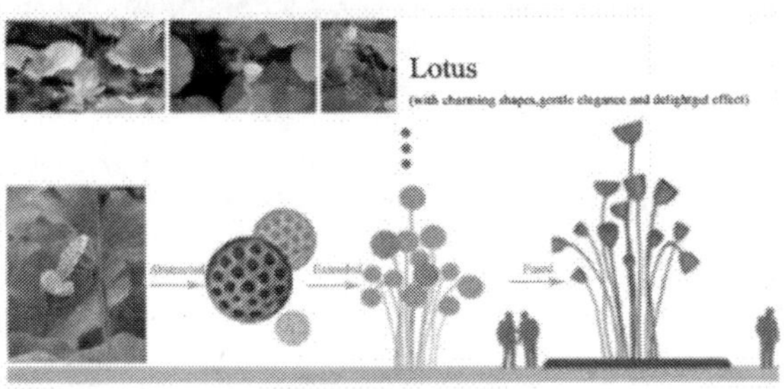

Figure 10. Lotus root shaped landscape feature

4.5 Selections of regional plants

Local plants are first choice when it comes to selection of plants in planning and design. Prior to this design, research and survey have already been conducted on existing aqua plants, plants living on the border of water and bank- solidifying plants. Vegetation that was damaged and contaminated is also restored. In addition, the growth of the individual plants and the stability of clusters must be taken into consideration when choosing the plant species for landscaping zone along the river banks. Plant species that are survival of the fittest in natural selection over a long term should be chosen while isolated or group planting are adopted based on the requirement of different microenvironment.

5 Summary

Landscape design is not merely rockeries and water design on the paper. More important agenda for the designers is to enable landscape to express local features, accurately interpret the loci and respect local culture in the context of accelerated urbanization. Many of the current landscape lacks in diversity of function usually with only visual effects achieved. Without local features and sense of loci, the landscape fails to interact with people and are reduced to negative space gradually, which leads to the eventual desolation of so called public open green space. This paper hopes waterfront landscape planning and design of Duohua River of Longli of Guizhou can set examples for future local landscaping projects and inspire more creative design concepts and more humane design thinking.

References

1. Man Lu, Chuanyue Wang, Landscape design of urban river channels based on regional culture–Case study of landscape design of Jiangdongzhi River in Xiaoshan of Zhejiang Province. J. *Architecture and culture*, **6**, 79(2013)
2. *Hamerton Philip.Landscape*, (Roberts Publication, Boston 1885)
3. Gao Yang, *Evaluation and Treatment Study on Ecological Situation of Rivers and Streams in Suburb Beijing*: doctoral dissertation submitted to Beijing Forestry University (2009)
4. Chendong Zhu, Preliminary exploitation of landscape planning of river channels in major cities. J. *Beijing Water Conservancy*, **3**, 7 (2002)
5. Anxia Wang, Hesheng Ye, Application of Regional Cultural Symbols in Urban Sculpture Design. J. *Ability and Wisdom*, **17**, 194 (2010)
6. Fengbin Yuan, The Excavation and Figure of the landscape Feature of the Cities' waterfront Area. D. Wuhan University (2005)

Monitoring of Daily Temperature Effect on Deck Deformation of Concrete Arch Bridge

Tang Yingying[1,2], Wang Yong'e[1], Niu Yanwei[1], Chen Hong[1] and Huang Pingming[1]

[1] *School of Highway, Chang'an University, 710064 Xi'an, China*
[2] *School of Science, Chang'an University, 710064 Xi'an, China*

Abstract. Temperature makes greatly differences on concrete in many aspects, including stress, strain and deformation, especially for arch concrete bridge structure. Some software can compute theoretical deformation, stress and bending moment of bridge structure due to temperature load, such as Midas Civil. In recent years, in order to learn about conditions of structure, many bridges have installed the healthy monitoring system. In this research, the monitoring data lasting approach a year was obtained from Haierwa Bridge, a concrete truss arch bridge, in Hebei province. The bridge belongs to Xuanda Highway, which undertook the main traffic flow of coal transportation. The proportion of heavy vehicles, exceeding 50t, ups to 30%, different with common highway. The objective of this paper is to monitor and analyse the deformation of concrete truss arch bridge due to air temperature change. Firstly, results show that the daily tendencies of temperature and bridge vertical deformation at mid-span, L/4 and arch foot in the winter and summer. The linear relationship was calculated between the temperature and deformation of critical sections based on the monitoring data. In addition, the finite element model was established to calculate the theoretical value, and further compared with practice values.

1 Introduction

Solar heat transfer Heat is divided into convection, conduction and radiation, whose effect can be replaced by temperature in some degree in practice. Temperature change makes greatly differences on concrete in many aspects, including stress, strain and deformation. Some software can compute deformation, stress and bending moment of bridge structure due to temperature load, such as Midas Civil. Many scholars have studied influences on the concrete box girder, tower and other components of bridges. Researches not only concentrated on the temperature distribution of bridge components, but also investigated the behaviour of the bridge structures due to temperature change. Recently, in order to further investigate behaviour of bridges, the healthy monitoring system is more and more popular. Generally, the system was composed by several subsystems to monitor global performance, local performance and surrounding environment conditions.

On the one hand, many researches focus on the temperature distribution of critical bridge component. J. Suzuki et al. 1 focus on the pre-stressed concrete box-girder bridge. They found that air temperature was higher than surface temperature during the night and opposite during the daytime. Lu Zhifang et al. 2 analysed environment temperature and relative humidity. The research provides a new reasonable and reliable method for creep effect analysis of steel-concrete composite continuous beam bridge and concrete bridge. Tian et al. 3 utilized the numerical method to get the quasi-static deformation of railway bridge resulted from temperature. Peiretti et al. 4 conducted experiment and found that temperature was uniform, linear variation and nonlinear distribution in concrete bridge deck based on a four-year temperature monitoring data. It is obvious that temperature makes great differences on stress, strain, bending moment, deformation and other behaviour of bridge structure.

On the other hand, temperature is an important and continuous input variable not only for bridge components but also for the entire structure. Shashi Moorty and Charles W. Roeder 5 analysed temperature-dependent bridge movement, suggesting that concrete bridges sometimes are designed for smaller temperature ranges and corresponding thermal movements than may be expected in practice. In Chinese, Zhou Guangdong and Yi Tinghua 6 proved that thermal load played a significant role in structural condition evaluation and bridge design procedure, especially for those statically indeterminate bridges and cable-supported bridges. Therefore, the temperature load has been taken into account in General Specifications for Design of Highway Bridges and Culverts 7. Other papers further studied the correlation between available thermal variables and deformation induced by thermal loading. Battista et al. 8 established a temperature-displacement model according to four type temperature data. It can be applied to

https://doi.org/10.1051/matecconf/201820601011

estimate the bridge girder displacement. Xia et al. 9 analysed the temperature and structure data of a long-span suspension bridge. There was a linear correlation between temperature, displacement and stresses. GD zhou 10 proposed a general approach for modeling closed-form thermal correlation of deformation applied in a long-span arch bridge according to long-term monitoring data.

It is certain that temperature changing plays a role in deformation of concrete bridges. However, the attention pay on the whole bridge structure was not enough for all kinds of bridge structure, such as concrete truss arch bridge. The objective of this paper is to analyse the effect of temperature on bridge deck deformation, based on the temperature and deformation monitoring data of Haierwa Bridge. Moreover, the finite element model was established to calculate the theoretical value, and further compared with practice values.

2 Bridge background

2.1 Bridge overview

Xuanda Highway, namely xuanhua - datong highway, is one of the main channels for transporting coal from Datong to other places in Hebei Province. The Highway is started from xuanhua, via a Shaohua Ying, YangYuan to shanxi provincial boundary, whose total length is 127.8 km. Xuanda Highway is the first mountainous area heavy highway in Hebei province. It is a two-way four-lane highway, with a speed limit range of 90 km/h to 110 km/h.

The Haierwa Bridge (as figure 1) is located in the Xuanda Highway in Hebei province, nearly by Xuanhua. The number of the starting mileage pile is K168+ 400, and the terminal mileage pile is K168+ 612. The total length of bridge is 191.6m, and the width is 24.5m. It was completed and opened to traffic in December 2000. Bridge design span is combination of 14m+138m+10m+2 ×8m, which design load is car-20, trailer-120. The main span is composite truss arch bridge with pre-stressed concrete structure. The bridge is two-way four-lane. Especially, lane 3 and lane 4, from Datong to Xuanhua direction, are heavy traffic lanes of the heavy highway, regarded as the research object of the paper.

Figure 1. Haierwa Bridge

2.2 Monitoring systems

The healthy monitoring system was conducted to monitor the behaviours of the Haierwa Bridge. The bridge monitoring system consists of monitoring center and five monitoring subsystems to monitor dynamic characteristics, environment, traffic, deformation and stress-strain. It is total of 101 monitoring points installed on Haierwa Bridge. There were 24 stress-strain sensors installed along the main arch ring, which can also monitor the structure temperature. The research main used data collected from 5 displacement sensors and a corresponding air temperature sensors. Five displacement sensors were installed at five places on bridge deck, respectively above arch foot, 1/4 span and mid-span. Sensors numbers are 771018, 771012, 771011, 771004, 769204 as figure 2. While the air temperature sensors were placed on corresponding positions at 174411 (as figure 2) to monitor the air temperature, replaced the effect of solar heat transfer. The monitoring data to analyse the relationship between deformation and solar heat transfer is from March to November 2015, lasting for three seasons. The writer selected one day in the summer and winter respectively to show the daily distribution of temperature and deformation.

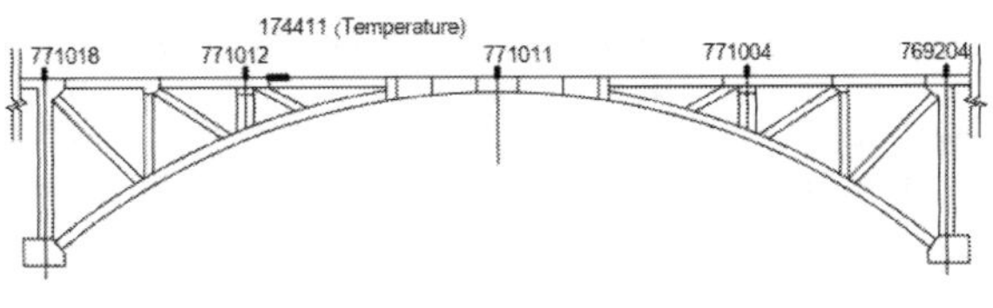

Figure 2. Sensors arrangement on Haierwa Bridge

The deck deformation sensors and air temperature sensors placement of Haierwa Bridge were as follows:

Table. 1 Sensors placement

Type of sensor	Number of sensor	Placement
Deformation sensors	771018	Arch foot on the side of Datong
	771012	L/4 on the side of Datong
	771011	Mid-span
	771004	L/4 on the side of Xuanhua
	799204	Arch foot on the side of Xuanhua
Air temperature sensors	174411	Bridge deck of L/4 on the side of Datong

3 Thermal actions and deck deformation

The air temperature sensor monitoring data was shown in figure 3, which presented the temperature tendency from March to October in 2015. The lowest temperature was -2 ℃ in March, and the highest temperature was 26.5 ℃ between June and August. Therefore, the temperature daily effect analysis used March 15th and August 5th monitoring data. The biggest different temperature value was 28.5, which can be regarded as temperature load of finite element model.

 https://doi.org/10.1051/matecconf/201820601011

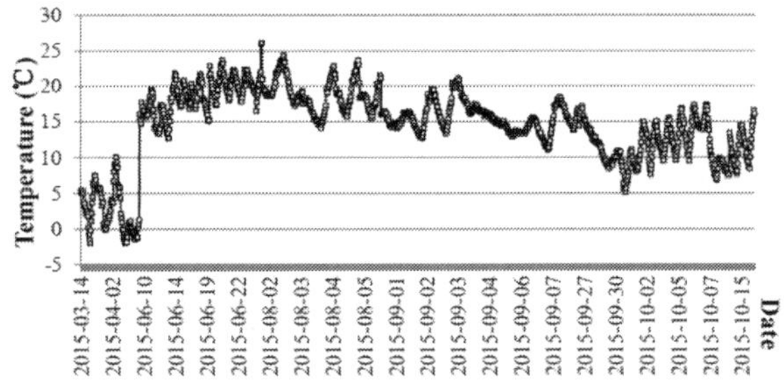

Figure 3. Temperature tendency in 2015

3.1 Thermal actions in the winter

Five critical positions were selected to analysis, including arch feet, 1/4 span and mid-span. It is symmetrical for the concrete truss arch bridge, so the following figures were presented by three critical sections. The deformation and air temperature conditions at mid-span was shown figure 4. The downward deflection is positive for all monitoring data. The curve of deflection is more stable than temperature.

The monitoring data on March 15th was selected to present the daily distribution in the winter. The picture shows that the deck deformation change of L/4 and arch foot was similar to the air temperature tendency. While the downward deflection of bridge mid-span decreased with temperature increasing in the winter, and increased with temperature declining. But there was a little delay between deformation and temperature. As for arch foot and 1/4 span, the deflection upward increased with temperature decreasing. The different deformation values were respectively 11mm, 6mm, 4mm for mid-span, L/4, arch foot, when temperature dropped about 12℃.

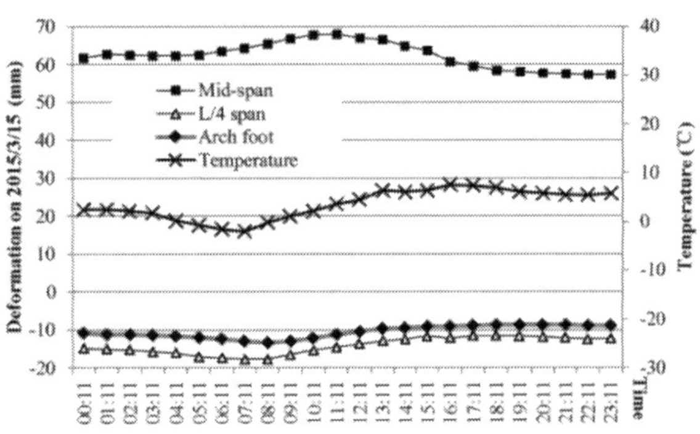

Figure 4. Deformation of arch bridge on 2015/3/15

3.2 Thermal actions in the summer

The monitoring data on August 5th was selected to present the daily distribution in the summer. The picture shows that the deck deformation change of L/4 and arch foot was similar to the air temperature tendency. However, the downward deflection of bridge mid-span decreased with temperature increasing in the summer. It needs to note that, the downward deformation is later than temperature decreased. As for arch foot and 1/4 span, the deflection upward increased with temperature increasing. The different deformation values were respectively 13mm, 8mm, 8mm for mid-span, L/4, arch foot, when temperature dropped about 10℃. In a word, the amplitude of temperature variation was smaller than that of winter, but the amplitude of deformation variation was larger in the summer.

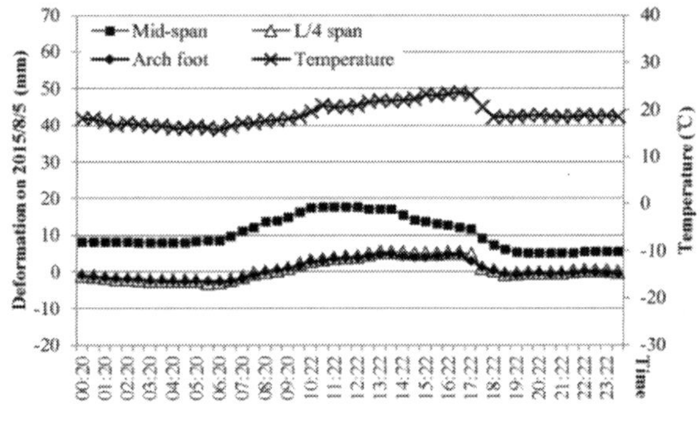

Figure 5. Deformation of arch bridge on 2015/8/5

3.3 Linear relationships between temperature and deformation

According to figure 4 and figure 5, there were strong relationship between temperature and deformation, which need to further investigation. All monitoring data in 2015 was spotted to regress the linear relationship. As shown in figure 6, the temperature effect on deck deformation was biggest at mid-span. Although the correlation coefficient was not enough, only was 0.75. As for arch foot and 1/4 span (figure 7 and 8), the deflection upward increased with temperature decreasing. The relationship between deflection and temperature change is linear, and the correlation coefficient can exceed 0.9 when the temperature higher than 10℃. In other words, the linear relationship was weak under 10 ℃. The slopes of temperature-deflection are 1.1, 1.3 and -2.3 at arch foot, 1/4 span and mid span, respectively. It proved that temperature effect on mid-span deformation was much more than L/4 and arch foot.

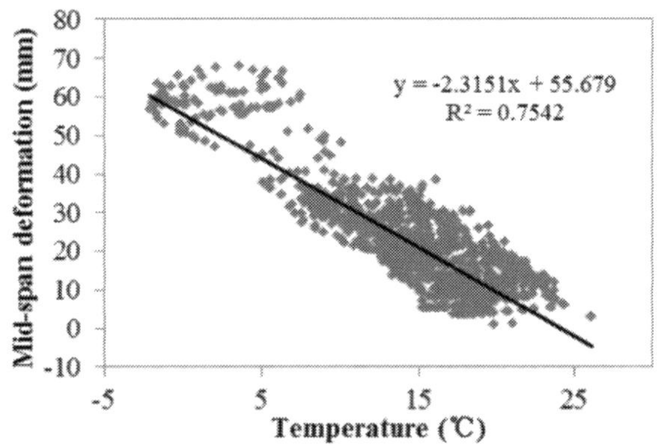

Figure 6. Relationship between deformation and temperature of mid-span

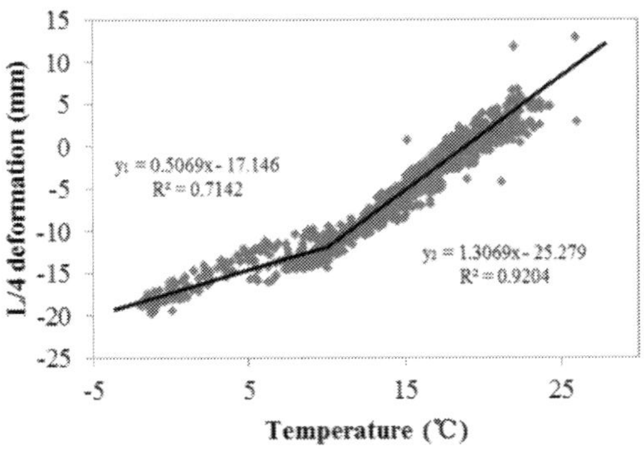

Figure 7. Relationship between deformation and temperature of L/4

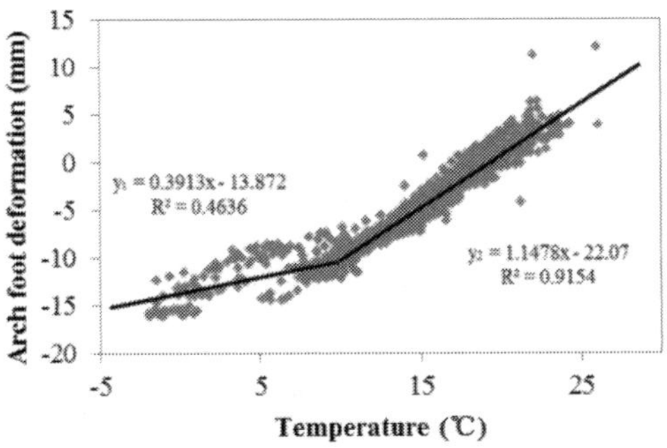

Figure 8. Relationship between deformation and temperature of arch foot

3.4 Verification of Finite Element

Haierwa Bridge is a concrete truss arch bridge. The main span is up to 138m. There are double columns on the main arch at L/4 span, and expansion joint is between the two columns, which is different from common truss arch bridges. In order to verify the rule got from monitoring data above, the write establish the finite element (FE) model of the bridge. The FE model has 408 nodes and 414 beam elements. The concrete selected C50 to simulate. At the same time, the corresponding temperature load was applied to Midas Civil to obtain theoretical deflection value as figure 9. Because the traffic load and self-weight of bridge are stable respectively and always existence, the write only take temperature load into the model. Compared with the monitoring data, most of the deformation of mid-span bridge is led by the change of temperature.

As table 2 shown, the deformation growth rate was obtained due to temperature changing based on regression equation. It is obvious that the deformation of mid-span is far more than L/4 span and arch foot. The conclusion is similar to the FE model. But the deformation growth rate is different. In all monitoring data, the maximum temperature is 26.5 ℃ and the minimum temperature is -2 ℃. Therefore, the biggest different value of air temperature can up to 28.5 ℃. According to regression equation of monitoring data, the deck of mid-span declined 6.598cm, while the theoretical was 8.623cm. The deflection is stable from arch foot to 1/4 span with temperature changing, which is not exceed 1 cm in the finite element model. However, the monitoring data were about 2~3cm, and the direction is different. In other words, when temperature dropping, the practical downward deformation will much over than the nominal data at L/4 and arch foot. However the mid-span different deformation is opposite. In conclusion, there were about 2~3.5cm different deformation value between theoretical and monitoring data. The different may be led by shrinkage, creep, material property, traffic conditions and others, which need to be further studied.

Table. 2 Comparison of monitoring data and theoretical value

Correlation of temperature and deformation		**Mid-span**	**L/4 span**	**Arch foot**
Monitoring data	t≧10℃	d=-2.3151Δt	d=1.3069Δt	d=1.1478Δt
	t<10℃		d=0.5069Δt	d=0.3913Δt
FE model		d=-3.028Δt	d=-0.2517Δt	d=-0.2567Δt
ΔT(℃)		-2~26.5	-2~26.5	-2~26.5
Monitoring value (cm)		6.598	-2.751	-2.300
FE theoretical value (cm)		8.623	0.717	0.732
Different value (cm)		2.027	3.468	2.080

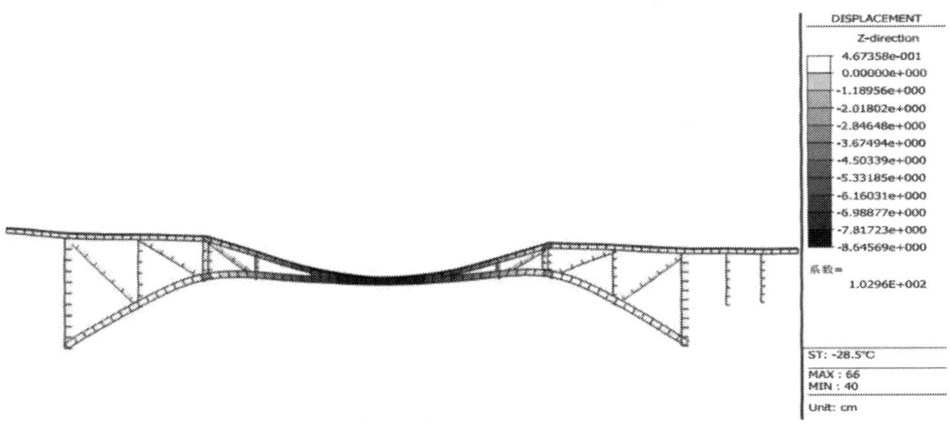

Figure 9. Deformation of structure when temperature dropped

4 Conclusions

In conclusion, air temperature change has influence on structure deformation in concrete arch bridge, especially for mid-span. There is a strong linear relationship between the structure deformation and temperature. Combined with the finite element model, conclusions were as follows:

Deformation curves were more stable than the temperature curve. For instance, some small fluctuations of temperature were not reflected at the same time.

The deck deformation varied with temperature. The deck deformation change of L/4 and arch foot was similar to the air temperature tendency, while there was a little delay between deformation and temperature. The downward deflection of bridge mid-span decreased with temperature increasing, and increased with temperature declining. In addition, the amplitude of temperature variation was smaller than that of winter, but the amplitude of deformation variation was larger in the summer.

There were strong linear relationships between air temperature and deck deformation at L/4 and arch foot of concrete truss arch bridge. As for mid-span, the linear relationship is weaker. However, the temperature effect on mid-span deformation was much more than L/4 and arch foot.

In addition, the finite element model was established to calculate the theoretical value, and further compared with practice values. There were about 2cm different deformation value between theoretical and monitoring data. It is essential to control deformation due to temperature change for concrete arch bridge, especial its middle span in long term operation.

Acknowledgments

This study is supported by the National Science Foundation of PR China (51208056), and the Fundamental Research Funds for the Central Universities (310821161013, 300102218213).

References

1. J. Suzuki, Y. Ohba, Y. Uchikawa, K. Hoshikawa and K. Kimura, Monitoring temperatures on a real box-girder bridge and energy budget analysis for basic information on bridge cooling and surface freezing, J. Bridge Eng., **12**, 7 (2007)
2. Z Lu, M Liu and Q Li, Creep effect analysis of steel-concrete composite bridge considering mutative temperature and relative humidity, J. Central South University, **46,** 2650-2657 (2015)
3. Y Tian, N Zhang and H Xia, Temperature effect on service performance of high-speed railway concrete bridges, Adv. Struct. Eng., **20,** (6) 865-883 (2017).
4. H. C. Peiretti, J. I. E. Parrotta, A. B. Oregui, A. P. Caldentey and F. A. Fernandez, Experimental study of thermal actions on a solid slab concrete deck bridge and comparison with Eurocode 1 J. Bridge Eng., **19,** 10 (2014)
5. S. Moorty and C. W. Roeder, Temperature-dependent bridge moment, J. Struct. Eng., **118**, 1090-1105 (1992)
6. GD Zhou and TH Yi, Thermal load in large-scale Bridges: a state-of-the-art review, Int. J. Distrib. Sens. N., **17** (2013)
7. CCCC Highway consultants CO. Ltd. *General Code for Design of Highway Bridges and Culverts* (China Communications Press, Beijing, 2004)
8. N. D. Battista, J. D. Brown, H. P. Tan and K. Y. Koo, Measuring and modelling the thermal performance of the Tamar Suspension Bridge using a wireless sensor network, Struct. Infrastruct. E., **11**, 176-193 (2015)
9. Q Xia, J Zhang, YD Tian and YF Zhang, Experimental study of thermal effects on a long-span suspension bridge, J. Bridge Eng., **22**, 7(2017)
10. GD Zhou, TH Yi, B Chen and X Chen, Modeling deformation induced by thermal loading using long-term bridge monitoring data, J. Perform. Constr. Facil., **32**, 3 (2018)

Variational Formulation of Interaction between Elastic Plate and Elastic Medium under the Influence of Surface Energy

S. Tirapat[1], T. Senjuntichai[1], J. Rungamornrat[1] and R. K. N. D. Rajapakse[2]

[1]Applied Mechanics and Structures Research Unit, Department of Civil Engineering, Faculty of Engineering, Chulalongkorn University, Bangkok 10330 Thailand
[2]School of Engineering Science, Simon Fraser University, Burnaby V5A 1S6 Canada

Abstract. A variational formulation of interaction problem is presented in this paper for the analysis of an elastic circular plate under axisymmetric vertical loading resting on an isotropic elastic half-space under the influence of surface energy. The Gurtin-Murdoch surface elasticity theory is adopted to take into account the surface energy effects. The contact surface between the plate and the half-space is assumed to be smooth, and the deflected shape of the plate is represented by a power series of the radial coordinate. The undetermined coefficients in the series are determined through the minimization of the total potential energy functional of the plate-half-space system. The accuracy of the present solution is verified by comparing with existing solutions, and selected numerical results are presented to portray the influence of surface energy effects on interaction between an elastic circular plate and an elastic half-space.

1. Introduction

Nano-scale materials are found to possess excellent mechanical properties when compared to conventional engineering materials. Besides conducting experiments, analytical investigation is also performed to study mechanical behaviors at such small scale. Two analytical approaches have commonly been employed, namely atomistic simulations and modified continuum-based models. The first approach yields more accurate results but requires a lot more computational efforts and resources. The latter one then becomes an attractive choice in obtaining first-approximation to nano-scale systems. In this method, the surface energy effects have to be considered due to the high surface to volume ratio at nano-scale level. To take into account the influence of surface energy, Gurtin and Murdoch [1-2] presented a mathematical framework based on continuum mechanics concepts. In their model, the elastic surface is formed as an idealized layer of zero thickness perfectly bonded to the underlying bulk material without slipping. The surface elastic constants can be determined by atomistic and molecular static simulations [3], whereas the bulk material is governed by the classical elasticity theory. The Gurtin-Murdoch continuum-based model has been employed to study a variety of problems with the consideration of surface energy effects in the past (e.g. see [4-14]).

The present study is concerned with the analysis of interaction problem between a circular nano-plate of radius a under axisymmetric vertical loading resting on an isotropic elastic half-space as shown in Figure 1 with the consideration of surface energy effects. This interaction problem is studied by employing a variational approach and the Gurtin-Murdoch surface elasticity theory [1-2]. The contact surface between the plate and half-space is assumed to be smooth, and the vertical displacement of the plate is represented by a power series of radial coordinate. The normal contact traction is expressed in terms of these generalized coordinates through the solutions of the flexibility equations based on the Green's functions for an elastic half-space under axisymmetric surface loading and surface energy influence, expressed in the form of Hankel integral transform [14]. Finally, the solution to the generalized coordinates can be obtained from the minimization of the total potential energy of the plate-half-space system. The accuracy of proposed solution scheme is confirmed through the comparison with exiting solutions. Selected numerical results for deflection and bending moment of the nano-plate are presented to demonstrate a significant influence of surface energy effects on the interaction problem.

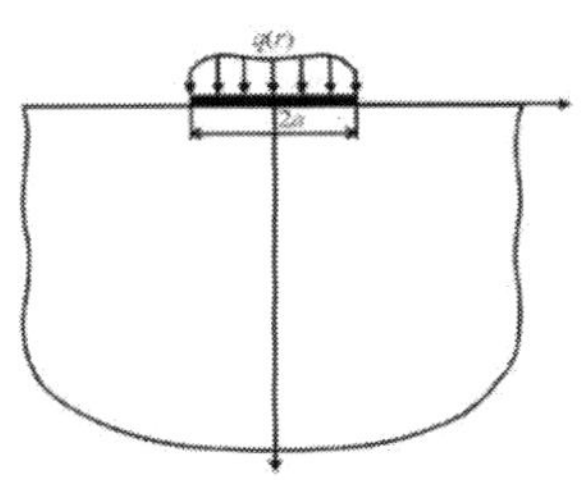

Figure 1. A circular nano-plate on an elastic half-space under axisymmetric vertical loading

2. Basic equations

Consider an isotropic elastic half-space undergoing axisymmetric deformations with the cylindrical coordinate system (r, θ, z) and surface energy effects as shown in Figure 1. According to the Gurtin-Murdoch continuum model [1-2], the surface is a zero-thickness layer perfectly bonded to the bulk material without slipping. The basic equations for the bulk follow directly from the classical linear elasticity theory. On the surface, the equilibrium equation, the constitutive relations and the strain-displacement relationship are respectively expressed for axisymmetric deformations as,

$$\frac{\partial \sigma_{rr}^{s}}{\partial r}+\frac{\sigma_{rr}^{s}-\sigma_{\theta\theta}^{s}}{r}+\sigma_{zr}+t_{r}^{0}=0;\quad \frac{\partial \sigma_{zr}^{s}}{\partial r}+\frac{\sigma_{zr}^{s}}{r}+\sigma_{zz}+t_{z}^{0}=0 \tag{1}$$

$$\sigma_{rr}^{s}=\tau^{s}+\left(2\mu^{s}+\lambda^{s}\right)\varepsilon_{rr}^{s}+\left(\lambda^{s}+\tau^{s}\right)\varepsilon_{\theta\theta}^{s};\ \sigma_{\theta\theta}^{s}=\tau^{s}+\left(2\mu^{s}+\lambda^{s}\right)\varepsilon_{\theta\theta}^{s}+\left(\lambda^{s}+\tau^{s}\right)\varepsilon_{rr}^{s};\sigma_{zr}^{s}=\tau^{s}\frac{du_{z}^{s}}{dr} \tag{2}$$

$$\varepsilon_{rr}^{s}=\frac{du_{r}^{s}}{dr};\quad \varepsilon_{\theta\theta}^{s}=\frac{u_{r}^{s}}{r} \tag{3}$$

where the superscript "s" is used to denote the quantities corresponding to the surface (or interface); μ^s and λ^s are surface Lamé constants; and τ^s is the residual surface stress (or surface tension) under unstrained condition. In addition, t_i^0 denotes the prescribed surface traction in the i-direction ($i = r, z$).

3. Variational formulation of interaction problem

3.1. Potential energy of elastic circular nano-plate

Consider an elastic circular nano-plate of radius a subjected to axisymmetric vertical loading as shown in Figure 1. The deflection of the plate, denoted by $w(r)$, can be represented in the following form:

$$w(r)=\sum_{n=0}^{N}\alpha_{n}r^{2n},\qquad 0\le r\le a \tag{4}$$

where α_n (n = 0,1,...,N) denotes a set of generalized coordinates.

In the classical thin plate theory, the out of plane stress σ_{zz} is neglected. According to Lim and He [15]; and Lu et al. [16], σ_{zz} is assumed to vary linearly through the plate thickness to satisfy the equilibrium conditions along the interface, and σ_{zz} can then be expressed as,

$$\sigma_{zz}=\frac{1}{2}\left(\sigma_{zz}^{+}+\sigma_{zz}^{-}\right)+\frac{z}{h}\left(\sigma_{zz}^{+}-\sigma_{zz}^{-}\right)=\frac{2\tau_{P}^{s}z}{h}\left(\frac{d^{2}w}{dr^{2}}+\frac{1}{r}\frac{dw}{dr}\right) \tag{5}$$

where σ_{zz}^{+} and σ_{zz}^{-} are the stresses at the top and bottom surfaces respectively of the bulk of the plate; h is the plate thickness; and τ_P^s is the residual surface stress (or surface tension) under unstrained condition of the plate. In addition, the other normal stresses in terms of generalized displacement, $w(r)$, can be expressed as,

$$\sigma_{rr}=\frac{E}{1+v}\left[\frac{1}{1-v}\left(-z\frac{d^{2}w}{dr^{2}}\right)+\frac{v}{1-v}\left(-\frac{z}{r}\frac{dw}{dr}\right)\right]+\frac{v}{1-v}\frac{2\tau_{P}^{s}z}{h}\left(\frac{d^{2}w}{dr^{2}}+\frac{1}{r}\frac{dw}{dr}\right) \tag{6}$$

$$\sigma_{\theta\theta}=\frac{E}{1+v}\left[\frac{1}{1-v}\left(-\frac{z}{r}\frac{dw}{dr}\right)+\frac{v}{1-v}\left(-z\frac{d^{2}w}{dr^{2}}\right)\right]+\frac{v}{1-v}\frac{2\tau_{P}^{s}z}{h}\left(\frac{d^{2}w}{dr^{2}}+\frac{1}{r}\frac{dw}{dr}\right) \tag{7}$$

where E and v is the Young's modulus and Poisson's ratio of the plate.

The strain energy of the plate contains two parts, i.e., the elastic strain energy stored in the bulk (U_B) and the elastic strain energy of the surface (U_S), which are expressed respectively as,

$$U_{B}=\frac{1}{2}\int_{V}\left(\sigma_{rr}\varepsilon_{rr}+\sigma_{\theta\theta}\varepsilon_{\theta\theta}\right)dV \tag{8}$$

$$U_{S}=\frac{1}{2}\int_{\Gamma}\left(\sigma_{rr}^{s}\varepsilon_{rr}^{s}+\sigma_{\theta\theta}^{s}\varepsilon_{\theta\theta}^{s}+2\sigma_{rz}^{s}\varepsilon_{rz}^{s}\right)d\Gamma \tag{9}$$

where V and Γ are the bulk volume and the surface area of the plate respectively.

Finally, the potential energy due to a vertical load $q(r)$ as shown in Figure 1 is written as,

$$W=-\int_{\Gamma}q(r)w(r)d\Gamma \tag{10}$$

In view of Eqs. (8) to (10), the total potential energy I_P of nano-plate is given by,

$$I_{P}=U_{B}+U_{S}+W \tag{11}$$

The minimization of I_P in Eq. (11) together with integration by parts leads to the governing equation for a circular nano-plate:

$$D\nabla^{4}w-2\tau_{P}^{s}\nabla^{2}w-q(r)=0 \tag{12}$$

where

$$D=\frac{Eh^{3}}{12\left(1-v^{2}\right)}+\left(\lambda_{P}^{s}+2\mu_{P}^{s}\right)\frac{h^{2}}{2}-\frac{\tau_{P}^{s}vh^{2}}{6\left(1-v\right)} \tag{13}$$

and

$$\nabla^{2}=\frac{1}{r}\frac{d}{dr}\left(r\frac{d}{dr}\right) \tag{14}$$

The shear force and bending moment of a circular nano-plate are given by,

$$Q_r = -D\frac{d}{dr}\nabla^2 w + 2\tau_P^s \frac{dw}{dr} \tag{15}$$

$$M_r = D\frac{d^2 w}{dr^2} + \frac{D_1}{r}\frac{dw}{dr} \tag{16}$$

where

$$D_1 = \frac{\nu E h^3}{12(1-\nu^2)} + \left(\lambda_P^s + \tau_P^s\right)\frac{h^2}{2} - \frac{\tau_P^s \nu h^2}{6(1-\nu)} \tag{17}$$

Equations (12) to (17) are identical to the governing equation for a nano-plate derived by Liu and Rajapakse [8] from the equilibrium of an infinitesimal plate element. This equation reduces to the classical Krichhoff equation when the surface energy effects is completely neglected (i.e., τ_P^s, λ_P^s, μ_P^s are zero). In addition, λ_P^s and μ_P^s are surface Lamé constants for the nano-plate.

3.2. Strain energy of elastic half-space

An elastic circular nano-plate under axisymmetric vertical loading is resisted by the normal contact traction acting on bottom surface of the plate. Let S denote the circular contact area between the plate and the supporting medium. The normal traction can be represented by a traction field $T_z(r)$ acting on the circular surface S. The strain energy U_h of the elastic half-space can be expressed in the following form:

$$U_h = \frac{1}{2}\int_0^a \left[2\pi r T_z(r) w(r)\right] dr \tag{18}$$

Note that $T_z(r)$ can be expressed in term of generalized coordinates α_n as follows:

$$T_z(r) = \sum_{n=0}^{N} \alpha_n T_{nz}(r) \tag{19}$$

where $T_{nz}(r)$ denotes the traction in the vertical direction applied over S such that the vertical displacement within S is equal to r^{2n}.

To determine the traction $T_z(r)$, the circular contact area S is discretized into a number of Ne annular ring elements. It is assumed that $T_{nz}(r)$ is constant within each ring element. The unknown contact traction $T_{nz}(r)$ is then evaluated by solving a flexibility equation based on the Green's functions of an elastic half-space with consideration of surface energy effects under axisymmetric vertical loading. These Green's functions, expressed in the form of Hankel integral transform, are given elsewhere [14].

In view of Eqs. (11) and (18), the total potential energy I of the plate-half-space system can be expressed as,

$$I = I_P + U_h \tag{20}$$

Finally, the generalized coordinates α_n ($n = 0,1,\ldots,N$) are determined by using the principle of minimum potential energy, which requires that

$$\frac{\partial I}{\partial \alpha_n} = 0, \qquad n = 0,1,\ldots,N \tag{21}$$

The solution of a linear simultaneous equation system given by Eq. (21) yields the solution of the generalized coordinates α_n ($n = 0,1,\ldots, N$) for a given plate-half-space system. Finally, the plate deflection, shear force, and bending moment can be obtained by back substituting the generalized coordinates into Eqs. (4), (15), and (16) respectively.

4. Numerical results and Discussion

4.1. Validation of numerical scheme

The accuracy of the present solution scheme is confirmed by comparing with existing solutions. Figure 2 shows the comparison of normalized deflection profiles of a circular nanoplate with a simply-supported edge under a uniform loading, q_0, between the present solution and the solution given by Liu and Rajapakse [8]. The material parameters employed in the comparison are $E = 90$ GPa and $\nu = 0.23$; and the surface material properties are τ_P^s = 0.5689 N/m, λ_P^s = 3.4939 N/m and μ_P^s = -5.4251 N/m. It is clearly seen that very good agreement between the two solutions is obtained with $N = 5$. Figure 3 (a) and (b) respectively show radial profiles of normalized contact traction and bending moment of a circular plate on an isotropic elastic half-space without surface stress effects, between the present solution with $N = 8$ and $Ne = 40$ and the solution proposed by Brown [17] for different values of relative plate stiffness, K_r, with $\nu = 0.3$. It is evident from Figure 3 that the present solution agrees very closely with Brown [17] for all values of K_r.

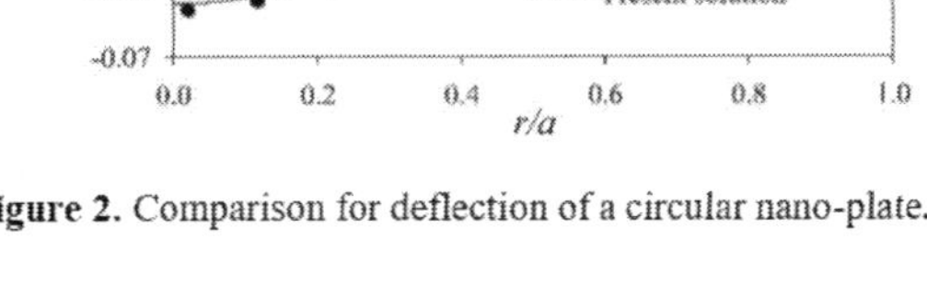

Figure 2. Comparison for deflection of a circular nano-plate.

 https://doi.org/10.1051/matecconf/201820601012

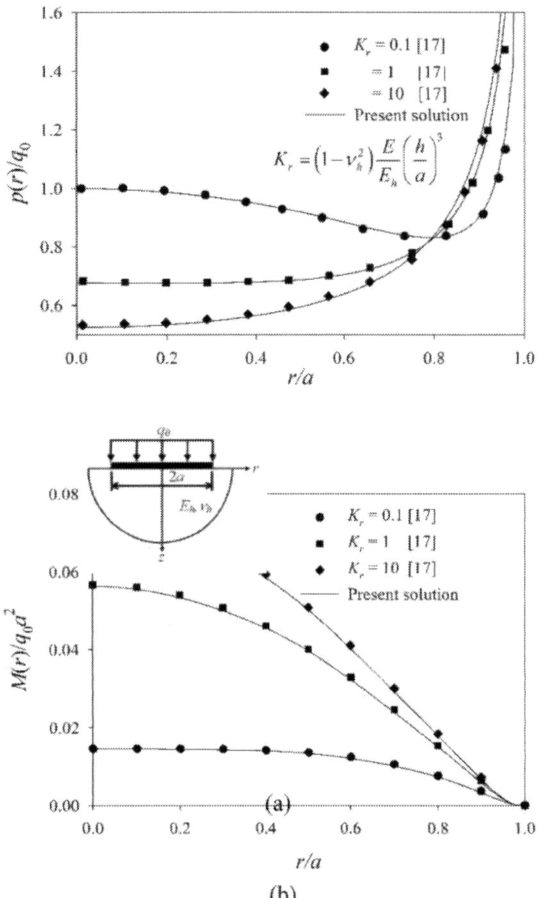

Figure 3. Comparison for: (a) contact pressure; and (b) bending moment of a circular plate on an elastic half-space without surface energy effects

4.2. Numerical results

The influence of surface energy effects on interaction between an elastic circular plate under axisymmetric vertical loading and an elastic half-space is investigated in Figures 4 to 6. The following material parameters are employed in the numerical study: E = 107 GPa and ν = 0.33 for the plate; and λ = 58.17 GPa and μ = 26.13 GPa for the half-space. In addition, τ_p^s = 0.6056 N/m, λ_p^s = 4.4939 N/m, and μ_p^s = 2.7779 N/m for the surface of the plate; and τ_h^s = 1 N/m, λ_h^s = 6.8511 N/m, and μ_h^s = -0.376 N/m for the top surface of the half-space. Note that the broken lines in all figures presented in this section denote the classical solutions where surface stress effects are ignored (i.e. $\tau^s = \lambda^s = \mu^s \approx 0$). Figure 4 and 5 respectively display radial profiles of normalized vertical displacements and bending moments of a circular nano-plate resting on an elastic half-space under a uniformly distributed vertical load of magnitude q_0 for different values of relative plate stiffness, i.e., K_r = 0.1, 1 and 10. In addition, $\bar{a}$ = 10 where $\bar{a}$ = a/Λ_h, and Λ_h = $(\lambda_h^s + 2\mu_h^s)(\lambda + 2\mu)/2\mu(\lambda+\mu)$. Numerical results in Figure 4 and 5 indicate a significant influence of surface energy effects on both deflection and bending moment. It can be seen from Figure 4 that the normalized deflection from the present study is lower than its classical counterpart for all values of relative stiffness K_r whereas the normalized bending moment under the influence of surface stresses shown in Figure 5 is always larger than the corresponding classical solution. Therefore, the presence of surface stresses renders the plate stiffer.

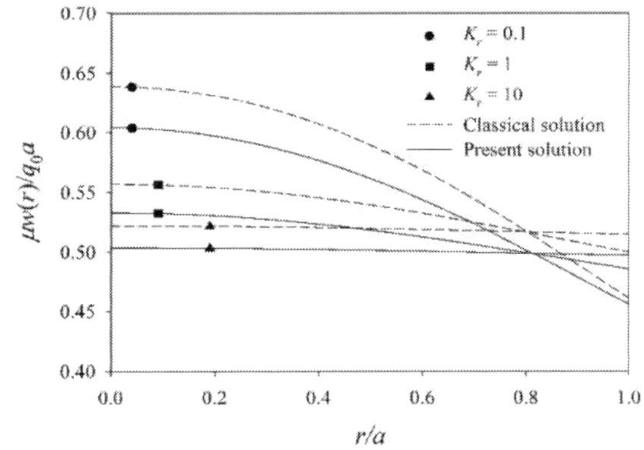

Figure 4. Radial profiles of normalized vertical deflection of a circular nano-plate resting on an elastic half-space under uniform vertical loading for different values of K_r with $\bar{a}$ =10.

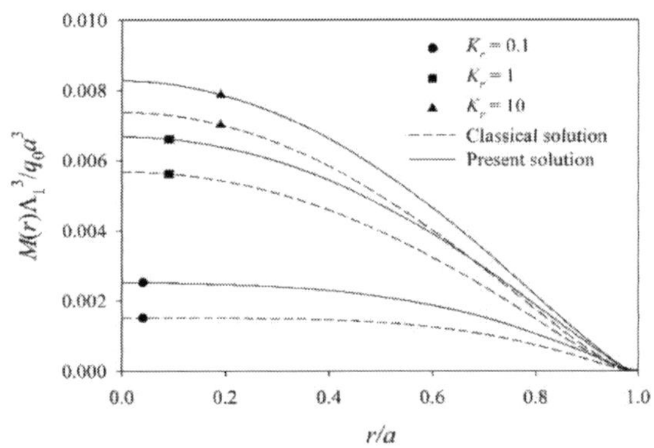

Figure 5. Radial profiles of normalized bending moment of a circular nano-plate resting on an elastic half-space under uniform vertical loading for different values of K_r with $\bar{a}$ =10.

Figure 6 shows variations of normalized central deflection of a circular nano-plate under the uniform vertical loading with normalized radius $\bar{a}$ for different values of relative stiffness K_r. It is evident from Figure 6 that under the influence of surface energy the normalized central deflection depends significantly on both normalized radius and relative stiffness of the plate. Size-dependent behaviour due to the presence of surface stresses is thus clearly observed in the present solution whereas the classical elasticity solution is size-independent. In addition, the influence of surface energy effects is reduced with increasing value of $\bar{a}$, and the present solution eventually converges to the classical solution.

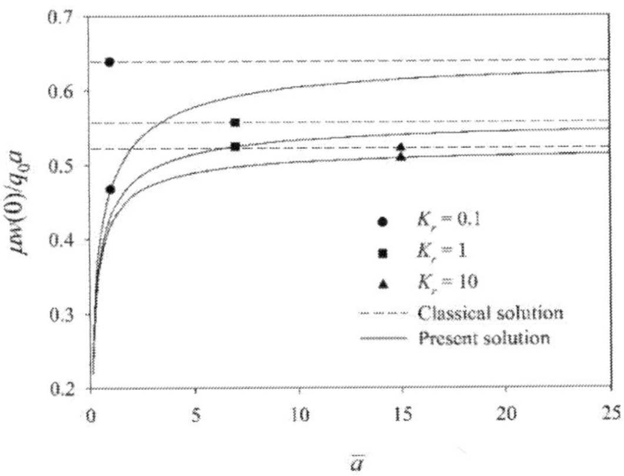

Figure 6. Variations of normalized central deflection with $\bar{a}$ of a circular nano-plate resting on an elastic half-space uniform vertical loading for different values of K_r.

5. Conclusions

A variational solution scheme is presented in this paper to study the interaction problem between an elastic circular plate and an isotropic elastic half-space under the influence of surface energy by adopting Gurtin-Murdoch surface elasticity theory. The contact surface between the plate and the half-space is assumed to be smooth, and the deflection shape of the plate is represented by a power series of the radial coordinate. The generalized coordinates in the series are obtained from the minimization of the total potential energy functional of the plate half-space system, which consists of the strain energies of the plate, the half-space, and the surface, together with the potential due to the applied loading. The accuracy of the proposed solution scheme is confirmed by comparing with exiting solutions. Numerical results presented in this paper indicate a significant influence of surface energy effects on the interaction problem, and the consideration of surface stresses renders the plate stiffer. In addition, the material behaviour becomes size-dependent when the surface stresses are taken into account. The proposed variational solution scheme can be used to study various interaction problems involving an elastic circular plate and an elastic medium with the presence of surface stresses for the applications related to nano-scale systems and soft elastic solids.

6. Acknowledgement

The work presented in this paper was supported by the Thailand Research Fund [grant numbers BRG 5880017 and PHD/0096/2554]. This support is gratefully acknowledged.

References

1. Gurtin M E and Murdoch A I 1975 Arch. Ration. Mech. An. 57 291-323.
2. Gurtin M E and Murdoch A I 1978 Int. J. Solids Struct. 14 431-440.
3. Shenoy V B 2005 Phys. Rev. B 71 094104.
4. He L H and Lim C W 2006 Int. J. Solids Struct. 43 132- 143.
5. Lü C F, Lim C W and Chen W Q 2009 Int. J. Solids Struct. 46 1176-1185.
6. Intarit P, Senjuntichai T and Rajapakse R K N D 2010 Eng. Fract. Mech. 77 3592-3603.
7. Sapsathiarn Y and Rajapakse R K N D 2012 IEEE T Nanotechnol. 11 247-254.
8. Liu C and Rajapakse R K N D 2013 IEEE T. Nanotechnol. 12 13-20.
9. Pinyochotiwong Y, Rungamornrat J and Senjuntichai T 2013 Int. J. Eng. Sci. 71 15-35.
10. Rungamornrat J, Tuttipongsawat P and Senjuntichai T 2016 App. Math. Model. 40 1532-1553.
11. Intarit P, Senjuntichai T, Rungamornrat J and Rajapakse R K N D 2017 Acta Mech. 228 617-630
12. Tirapat S, Senjuntichai T and Rungamornrat J 2017 Adv. Mater. Sci. Eng. 2017 7530936
13. Intarit P, Senjuntichai T and Rungamornrat J 2018 Z. Angew. Math. Phys. 69 29
14. Intarit P 2012 Solutions of Elastic Medium with Surface Stress Effects. Ph.D. Thesis, Chulalongkorn University
15. Lim C W and He L H 2004 Int. J. Mech. Sci. 46 1715-1726.
16. Lu P, He L H, Lee H P and Lu C 2006 Int. J. Solid Struc. 43 4631-4647.
17. Brown 1969 Geotechnique 19 399-404

Fuzzy Based Bridge Structural Health Rating of Existing Bridges using Accelerometer Sensors: Case of Zamora Bridge

John Victor J. Lau, Charmaine Aila Mai L. Bassig , Earl Justin C. Celino, and John Paul D. Carreon

Mapúa University, Muralla St., Intramuros, Manila, PHILIPPINES

Abstract. This study presents the development of a fuzzy-based procedure for condition assessment and bridge rating in structural health monitoring of existing bridges. A sample bridge, Zamora Bridge, was assessed to determine its structural health through fuzzy-based analysis utilizing accelerometer sensors installed in the bridge. Sensors were installed on superstructure and substructure components to record the three-dimensional accelerations of each element. Accelerations were then compared to the designed acceleration of the member to obtain structural ratings. Using fuzzy logic, the resultant rating set for the bridge has been evaluated based on specific ratings and importance factor for elements of the bridge. The final rating for the bridge can be evaluated by converting the collective fuzzy rating that defines the membership value. Comparison between the data of fuzzy based analysis for structural health assessment and visual bridge inspection for condition rating verified the reliability of the findings of this study.

1 Introduction

There are 8,131 permanent bridges in the sixteen regions of the Philippines according to the Department of Public Works and Highway (DPWH) as of December 2015. Bridges connect barangays, cities and regions to provide goods and mobility thus should be always functional especially during critical conditions in the area. Visual inspection is employed by DPWH to assess the structural integrity of existing bridges. This takes time to complete the assessment of the existing bridges. As indicated by various codes, assessment of existing bridges was developed; deterioration and performance were accounted [1].

In one of the earliest accounts in bridge condition assessment, bridge elements were considered from substructure, superstructure and deck [2]. Assessment was carried out not only to identify the structural health but also detect damages for failure prevention [3]. With the importance of the bridge in mobility, bridges have to be maintained and assessed frequently with reliability. This study aimed to develop a fuzzy-based structural health assessment of bridges utilizing readings from sensors installed at key structural members.

The study covered the development of a systematic process and program that will assess structural health of bridges using fuzzy based analysis. As such, this study was limited to the analysis and interpretation of data collected from installed accelerometer sensors on and the design specifications of Zamora bridge; the establishment of membership functions; the incorporation of relative importance factors; and, the development of a program for the structural health monitoring.

The study covered the development of a systematic process and program that will assess structural health of bridges using fuzzy based analysis. As such, this study was limited to the analysis and interpretation of data collected from installed accelerometer sensors on and the design specifications of Zamora bridge; the establishment of membership functions; the incorporation of relative importance factors; and, the development of a program for the structural health monitoring.

2 Fuzzy Logic

Visual inspection techniques were the primary methods used to evaluate the condition of the majority of bridges. These subjective assessments cause impact on the safety and management of a bridge [4].

The bridge evaluation and remaining service life prediction processes were conducted in order to efficiently maintain existing bridges, it was based on the results of periodical visual inspections [5]. Conventional methods for crack detection were applied by experienced inspectors who mark the cracks manually and read the width of the cracks with their naked eyes [6,7].

3 Fuzzy Logic

Fuzzy logic covers the shortcomings of traditional Boolean logic which is currently used for visual bridge condition assessment, and therefore can provide more accurate results. Fuzzy logic is an evaluation process wherein statements or values are represented in varying degrees of truthfulness [8] as opposed to Boolean Logic which is confined to either a true *(1)* or a false *(0)* [9]. In Fuzzy logic, there exist degrees such as partially true, slightly false, which can be represented as real numbers between 1 and 0.

Fuzzy logic is based on the concept of the fuzzy set. These are sets without a clearly defined boundary [10]. Contradictory to a fuzzy set, a classical set is one with clearly defined boundaries. These cases cannot be possible in classical sets, therefore, to define such cases, the concept of fuzzy sets was introduced.

4 Bridge Condition Assessment

Utilizing fuzzy logic concepts, a condition rating which evaluates a reinforced concrete bridge's strength and serviceability condition can be formulated [11]. Sasmal et al. proposed a systematic approach for the condition rating of existing bridges through combining fuzzy mathematics with techniques from eigenvector-based priority-setting. In this study however, a simplified fuzzy concept was applied using fuzzy weighted average approach.

5 Research Setting and Design

The study considered the structural components of the bridge where accelerometer sensors were installed, these locations shall be referred to as nodes. The structural components of the bridge, its National Bridge Elements (NBEs), is divided into three major components, namely: deck, superstructure, and substructure. Each component was further divided into several elements. For the case study, analysis covered select girders in the superstructure, and select piers in the substructure of the chosen sample bridge, Zamora bridge.

The Padre Jacinto Zamora Bridge spans about 410 meters, linking Paco and Pandacan, Manila, Philippines and is a reinforced concrete bridge built in the 1960's and became operational in 1963. Periodic inspection was done on the bridge through the years by bridge inspectors of the DPWH.

Through analysis of readings of installed accelerometer sensors on the bridge, the structural health of the Zamora Bridge is determined. The sensors provide real-time structural assessment of Zamora bridge by translating sensor acceleration outputs into more comprehensible standards used by the DPWH, to hasten bridge maintenance proceedings and avoid greater losses in the case of failure. Data necessary for the study were taken from the SMART Bridge Project and the latest available bridge inspection data from the DPWH. Statistical analyses and quantitative comparison of values were utilized to produce a well- grounded basis for the structural health monitoring system of Zamora bridge.

The Smart Bridge sensors installed in Zamora bridge were placed on the columns and on the interior face of the concrete girders of the bridge for the reason that these members are easily accessible and are subject to deflection.

Each interior girder was installed with three (3) sensors: one at the midspan of the girder, and two sensors placed 2H from each support, wherein H was the overall depth of the girder. Sensors were placed on the midspan since it is where the highest deflection was most likely to occur and was the theoretical weakest point of the member.

The sensors record 100 samples per second of acceleration of the member with respect to the x, y, and z- axes respectively of each sensor.

Adopted from a study done in Indiana and its neighboring states, where a survey was conducted with bridge engineers and inspectors, the structural importance for each membership function was established [12]. In this study, the scale used was adjusted into four classifications: good, fair, poor and bad, which are the membership functions used in this study.

Table 1 shows the converted importance factors via interpolation for each classification to be multiplied to the membership functions for each node.

Table 1. Element Importance Factors

	Bad	Poor	Fair	Good
Pier	1.0000	0.8700	0.7000	0.4300
Girder	1.0000	0.9500	0.8500	0.5800
Superstructure	1.0000	0.8867	0.7633	0.4700
Substructure	1.0000	0.6333	0.3167	0.1000

The designed acceleration of the members taken from STAAD is converted to LSB to correspond for each of the membership function automatically.Fuzzified data served as inputs for the fuzzy based mathematical analysis to determine the fuzzy rating of the entire bridge. The fuzzified data of each of the member had a single value for every membership function. Thus, the resulting data was a one-by-four matrix for each of the axes.

A 1 x 4 matrix was formed in Eq. 1 when the mean acceleration in the X-axis is determined in the different MFs. The fuzzified data for the 3-dimensional axes were then combined into a larger matrix that has a dimension of 3 x 4 as shown in Eq. 2.

$$[0.7445 \quad 0.255 \quad 0 \quad 0] \tag{1}$$

$$\begin{vmatrix} 0.7445 & 0.2555 & 0 & 0 \\ 0 & 0 & 0.7995 & 0.2005 \\ 0.4322 & 0.5678 & 0 & 0 \end{vmatrix} \tag{2}$$

Eq. 2 was multiplied with a weight factor. Each element of the matrix represents the relative importance of each axis. The matrix was normalized resulting to Eq. 3 which shows the membership values for all the MFs considering actual acceleration of the member on 3D analysis.

$$[0.3992 \quad 0.2744 \quad 0.2665 \quad 0.0668] \tag{3}$$

The relative importance factors were adopted [8]. The members were then classified to either superstructure elements or substructure elements. Piers were considered as substructure elements while girders were considered as superstructure elements which result to a 1 x 5 matrix.

For the structural health rating of Zamora bridge, a scale of 0% to 100% was used. A rating of 100% would mean that the bridge was in the perfect condition; 0% in the worst condition. The rating of good falls under 75% to 100%, fair for 50% to 75%, poor for 25% to 50%, bad for lower than 25%.

The resulting overall rating for the Zamora bridge is 0.538433 or 53.84% which is in the range of 50% to 75% that corresponds to a fair condition rating for Zamora bridge. The results were then compared to the latest Bridge inspection results of the Zamora Bridge performed by the DPWH showed that the Zamora bridge was given a fair condition rating.

Although the results obtained from the program developed were the same with the result of the bridge inspection, further statistical analysis for the percentiles to be used in the ratings to optimize accuracy is necessary.

Simulations were done with the program using varying percentile ranges for the membership functions. Including the membership function established in the study, mean percent increase or decrease in the magnitudes of the accelerations were simulated to determine the percent change in acceleration for the rating of Zamora bridge to change. Table 2 illustrates the bridge structural health rating of Zamora bridge using varying percentiles for the membership functions using the same sample data taken from the SMART Bridge sensors.

Table 2. Zamora Bridge Structural Health Rating using Membership Functions (MF)

MF_{Good} (LSB)	MF_{Fair} (LSB)	MF_{Poor} (LSB)	MF_{Bad} (LSB)	Structural Health Rating
0%	90%	99%	100%	59.3942% (Fair)
0%	80%	98%	100%	53.8433% (Fair)
0%	70%	97%	100%	50.2792% (Fair)

The peak values for the membership functions were adjusted to different percentiles of the design accelerations based on the STAAD file of Zamora bridge. In the study, the 80th and 98th percentiles were used, which results to a 53.84% rating for the bridge. Adjusting the percentile ranges for the fair and poor membership functions to the 90th and 99th percentile, respectively, results to a rating of 59.39%.

This rating still falls under the fair condition rating region of 50% to 75%, but was higher compared to the previous membership functions used. Considering that the same set of data was used for the simulation, this portrays that increasing the percentile ranges used in the establishment of membership functions results to a less conservative bridge structural health rating. This means that damages on the bridge had less impacts to the overall rating, and therefore yielded a higher structural health rating for the bridge.

Consequently, adjusting the rating lower, with the fair and poor membership functions at the 70th and 97th percentile, respectively, would result to a rating of 50.28%. This also falls under the fair condition rating range of 50% to 75%, but was lower compared to the first membership functions used.

Using membership functions with higher percentiles used for the fair and poor membership functions would result to a less conservative structural health rating and vice versa. This was illustrated in the simulations done, whereas the increase or decrease in the actual acceleration readings to yield the same structural health rating at the 25%, 50% and 75% mark were higher for the 90th and 99th percentiles for fair and poor membership functions compared to the 80th/98th and 70th/97th percentiles for fair and poor membership functions.

6 Conclusion

The study successfully developed a means to assess structural health rating of Zamora Bridge, a reinforced concrete bridge, through fuzzy-based analysis utilizing accelerometer sensors installed in bridge girders and piers.

The structural health monitoring program is based on MATLAB software, using readings from the accelerometer sensors in terms of LSB, thus removing the need of unit conversion, as input. Assessment was based on the design and specifications of the bridge, analyzed through STAAD software, processed through fuzzy based analysis. This eliminates the subjective nature of visual inspection to assess bridge conditions. Consequently, the study also provides real-time bridge health ratings to improve currently used structural health monitoring programs.

Furthermore, the structural health monitoring program developed can be adjusted to produce more conservative assessments of bridge condition. Peaks for good, fair, poor and bad criteria can be revised through statistical analysis of projected design accelerations. Placing peak values for each criterion closer to the central value produces a more conservative analysis, whereas placing them farther from the central value produces less conservative analysis. The conservative design for the structural health monitoring program was aimed to detect damage early on so that maintenance can be conducted in advance.

7 Acknowledgement

The authors acknowledge the Department of Science and Technology (DOST) for funding the SMART Bridge Project where the data in this study were obtained. This project was proposed by Mapúa University. Also, the authors recognized the support of the Department of Public Works and Highways in assisting in the installation of the sensors.

References

1. S. Sasmal, K. Ramanjaneyulu, S. Gopalakrishnan and N. Lakshmanan, "Fuzzy Logic Based Condition Rating of Existing Reinforced Concrete Bridges," 2006.
2. A. B. Tee, "The application of fuzzy m athem atics to bridge condition.," 1988.
3. Y W. R. Wickramasinghe, D. P. Thambiratnam, T. H. Chan and T. Nguyen, "Vibration characteristics and damage detection in a suspension bridge," Journal of Sound and Vibration, pp. 254-274, 2016.
4. Tarighat, A., & Miyamoto, A. (2009). Fuzzy concrete bridge deck condition rating method for practical bridge management system. Expert Systems with Applications, 36, 12077-12085.
5. Emoto, H., Takahashi, J., Widyawati, R., & Miyamoto, A. (2014). Performance evaluation and remaining life prediction of an aged bridge by J-BMS. Procedia Engineering, 95, 65-74.
6. Dilena, M., & Morassi, A. (2011). Dynamic testing of a damaged bridge. Mechanical Systems and Signal Processing , 25, 1485-1507.
7. Dingle, N. (2011, November 4). Artificial Intelligence: Fuzzy Logic Explained. Retrieved from Control Engineering: http://www.controleng.com/single- article/artificial-intelligence-fuzzy-logic-explained/8f3478c13384a2771ddb7e93a2b6243d.html
8. Sheposh, R. (2017, January). Salem Press Encyclopedia of Science. Ipswich: EBSCO Publishing.
9. Fairhead, H. (2015, November 2). Introduction to Boolean Logic. Retrieved from I Programmer: http://www.i- programmer.info/babbages-bag/235-logic-logic-everything-is-logic.html
10. Ross, T. J. (2010). Fuzzy Logic with Engineering Applications. West Sussex: John Wiley & Sons.
11. Sasmal, S., Ramanjaneyulu, K., Gopalakrishnan, S., & Lakshmanan, N. (2006). Fuzzy Logic Based Condition Rating of Existing Reinforced Concrete Bridges. Americal Society of Civil Engineers, 261-273.
12. Tee, A., & Bowman, M. (1988). Bridge condition assessment using fuzzy weighted averages. A fuzzy mathematical approach for bridge condition evaluation, 49-57.

The Numerical Investigation of Aluminum Foam-protected Reinforced Concrete Slabs under Blast Loading Using AUTODYN-3D

Ahmed K. Taha, Zhengguo Gao and Dahai Huang

school of transportation science and engineering, Beihang university (Beijing university of aeronautics and astronautics), Beijing, P.R.C., 100191

Abstract. Aluminum foam is a lightweight material with high energy absorption capacity. In this study A Nonlinear three-dimensional hydrocode numerical simulation was carried out using *autodyn-3d*, which is an extensive code dealing with explosion problems. In this simulation, a high explosive material (comp B) is blasted against several concrete panels. The model was first validated using experimental tests carried out by Chengqing and has shown good results. Several numerical tests were carried out to study two parameters that affect the deflection of reinforced concrete panels. The parameters included are the thickness of concrete target and the thickness of steel plate. The results showed that increasing the thickness of the steel plate has an insignificant effect on the deflection of the reinforced concrete target while increasing the thickness of the concrete panel has a significant effect on the deflection of the concrete target.

1 Introduction

Structural buildings can be exposed to blast loadings including accidental gas explosions or terrorist attacks during their service life, which might cause heavy human and economic losses. Over the last several decades, there has been great interest shown by the military and other governmental agencies in designing structures to withstand blast loadings. A variety of protective techniques are developed against increasing accidental explosion and terrorist attacks [1,2].

The entire chain involved in the design and construction of structures has shown keen interests in the performance assessment of existing as well as new civilian structures when subjected to blast loading due to the increase in the number of intentional and accidental blast events throughout the world.

Blast events produce short duration high magnitude loadings which can greatly influence structural blast response. When compared to conventional loads, the frequencies of explosive loads are usually much higher. The action applied to a structural member when subjected to an explosive detonation is in the form of an impact shock wave. The parameters that define the shock wave are the reflected pressure and the duration of the positive phase. The time integral of the reflected pressure over the duration of the positive phase is known as reflected impulse. Public buildings, such as railway stations, embassies and airports, should be designed to ensure as much safety of the occupants as possible.

Furthermore, counter-measures need to be taken to reduce the severity of the explosion, indirect means such as using blast barrier to protect important infrastructures and people inside them are widely used. However, several recent terrorist attacks have proven that such indirect methods cannot effectively prevent attacks initiated by suicide bombers or suitcase bombs. Therefore, there is an urgent need to directly enhance the blast-resistance of important structures through using new structural types or new materials.

Newly developed mobile and lightweight materials such as aluminum foams are very attractive for use as protective layers in many applications because they have lightweight, are cheaper, and have greater energy absorption capacities than traditional technologies[3-7].

Metallic foam consists of a matrix of aluminum filled with pockets of air. Because of its long, plastic plateau in compression, metallic foam allows high energy absorption at a nearly constant stress level, making it an ideal material for mitigating the effects of explosive loads on a structural system.

When an explosion occurs near a foam cladded reinforced concrete member, the foam layer compresses and absorbs a large amount of energy and offers protection for the members against blast loads [8].

Studies conducted by Schenker [9] have shown that foam protected specimens efficiently mitigated the effects of the blast loading on reinforced concrete walls. Schenker also conducted two full-scale blasts on foam-protected reinforced concrete walls [10], which have shown the effect of foam on protecting concrete from blast loads.

Results from tests conducted by Hanssen have shown that when field blast tests have been conducted with a ballistic pendulum, the aluminum foam layers added as a protective layer have negative effect on energy and impulse. Where the energy and impulse have increased instead of decreasing as expected [11].

2 Numerical Simulations

In the current study, a numerical model was made to simulate the effect of blast wave on different configurations of concrete slab covered with aluminum foam and steel plate.

2.1 Numerical Tool: Hydrocode

A computer program that is capable of computing strains, stresses, velocities and propagation of shock waves as a function of time and position is known as a hydrocode. In a hydrocode simulation, the response of a continuous media subjected to dynamic loading is governed by the conservation of mass, momentum and energy, and also the equation-of-state and constitutive relation of the media. The equation-of-state takes into account the effects of compressibility of the continuous media and is a function of internal energy and density, whereas the constitutive relation represents the media's resistance to shear. In this study the hydrocode simulations on the explosion effect on a concrete target are performed using AUTODYN-3D [12], a fully integrated and interactive hydrocode.

In AUTODYN-3D the fundamental equations together with the initial and boundary conditions are solved using a finite difference scheme.

2.2 Model Validation

This paper presents a 3D hydrocode simulations using autodyn-3d [13] on the effect of explosion on a reinforced concrete target as shown in Fig.1. The set of experimental data published by Chengqing Wu [14] for concrete slab target covered with an aluminum foam sheet and a steel plate is used for validation.

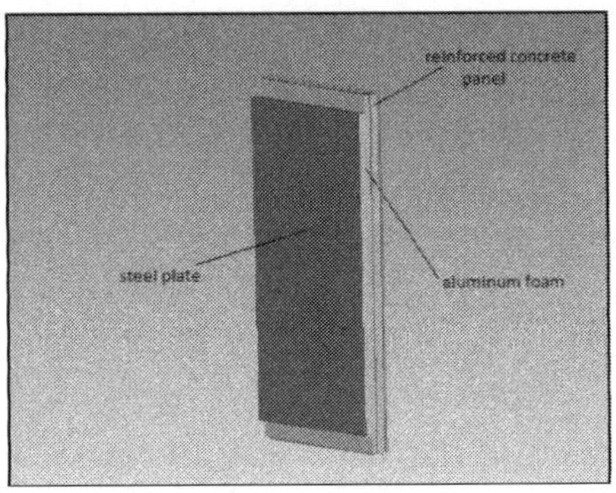

Figure. 1 Structural configuration of the proposed model

These experiments aimed at testing the performance of aluminum foam-protected RC slabs. Five foam protected slabs were tested against blast loads with two different types of aluminum foams and different weights of explosive charge.

The dimensions of the tested concrete slab was 2000×1000×100 mm. the slabs were reinforced both in the tension and compression faces with a 12mm diameter mesh that was spaced at 100mm in the major bending plane and at 200mm in the minor plane as shown in Fig. 2. The concrete slab's cover was 10mm. The concrete's compressive strength was 39.5 MPa , tensile strength was 8.2 MPa and young's modulus was 28.3 GPa. The reinforcement had yield strength of 600 MPa and young's modulus of 200 GPa.

A 1.15 mm thick steel plate sheet was attached to the outer face of the aluminum foam. The explosive used in the air blast was of weight 8.05 kg and 14.0 kg.

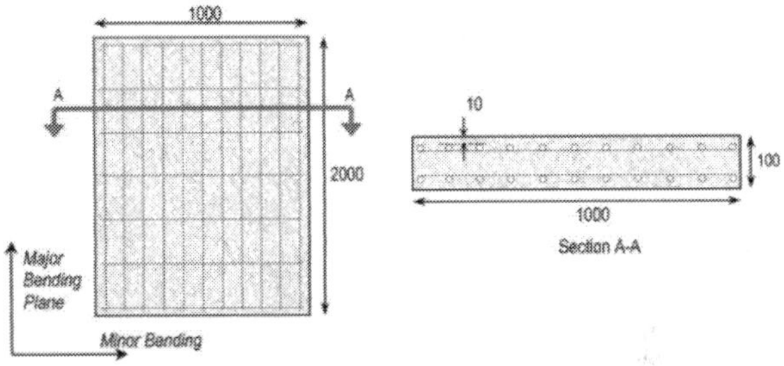

Figure. 2 Reinforced concrete target

2.2.1 Finite Element Model

In this model the concrete target is modeled with 7956 nodes and 6250 element, the aluminum foam part is modeled with 7956 nodes and 6250 elements, the steel plate part is modeled with 2652 nodes and 1250 elements while the air domain is modeled with 574488 nodes and 553800 elements.

Remap technique was used during numerical simulation [15]. The initial detonation and blast wave propagation of the explosive in free air were first calculated in a 2D domain, the result was then remapped into a 3D space as shown in Fig.3. This technique is used to save time and make the model more efficient.

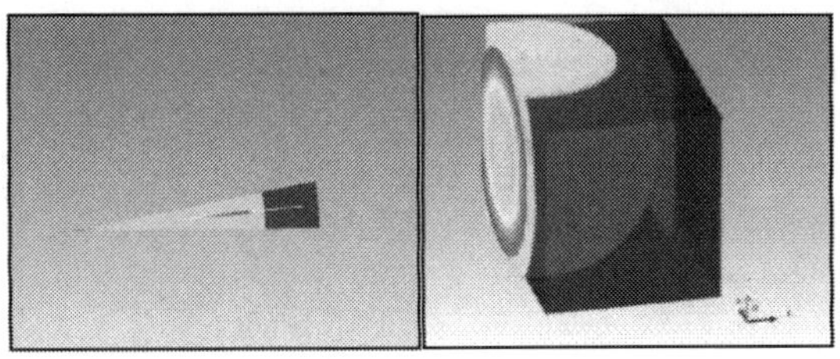

Figure.3 Remapping technique

Flow-out boundary conditions were applied on the outer surfaces of the air domain to allow the pressure of the blast to be dissipated outside the air domain without reflecting and affecting the concrete target.

Boundary conditions are applied on the concrete target preventing the motion of the concrete target in x and y direction and making the effective span of the concrete target equal to 1.8 m according to the details of the experimental work as shown in Fig.4.

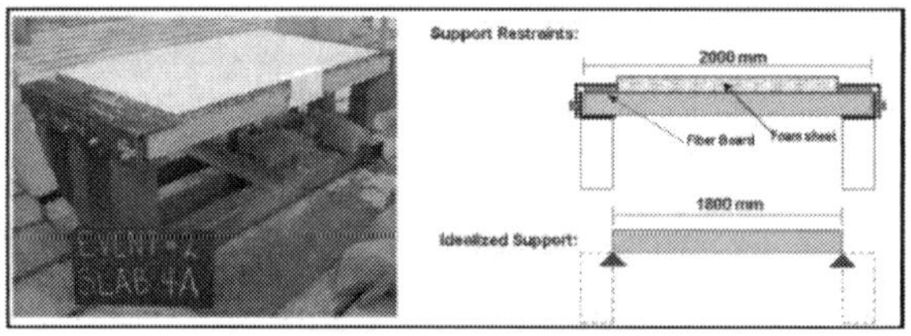

a) Experimental configuration

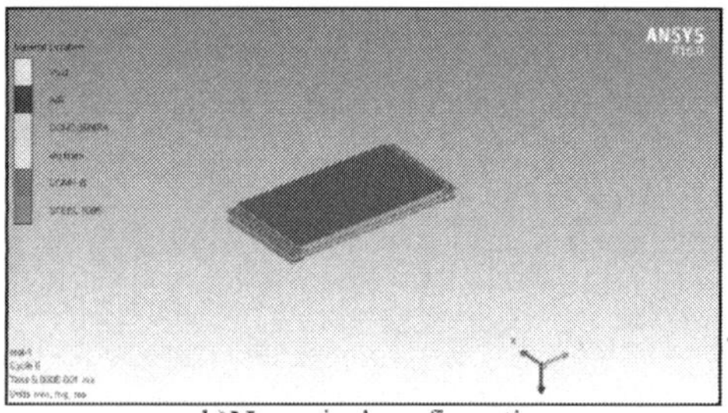

b)Numerical configuration

Figure.4 Boundary conditions

The concrete target is described with the Lagrange solver and the reinforcement bars are defined as beam element. Comp B is modeled using Jones-Wilkins-Lee equation of state which models the pressure generated by chemical energy in an explosion. Air was modeled by an ideal gas equation of state, which is one of the simplest forms of equation of state.

The first specimen used for verification is specimen 2B where as shown in Fig. 5 the overpressure for specimen 2B measured from experimental work is compared to the overpressure measured from numerical model. Specimen 2B is a reinforced concrete slab covered with 25mm aluminum foam layer where the explosive used is of weight 8.070 kg at standoff distance of 1.47m.

The overpressure measured from experimental work was 1420 kPa while the one measured from the numerical model was 1169 kPa with a percentage of error of 17.6%.

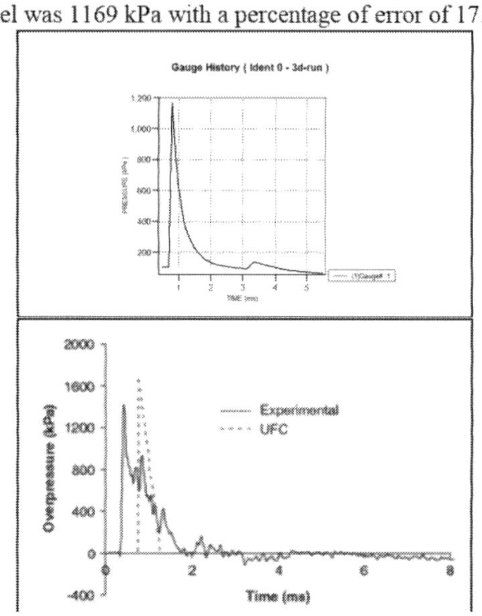

Figure. 5 Overpressure for specimen 2B

Specimen 4A is also used for verification, the displacement of specimen 4A measured from experimental work is compared to the displacement from the numerical model as shown for Fig.6.

Specimen 4A is a reinforced concrete slab covered with 43.2mm aluminum foam layer where the explosive used is of weight 8.084 kg at standoff distance of 1.50m.

The displacement measured from experimental work was 31mm while the one measured from the numerical model was 26.8mm with a percentage of error of 13.5 % as shown for Fig.6.

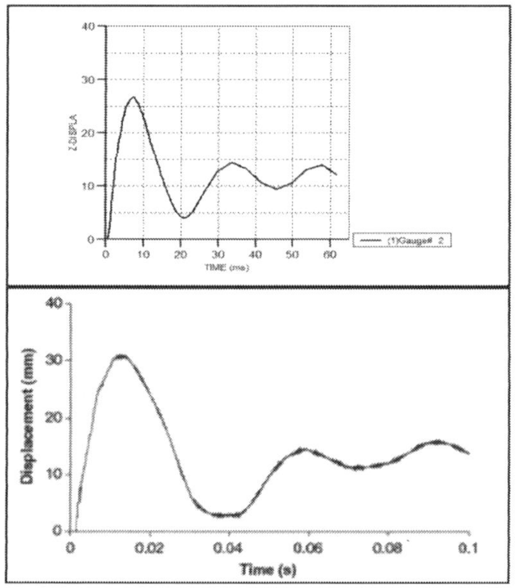

Figure. 6 Displacement of specimen 4A

3 Parametric studies and results

The effect of two factors on the displacement in the middle point of the reinforced concrete target is studied as follows:

3.1. Thickness of concrete target

In this section the effect of the thickness of the concrete target is studied. The thickness of the concrete slab in the experimental tests and the validated numerical model was constant (100mm).

Two numerical models were made with two different thicknesses of concrete target (150 &200 mm) to investigate the effect of increasing the concrete slab thickness on the deflection of it. The results were as shown in table 1.

Table 1. The effect of thickness of concrete target on deflection

Thickness of steel plate(mm)	Deflection (mm)
1.15	26.8
5	25.68
10	25.48

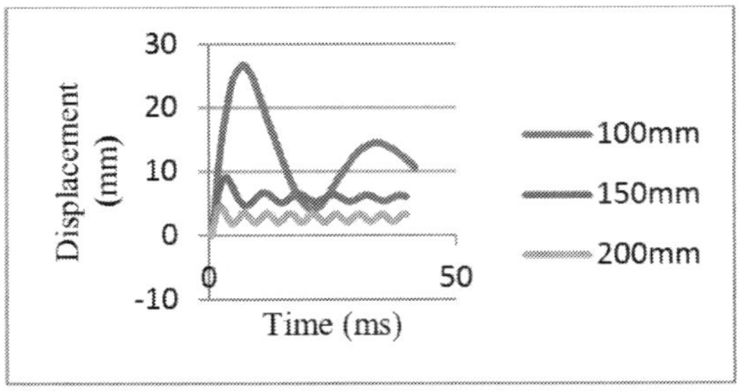

Figure.7 Relation between thickness of concrete target and deflection

The increase in the thickness of the concrete target greatly influences the amount of deflection in it as shown in Fig.7 as when the thickness of the concrete target increased from 100mm to 150mm, the deflection decreased from 26.65mm to 9 mm and when the thickness increased again from 150mm to 200mm, the deflection decreased once more to reach 4.5mm.

3.2. Thickness of steel plate

The influence of increasing the thickness of steel plate was also studied where the thickness of steel plate in the experimental work and in the validated numerical model was 1.15mm.

Two numerical models were made with two different thicknesses of steel plate to investigate the effect of increasing the steel plate thickness on the deflection of the concrete target. The results were as shown in table 2.

Table 2. The effect of the thickness of steel plate on deflection

Thickness of conc. Target (mm)	Deflection (mm)
100	26.8
150	9
200	4.5

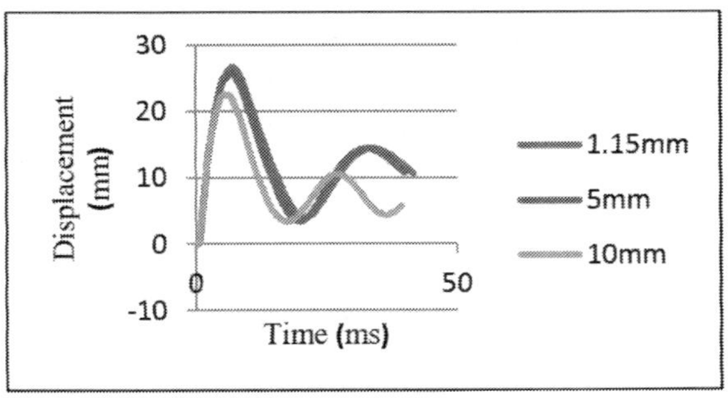

Figure.8 Relation between thickness of steel plate and deflection

The increase in the thickness of the steel plate has insignificant effect on the deflection of the mid-span of the concrete target as shown in Fig.8 as when we increased the thickness of the steel plate from 1.15mm to 5mm the deflection decreased to 25.68mm and when we increased the thickness again to 10mm the deflection decreased to 25.48mm.

4 Conclusion

The present study investigates the dynamic response of composite panels containing aluminum foam against blast waves and a non-linear 3d model is used to model the proposed configuration. The effect of increasing the thickness of concrete panel and the thickness of steel plate is examined. The main findings of this paper are as follows:

The numerical model presented has a good agreement with the experimental work. Using numerical simulation in studying the effect of blasting loads is a very effective way in studying different parameters. Increasing the thickness of the concrete target has a great influence on the deflection of the concrete target. Increasing the thickness of the steel plate has a slight influence on the deflection of the concrete target.

Some future researches can be made about using some other new materials instead of using aluminum foam and studying the response of a complete structural member against blasting.

5 References

1. Chengqing Wu,"Research development on protection of structures blast loading at university of Adelaide" Australian journal of structural engineering(2012) **13**,No.1.
2. Inderpal Singh Sanhu, et al. "Study of blast wave pressure modification through rubber foam" 11th International symposium on plasticity and impact mechanics 173(2017) 570-576.
3. Sanam Aghdamy, Chengqing Wu and Michael Griffith , "Simulation of Retrofitted Unreinforced Concrete Masonry Unit Walls under Blast Loading" International journal of protective structures(2013)**4** No. 1.
4. Mohd Fadzli Ismail ,et al." Investigation on energy absorption of aluminium foam-CFRP sandwich panel subjected to impact loading" JurnalTeknologi (Sciences & Engineering) 75:8 (2015) 113–116.
5. Ye Xia, etal.,"Numerical simulation of foam-protected RC members under blast loads " , International Journal of Protective Structures(2014) **5**,No.4.
6. Feng Zhu et al.," Structural response and energy absorption of sandwich panels with an aluminium foam core under blast loading", Advances in Structural Engineering ,**11** (2008), No. 5.
7. Li Shiqiang et al.," Dynamic response of sandwich spherical shell with graded metallic foam cores subjected to blast loading", Composites: Part A56(2014) 262-271.
8. C. Jayarami Reddy,V. Madhu," Dynamic behavior of foams and sandwich panels under shock wave loading" 11th International symposium on plasticity and impact mechanics 173(2017),1627-1634.

9. Schenker, A., et al. "Foam-protected reinforced concrete structures under impact: Experimental and numerical studies" J. Struct. Eng., **131**(2005), 1233–1242.
10. Schenker, A., et al. "Full-scale field tests of concrete slabs subjected to blast loads." Int. J. Impact Eng., **35**,(2008), 184–198
11. Hanssen, A. G., Enstock, L., and Langseth, M. (2002). "Close-range blast loading of aluminum foam panels." Int. J. Impact Eng., 27(6), 593–618.
12. AUTODYN, Theory Manual, Revision 4.0, Century Dynamics Inc., 1998.
13. AUTODYN, Theory Manual, Revision 4.0, Century Dynamics Inc., (1998).
14. Chengqing Wu, Liang Huang and Deric John Oehlers " Blast Testing of Aluminum Foam–Protected Reinforced Concrete Slabs" journal of performance of constructed facilities , **25** (2011) :464-474.
15. AUTODYN. Remapping tutorial. California: Century dynamics; (2005) pp. 258

Research on Risk Management of Gongbei Tunnel Engineering Excavation Team

Dongchen Guo

China Academy of Transportation Sciences, Beijing, China

Abstract. Gongbei tunnel is the key project of Hong kong-Zhuhai-macao bridge Zhuhai connecting line. The tunnel adopts frozen Curve pipe curtain digging method. In order to ensure the safety of Gongbei tunnel engineering, the rsik management of tunnel excavation team has studied. It has studied on the relevancy between on the organizational structure, rules and regulations, education and training, assessment of rewards and punishments of construction team. Then it make sure the best relevancy risk to plan the risk prevention and controlling measures. Finally, the safety management strategy has targeted to the Excavation team. After practical application, the excavation team of Gongbei tunnel has achieved all-round and focused risk management, which has guided the implementation of the excavation work.

1 Preface

Gongbei tunnel is the key control project of Hong kong-Zhuhai-Macao bridge Zhuhai connecting line. The tunnel's starting point is located in south of Zhuhai Gongbei bay and east of the Macau peninsula waters, the terminal is located in the fifth branch management area of Guangdong provincial public security frontier. It picks up the Gongbei bay bridge after artificial island of the Zhuhai-Macau port, and traverses Gongbei port, which is the largest land road port of China.

Tunnel excavation adopts the 255m curve pipe curtain and frozen excavation method, which is so complicated and high risk. This requires the construction team to work closely together. Considering the particularity of its construction technology, it is necessary to optimize the configuration of the crew, and then put forward the targeted control measures and requirements.

2 Project overview and research ideas

2.1 Overview of Gongbei tunnel

The construction conditions of tunnel excavation are extremely complicated.

First, geological conditions are complex. The tunnel is built in the marine, marine-terrigenous and terrigenous facies strata. The average thickness is about 28~35m, and the soft soil layer is dominated by Marine sediments. It has multi-layer, large thickness, wide distribution, high compressibility, high sensitivity, high moisture content, high porosity ratio, low strength, etc. These factors make the tunnel have many disadvantages in the stability of surrounding rock, design of support and construction.

Second, it has high political sensitivity. It needs to traverse the first land port that have the daily passenger flow of 250,000 passengers (the peak of 300,000) and the average daily flow of 7,000 vehicles (the peak of 10,000 vehicles).

Third, the surface buildings are dense and important. The outer edge of the tube was only 0.75m from the pile foundation of the Macau joint inspection building. The minimum distance of the Zhuhai Gongbei station is less than 24m. These are very important public buildings.

Fourth, the power, telecommunication, water supply and drainage network are numerous in the area of tunnel line.

Therefore, the tunnel construction adopts curvilinear curtain to freeze soil construction method.

2.3 Research approach

• Through the investigation of the basic situation of the excavation team, the factor M affecting the safety management of the construction crew were analyzed.

• Through the analysis of the excavation operation process, the risk analysis is carried out to form the construction safety factor N.

•Through correlation analysis between factor M and factor N, to carry out the, the key safety management factors that influence the excavation of Gongbei tunnel are determined. Furthermore, the related management countermeasures are developed to guide the safety operation of the team.

3 Construction personnel influence factor analysis (factor M)

3.1 Tunnel excavation personnel analysis

Based on the basic information of 375 tunnel construction personnel, the analysis of the demographic characteristics of the construction crew of tunnel engineering was concluded [1], [2].

Gender: in tunnel engineering construction group, mainly male, female mainly engaged in logistics. About 96.3% is men and 3.7% is women. The gender ratio is about 26:1.

Age: among the tunnel builders, the maximum number of employees aged 40 to 49 was 137, accounting for 36.5% of the total. A total of 78 persons under 30 years of age accounted for 20.8%. Five people over the age of 60 accounted for 1.3%. The age structure is reasonable.

Education background: The construction personnel of high school and above are responsible for 11.5%. The construction personnel of the junior high school and the following qualifications account for 88.5%. The ratio of high school and above education to junior high school and below is 1:7.7.

Construction experience: the tunnel construction personnel have rich construction experience. People with more than five years of experience have a ratio of 82.4%.

Staff mobility: tunnel builders who work less 6 months accounted for about 60.7% in a single project. The staff mobility is frequent.

3.2 Analysis of safety factors

The factors that affect the safety on construction team are multifaceted. According to the four elements of safety management that include human, objects, environment and management, the analysis of influencing factors was carried out.

Human factors. The construction group is the smallest unit in the construction organization. The team leader is responsible for the daily management and safety management of the team. Therefore, the selection of group leader is the key of group organization and safety management. n addition, the workers' skill level and safety ability determine whether the construction site is safe. Therefore, construction personnel ability is an important factor in the safety management of construction crew.

Objective factors. The main items involved in construction team include mechanical equipment, tools, construction materials, and so on. The main factors affecting the safety of construction are the lack of safety equipment, failure, unqualified tools and construction materials.

Environmental factors. The environment of the tunnel excavation group is complex and changeable. The excavation environment of Gongbei tunnel is mainly the enclosed space formed by the pipe curtain and the support after the frozen soil. The main influencing factors are the freezing injury caused by the curtain, the construction injuries in the construction cross operation environment, such as mechanical injury, object strike, high fall, etc.

Managerial Factors. The safety management system of the construction project department is relatively perfect, but for the construction team, the management system is imperfect and the target is poor, and many systems are not in place.

In the above influencing factors, the objective factors can be eliminated by inspection and acceptance. Other factors, such as human factors (include team leader, construction personnel), environmental factors (include low temperature environment, cross-operation environment) and management factors (include responsibility system, education training, assessment, rewards and punishments), are all factors affecting the safety of Gongbei tunnel excavation team.

4 Influence factors of tunnel excavation process (factor N)

4.1 Decomposition of excavation process

There are three main stages.

• Subsurface excavation and secondary lining. It includes the soil grouting reinforcement in the hole, the excavation of the underground continuous wall, the excavation of the hole, the water stop between the pipes, frozen exterior-protected construction, the initial support, the temporary support, the secondary lining, etc.

• Third lining. It includes the dismantling of the temporary support, the third lining in sections, the thawing settlement grouting, and the filling of water mud in even tube, etc.

• Ancillary and traffic engineering in the cave.

The construction paragraph of tunnel excavation in Gongbei tunnel is shown in Figure 1.

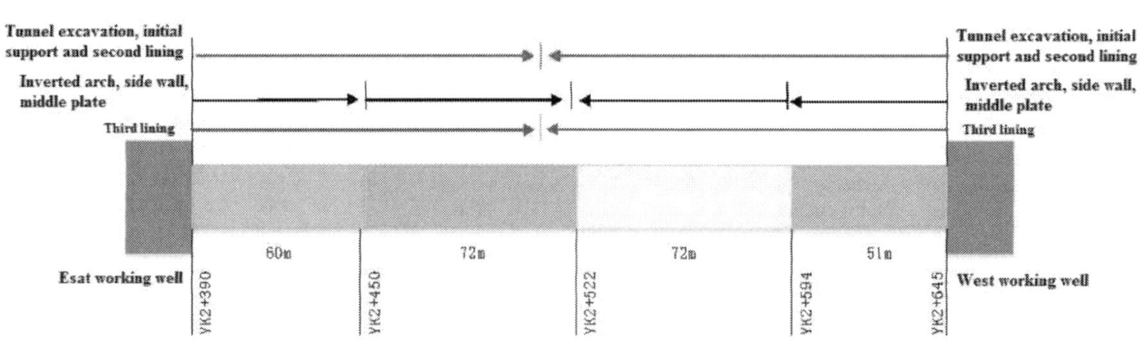

Figure 1. The section of the excavation works of Gongbei tunnel construction.

4.2 Analysis of safety factors

Based on the tunnel construction process, the paper analyzes the risk source, the risk factors and the types of accidents that may occur, and puts forward the corresponding risk control measures. It is shown in "Table 1".

Table 1. Safety risk identification and control measures during excavation of Gongbei tunnel.

SN	Content	Risk Factor	Accident Pattern	Control Measures
1	Clean the roof loose soil	Top cleaning is not thorough; Exposed soil is instability; Loose stone.	Falling accidents; object strike.	Set warning area; Remove loose stone and concrete blocks before work; Provide adequate lighting for the construction area.
2	supporting erection	Use the loader as the construction platform; The steel arch is not strong and the spacing is too large; The arch foot is suspended or placed on waste residue.	Falling accidents; collapse accident.	Use a safe and reliable working platform; The steel arch should be connected in time to prevent dumping; The suspended part of the arch and the wall must be filled with concrete.
3	Frozen earth excavation (artificial mating machinery)	Dust; Knocked over; Noise damage and visual impairment; Continue drilling in the rejected hole.	Falling accidents; collapse accident; water and mud burst; explosive accident.	Choose the safe construction site and check the safety condition of the drill before drilling; The construction site safety work protection is done ; Wear labor protection articles.
4	preliminary bracing	Not wearing protective equipment or incorrect use; Dust; The concrete rebound.	object strike; Falling accidents.	Wear personal protective equipment; Do not enter the construction area, before the nozzle of construction is strictly prohibited; Use a working platform.
5	dregs transportation	Mechanical and human cross-work; Collapse of surrounding rock; The vehicle slidng, overturn and collision; Dust exceeded.	vehicle injuries.	Setting up a warning zone in a dregs area; The loading and unloading operation should be directed by special persons; Loose rocks should be dealt with first; It is not allowed to mix; It is strictly forbidden to overload, exceed limit and speed; Set tunnel pedestrian and vehicle access.
6	cross-operation	The construction sequence is unreasonable; Excavators and mechanical arrangements do not meet the flow of water; The layout of excavation is not reasonable; The ramp between the steps is too steep.	object strike; vehicle injuries.	Construction machinery and vehicles should be covered with reflective film outside the body; The speed of construction machinery and vehicles must not exceed 5km/h; The construction machinery and vehicles should turn on headlights when driving; Construction personnel should wear safety reflective helmet and reflective vest; A movable reflective warning sign should be placed on the upper and lower steps to alert drivers.

Through risk analysis, we can draw the influence factors of affect the process of construction safety, mainly has to clean the loose roof soil (include soil instability, top cleaning), the erection of support (include mechanical work, support interval, unstable arch foot), frozen soil excavation (include dust, noise, drilling, mechanical operation), primary support (include individual protection, dust, and concrete), sediment transport (include crossover operation, the instability of surrounding rock, slag discharge car, dust), crossover operation sequence (include construction sequence, builders, machinery, layout) and other factors.

5 Correlation analysis of influencing factors

For the above factors, we designed a questionnaire for the front-line construction personnel. A total of 1214 valid questionnaires were collected.

Through questionnaire survey, the relevant information of each influencing factor was calculated.

Then using SPSS software to carry out data analysis, the related coefficients of each influencing factor were obtained [3]-[8]. This is shown in "Table 2".

Table 2. The related coefficient between the safety factors of Gongbei tunnel excavation process.

Related coefficient / Factor M / Factor N		Human factors		Objective factors		Managerial Factors			
		Group leader	*Front-line constructors*	*Low temperature*	*Cross operation*	*Responsibility system*	*Educational training*	*Assessment*	*Rewards and punishments*
Clean the roof loose soil	***Soil instability***	0.912	0.812	0	0	0.856	0.658	0.328	0.786
	Top cleaning	0.911	0.885	0	0	0.754	0.645	0.355	0.812
Supporting erection	***Mechanical work***	0.523	0.964	0	0.875	0.823	0.765	0.576	0.825
	Support interval	0.825	0.967	0.512	0	0.867	0.945	0.465	0.764
	Unstable arch foot	0.852	0.982	0	0	0.853	0.854	0.532	0.756
Frozen earth excavation	***Dust***	0.545	0.565	0	0	0.546	0.589	0.532	0.465
	Noise	0.456	0.478	0	0.125	0.125	0.156	0.214	0.323
	Mechanical work	0.512	0.716	0	0.256	0.865	0.786	0.643	0.564
	Drilling work	0.752	0.867	0.216	0.126	0.764	0.574	0.649	0.598
Preliminary support	***Dust***	0.547	0.457	0	0	0.235	0.325	0.453	0.534
	individual protection	0.756	0.985	0.564	0.876	0.654	0.867	0.912	0.954
	Concrete	0.451	0.549	0	0.354	0.239	0.247	0.258	0.387

Dregs transportation	***Cross-operation***	0.765	0.778	0	1	0.766	0.656	0.812	0.885
	Surrounding rock failure	0.421	0.873	0	0	0.148	0.471	0.232	0.321
	Vehicle	0.321	0.857	0	0	0.788	0.786	0.789	0.879
	Dust	0.347	0.657	0	0	0.235	0.325	0.453	0.534
Cross-operation	***Construction sequence***	0.952	0.912	0	1	0.953	0.904	0.923	0.954
	Builders	1	1	0	1	0.852	0.886	0.876	0.907
	Machinery	0.765	0.965	0	1	0.854	0.868	0.829	0.867
	Layout	0.945	0.911	0	1	0.876	0.852	0.782	0.813

Through correlation analysis, it can be concluded that there are important factors related to the construction safety of Gongbei tunnel excavation. It mainly refers to the group leader, front-line constructors, cross-operation, construction sequence, responsibility system, education training, rewards and punishments.

6 Safety management countermeasures of tunnel excavation team

6.1 Technical strategies

According to the correlation of safety influence factors, it is suggested to select small sections and determine the reasonable excavation steps, spacing, support and types.

6.1.1 Excavation steps

The tunnel was eventually determined to be excavated in fourteen parts. It is divided into 5 steps: A, B, C, D and E. The steps are about 3.8 ~ 5.0 m. It is shown in Figure 2.

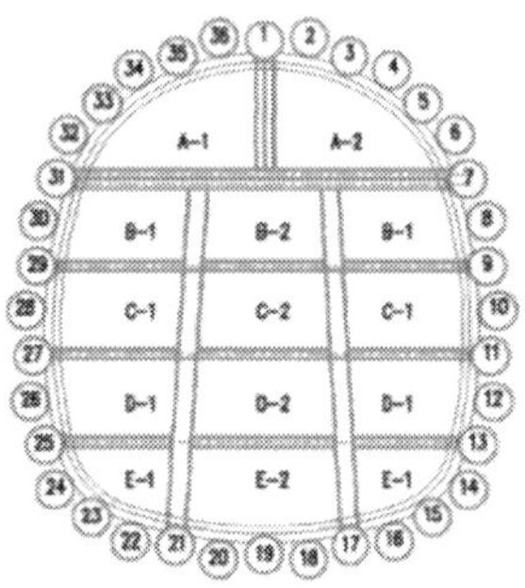

Figure 2. Tunnel excavation section layout.

6.1.2 Excavation space

a) The first step is to excavate a-2 guide pit first, then excavate a-1 guide pit, and the a-1 guide pit lags behind the a-2 guide pit about 5m.

b) The second step is separated from the first step by 20m.

c) The third step and the second step are separated by 15m.

d) The fourth step and the third step are separated by 15m.

e) The fifth and fourth steps are separated by 15m.

6.1.3 Support

a) The first step: for every 1 ~ 2 steel spacing (0.4m spacing), the initial support and temporary support structure should be followed. After the first step of excavation is completed, the instant support molds are cast concrete and the second lining.

b) The other steps: start to make a second line when the distance is more than 10m.

6.1.4 Excavation types

a) The first step: all the holes are excavated with PC55 small excavators, with small auxiliary tools such as pneumatic picks. The residual soil is transported with the self-unloading vehicle.

b) The other steps: first, the two side headings should be opened, then the center heading to be excavated. The center heading is spaced about 5m between the two side headings.

6.2 Construction strategies

6.2.1 Before work

a) The special construction plan should be prepared before the tunnel construction, and the on-site disposal plan of the construction site shall be improved, and the practice shall be organized regularly. In the construction, there shall be sufficient emergency and emergency supplies.

b) The tunnel excavation shall be adjusted according to its geological conditions, section size, suitable construction equipment, proper operation area and adjustment in the construction process.

c) Prior support should be checked before excavation, and whether the thickness of freezing meets the design requirements.

d) Before the operation of each construction process, the operation risk and safety measures shall be informed to all team members by the safety officer through the form of pre-class safety speech.

e) Before the excavation of the next cycle, it is necessary to check the initial support to ensure the safety of the construction work.

f) The roof cleaning must be directed by a specialist, and lighting should have adequate illumination.Safety confirmation must be carried out after clean the top. After qualified, other workers can enter the excavation work.

6.2.2 In the work

a) The excavation length must be controled strictly. Meanwhile, it shall be reduced the perturbation of surrounding rock, and do not damage the initial support, lining structure and construction equipment.

b) After excavation, the initial support and temporary support should be made to form a full - section circular closure. It is necessary to make the second lining structure as soon as possible. In the construction, the stability of the tunnel face and the ground of the temporary vertical support are always concerned.

c) The initial support and temporary support are followed by the excavation of the guide hole to complete the structure of the supporting body. When excavating, the special person should be appointed to conduct the excavation operation, Strictly limit the mechanical operation boundary to prevent collision steel frame.

d) When using mechanical excavation, the machine should be selected according to the tunnel's section and environment, and the safe operation area shall be defined and the warning signs shall be set. Non-operating personnel are not allowed in.

e) In the case of manual excavation, the workers shall maintain the necessary safety distance and set up a special conductor.

f) When the deformation rate of surrounding rock and support system is found to be abnormal, effective measures should be taken immediately and all personnel should be evacuated from the dangerous area when the situation is serious.

6.2.3 After the work

a) At the end of the day, the team leader and the safety officer should count the personnel. The team leader should review the work area and find out the security risks.

b) At the time of the transfer, the situation of work completion, the risk point of the construction process, and possible safety hazards are explained to the next group. The signing procedures of the succession shall be performed.

c) The tunnel construction plan should be adjusted in time according to the geological prediction.

6.3 Management countermeasures

6.3.1 Team organization

a) Group leader selection: The group leader of Gongbei tunnel should select a person with more than 10 years of experience in construction and should not be older than 55.

b) Construction personnel requirements: A construction worker with more than 5 years experience can bring one worker who has one year working experience. People without experience in tunnel construction are not allowed to enter the excavation team.

6.3.2 Team safety management system

The responsibility of the post should be clearly defined, the responsibility system should be implemented, and relevant security system (such as safety education training system, safety assessment and punishment system, etc.) should be formulated, and the relevant system should be strictly implemented. We will strengthen the management of the team from the three aspects of system construction, responsibility implementation, rewards and punishments.

7 Conclusion

Based on information research and questionnaire survey to Gongbei tunnel construction personnel, The basic data analysis of tunnel construction personnel and the risk analysis of excavation process are carried out. Using SPSS software to carry out correlation analysis, the important factors that influence the tunnel's construction safety are obtained. Finally, the countermeasures of the construction team has been proposed. The following conclusions are obtained.

a) The important factors related to the construction safety of the tunnel excavation are obtained. There are team leader, front-line construction personnel, construction step and spacing, cross operation, responsibility system, education training, rewards and punishments.

b) In the three aspects of technical strategies, construction strategies and management countermeasures, the safety management measures of the excavation team are formulated.

c) Through the application of relevant research results, the construction quality of tunnel engineering is effectively guaranteed, and the zero casualties target of the excavation team has realized.

This paper discusses the safety management of the tunnel excavation of Gongbei tunnel in order to provide reference for the safety management of other similar projects.

References

1. GUO Dongchen, SU Xinguo, HUANG Xuewen, ZHANG Weijun, WU Linsong, "Safety Management Mode of Bridge & Tunnel Construction Team Based on Demographic Characteristics," Transport Research, vol. 1, pp. 66–74, August 2015.
2. State Council Research Office, "Chinese migrant workers survey report," Yanshi Press. Beijing: 2006.
3. JIANG Yizhou, LI Zhiqiang, ZHANG Weijun, WU Linsong, "Influence Analysis of Psychological Factors of Bridge & Tunnel Builders on Operation Safety Based on SAS/SDS," Safety and Environmental Engineering, vol. 23, pp. 96–101, May 2016.
4. LIANG Zhendong, "Study on relationship of demographic characteristics with unsafe behavior and unsafe behavioral intention," Journal of Safety Science and Technology, vol. 10, pp. 46–52, January 2014.

https://doi.org/10.1051/matecconf/201820601015

5. ZHAO Jun, CAI Xiang, "Demographic Characteristics in Workplace Deviance," Soft Science, vol.28, pp. 112–116, August 2014.
6. Zhang Guiqun, Fang Weihua, "Status Survey and Variance Analysis on the Beijing Scientific and Technological Workers' Job Satisfaction Index:Based on Demographic Characteristics," Journal of Beijing University of Aeronautics and Astronautics(Social Sciences Edition), vol.26, pp. 1–6, November 2013.
7. Zhang Guiqun, Fang Weihua, "Status Survey and Variance Analysis on the Beijing Scientific and Technological Workers' Job Satisfaction Index:Based on Demographic Characteristics," Journal of Beijing University of Aeronautics and Astronautics(Social Sciences Edition), vol.26, pp. 1–6, November 2013.
8. WANG Panpan, LI Qiming, DENG Xiaopeng, "Study on Safety Competency Model of Construction Workers," China Safety Science Journal, vol.19, pp. 40–45, Augest 2009.

Study on the Calculation of the Post-construction Settlement after the Secondary Preloading of Sand Pile Composite Foundation of the High-speed Railway

ZUO Shen[1], LI Jin[2] and ZHANG Li[3]
1 Shandong jiaotong university, Jinan, China
2 Shandong jiaotong university, Jinan, China
3 Shandong jiaotong university, Jinan, China

Abstract. To complete the technology upgrade of the Kunshan section of the Beijing-Shanghai Railway, the secondary preloading of the original sand pile foundation was carried out in five years after the completion of the railway. Whether the control requirements of the post-construction settlement of the upgraded railway could be met was the key. Therefore, in this paper, the author analyses the consolidation deformation law of the sand pile foundation, sets up the membership function to describe the conversion from the drainage by consolidation to the composite foundation based on the Ascend Semi-cauchy Distribution Function of the Fuzzy Mathematics Theory, puts forward the time expression of the consolidation degree, verifies the computational accuracy of the expression by measured data and confirms that the e-lop curve method which considers the historical influence of the soil stress is more accurate in calculating the total settlement of the secondary preloading after comparison and calculation because this method can achieve the precise calculation of the post-construction settlement with the expression of degree of consolidation mentioned above.

1 Introduction

Please Sand pile composite foundation is a kind of bulk solid pile composite foundation which has been applied earlier. It is applicable to the compaction of the loose sand, silt, cohesive soil, plain fill, miscellaneous fill and other foundation treatment and is widely used in highways, railways, housing building and other infrastructure construction.

Therefore, the study on sand pile is significant in construction technology [1], [2], reinforcement mechanism [3]-[5], and design calculation [6]-[8], which has relatively mature theories and practical experience. However, applied research in railways is still rare. Because of the extremely strict control requirements of the post-construction settlement of the high-speed railways, the rigid pile which is safer and more conservative is used more in foundation treatment. Since the sand pile has a low bearing capacity and the consolidation deformation takes a long time, it often cannot meet the demand of the subgrade construction of the railways.

The testing section of Kunshan of the Beijing-Shanghai Railway was originally planned to adopt the ballast track. The control standard of the post-construction settlement of the general subgrade was no more than 10cm. Sand pile-net composite foundation and surcharge preloading were used for reinforcement. After that, ballastless track was used in the subgrade. The control standard of the post-construction settlement was raised to 15mm or less than 15mm. Thus, the surcharge preloading was carried out in the original testing section for the second time without changing the reinforcement measures relying on the sand pile-net composite foundation. Whether the control requirements of the post-construction settlement of the railway could be met became the key. Therefore, the calculation of the post-construction settlement after the secondary preloading of sand pile composite foundation which has been put aside for five years is demanded to be highly accurate. Nowadays, the settlement calculation method of the sand pile is mainly applied to harbors, highways, housing building and other fields. It is difficult to satisfy the strict standard of no more than 15mm. Therefore, in this paper, the author analyses the consolidation deformation law of the sand pile foundation, considers the historical influence that the secondary preloading exerts on the soil stress, puts forward the degree of consolidation and the calculation of the total settlement, which provides suggestions and help to the sand pile foundation in the field of railways.

2 General introduction to the test

The foundation of the testing section belongs to the quaternary holocene alluvia (Q4al). The surface layer is the yellowish clay which is in soft-hard plastic state and is 0.76-3.60m thick. Below is the muddy silty clay which is dark gray and in fluidity plastic state. It contains a small amount of humus and thin silt on the local scale. It is of high compressibility and thixotropy and low strength. The thickness ranges between 3.2 meters to 16.5 meters. The substratum is made up of clay, silt and silty clay and it is in soft-hard plastic state. The foundation of the testing section is 13.8-meter-wide. In consideration of the settlement of the embankment and the post-construction settlement of foundation, 0.2 meters have been widened on each side and the distance between centers of tracks made 5.0 meters. The height of the embankment filling ranges between 4.2 meters to 5.5 meters. The slope ratio of the side slope of the embankment is 1:1.5 The slope surface of the side slope is defended with geosynthetic fiber mattress to enforce the soil and protected by planting grasses by spraying sowing. During filling the embankment, geogrids have been placed every 0.3 meters within 2.5 meters of the side scope. The reinforced length of the sand pile-net composite foundation is 124.7 meters and the reinforcement measures are shown in Table 1. Preloading of the sand pile-net composite foundation of the Kunshan section has been carried out for two times, which is shown in Table 2.

Table 1. The reinforcement measures of the sand pile-net composite foundation.

Reinforce-ment measures	Pile spaci-ng /m	Pile Dia-meter /m	Reinforced depth/m	Pile arrange-ment	Notes
Sand pile-net composite foundation	2.0	0.4	15.0	Quincunx	0.6-meter-thick sand cushion, a layer of geogrids

Table 2. Two preloading of the sand pile-net composite foundation.

No.	The time of beginning loading	The time of stopping loading	The height of loading	Obse-rvation time	The time of unloading	The height of over-loading
1	2003-5-19	2003-10-17	6.29m	379d	/	0
2	2008-9-30	2008-10-16	2m	380d	2009-7-12	2m

3 Consolidatio calculation of sand pile composite foundation

The sand pile-net composite foundation meets the requirement for post-construction settlement of the ballast track subgrade after the second preloading in September 30th, 2008, 1810 days after the first preloading in October 17th, 2003. How the settlement of the sand pile-net composite foundation develops in five years after the first preloading, whether the secondary preloading is necessary or not, how the calculation method of the total settlement after the secondary preloading influences the preloading measures of the sand pile-net composite foundation with low and medium compressibility are of great significance.

3.1 Introduction of the consolidation deformation law

The law of how the settlement of the sand pile-net composite foundation develops in five years after the first preloading is the variation characteristic of the consolidation. Based on the consolidation theory of the sand drained foundation and the composite foundation, and the Ascend Demi-cauchy Distribution Function of the Fuzzy Mathematics Theory which describes the degree of how the sand pile-net composite foundation is attached to the vertical drains foundation and the composite foundation at some point, the time expression of the consolidation degree of the sand pile-net composite foundation is proposed. The key to this expression is the membership function of the Fuzzy Theory. In this paper, β, which is in $U = 1 - \alpha \cdot e^{-\beta t}$, expresses how the sand pile foundation is attached to the vertical drains foundation and the composite foundation. According to the measured value of the settlement in equal time intervals, how β changes along with the change of t is shown in Figure 1. The expression is shown as follows:

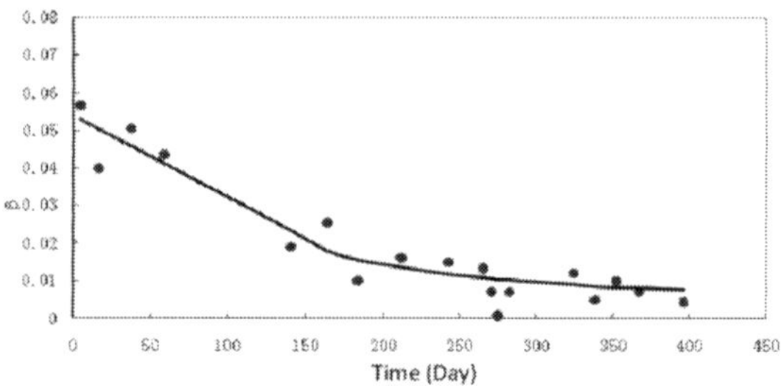

Figure 1. Trend graph on β-t of the sand pile-net composite foundation.

β of the sand drained foundation generally ranges between 0.04 and 0.07 (1 per day) and that of the cement-soil pile composite foundation ranges between 0.001 and 0.003 (1 per day). The independent variable of the Ascend Demi-cauchy Distribution Function, however, mainly ranges between 1 and 60. Given this, in this paper, β'(1000β) is the independent variable. Then the function that describes the degree of how the sand pile foundation is attached to the vertical drains foundation is shown as follows:

$$\beta(t) = \begin{cases} -0.00022t + 0.0549 & 0 \le t \le 164 \\ 0.3 \times (t - 83.369)^{-0.6419} & t > 164 \end{cases} \tag{1}$$

$$\mu_A(\beta') = \begin{cases} 0 & (0 \le \beta' \le 1) \\ \dfrac{0.03(\beta'-1)^2}{1+0.03(\beta'-1)^2} & (1 < \beta' < 60) \\ 1 & (\beta' \ge 60) \end{cases} \tag{2}$$

According to related parameters of the fracture surface, the time expressions of the sand drained foundation and composite foundation are respectively:

$$U_{\text{井}}(t) = 1 - 0.469 \times \exp(-t/12.45) - 0.323 \times \exp(-t/80.58) \tag{3}$$

$$U_{复}(t) = 1 - 0.7778 \times \exp(-t/370.81) \quad (4)$$

According to (2) (4) and (5), the expression of the consolidation degree of the sand pile-net composite foundation at time t can be worked out as follows:

$$U(t) = \mu_A(\beta') \cdot U_{井}(t) + [(1-\mu_A(\beta'))] \cdot U_{复}(t) \quad (5)$$

In the expression:

U(t): The consolidation degree of soft foundation reinforced by sand pile at time t.

U 井(t): The consolidation degree of foundation at time t according to the vertical drains theory.

U 复(t): The consolidation degree of foundation at time t according to the consolidation theory of composite foundation.

$\mu_A(\beta')$: The membership function that describes to what extent the sand pile foundation can be regarded as the sand drained foundation. In the function, β' is equal to 1000β.

3.2 Application of consolidation deformation law

According to Table 3 and Table 4, the estimated value of the post-construction settlement of the foundation center when t is 366d is 64.5 mm while the predicted value is 57 mm. It shows that the estimated value based on the Ascend Demi-cauchy Distribution Function of the Fuzzy Theory is slightly larger than the predicted value, which is much safer. The relative error between them is 13.1%, so it can be used in the calculation of post-construction settlement.

Table 3. The estimated value of consolidation degree based on the Ascend Demi-cauchy Distribution Function of the Fuzzy Theory.

Time/d	0	150	366	500	1000	1500	1963
Degree of consolidation/%	0	68.1	82.6	83.5	91.4	95.0	95.9

Table 4. The predicted value of the post-construction settlement after the first preloading (Unit:mm).

Location		Three-point method	Hyperbolic method	Asaoka Method	GM(1,1) Method
The foundation center (314)	The estimated value of the total settlement	359	371	351	351
	Post-construction settlement	45	57	37	37
The left observation pile (283)	The estimated value of the total settlement	310	331	309	308
	Post-construction settlement	27	48	26	25
The right observation pile (265)	The estimated value of the total settlement	292	308	288	289
	Post-construction settlement	27	43	23	24

Up to September 30, 2008 (when t is 1963d), the estimated value of the post-construction settlement based on the Ascend Demi-cauchy Distribution Function of the Fuzzy Theory is 15.25mm, which does not meet the control standard of the post-construction settlement of the ballast track subgrade. Therefore, the secondary preloading has been carried out to the sand pile-net composite foundation in the primary test.

The settlement from the first preloading on May 19, 2003 to the end of the second unloading on October 15, 2009 is shown in Figure 2. The deadline of observing the settlement during the first preloading is when t is 366d. The settlement from the deadline to the secondary preloading (when t is 1963d) is estimated based on the Ascend Demi-cauchy Distribution Function of the Fuzzy Theory.

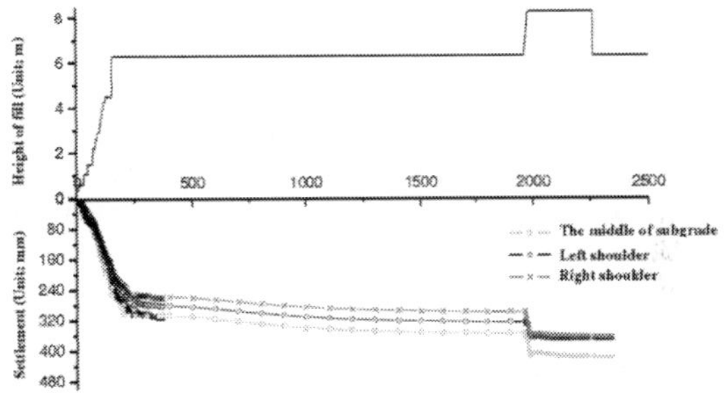

Figure 2. The settlement of the first and secondary preloading of testing section of Kunshan.

According to Table 5, the predicted value of the total settlement of the foundation center after the second preloading is 68.60mm while that of the residual settlement after the first preloading is 15.25mm, so the value of the total settlement after the second preloading is 53.35mm.

As is shown in Table 5, the completion rate of the settlement after the end of the secondary filling is very high. 30 days later, the rate is 76%-89.2%; 180 days later, the rate is 91.7%-95.3%; 320 days later, the rate is 93.1%-99.6%. That is, from the end of filling to the stabilization of the settlement, the completion rate can be regarded as 91.7%-93.1%. Therefore, the post-construction settlement can be figured out based on the completion rate after the filling in a certain time and the total settlement.

Table 5. The completion rate of the settlement after the secondary preloading.

Location		The foundation center	The left observation pile	The right observatio n pile
The predicted value of the total settlement/mm		68.60	46.01	66.78
At the end of the secondary filling	The measured value of the settlement/mm	48.1	27.97	50.17
	The completion rate of the settlement%	70.1	60.8	75.1
30 days after the end of the secondary	The measured value of the settlement/mm	52.43	40.05	52.43

 https://doi.org/10.1051/matecconf/201820601016

filling	The completion rate of the settlement%	76.4	87.0	89.2
180 days after the end of the secondary filling	The measured value of the settlement/mm	62.9	46.04	63.66
	The completion rate of the settlement%	91.7	100	95.3
320 days after the end of the secondary filling	The measured value of the settlement/mm	63.81	45.84	64.85
	The completion rate of the settlement%	93.1	99.6	97.1

3.3 Calculation of total settlement after the secondary preloading of the sand pile foundation

The total settlement is composed by instantaneous settlement, primary consolidation settlement and secondary consolidation settlement. The instantaneous settlement is calculated according to the elastic theory. Based on the layer-wise summation method, the primary consolidation settlement is calculated according to the coefficient of compression, modulus of compression, e-p or e-lgp compression curve. In this paper, the results according to e-p or e-lgp compression curve are compared and the calculation method of the total settlement is finally determined.

(1)E-p Curve

Hyperbolic equation describes the test curve of the relationship between the soil's stress and strain. As the research suggests, the relationship between compressive strain (ε) and effective consolidation stress (p) also satisfy the equation.

$$\varepsilon = \frac{p}{A+Bp} \quad \text{or} \quad \frac{1}{\varepsilon} = \frac{A}{p} + B \tag{6}$$

ε -the compressive strain; p-the compressive stress; A and B -undetermined coefficients.

According to the compression test, the relationship between compressive strain and void ratio is:

$$\varepsilon = \frac{e_0 - e}{1+e_0} \tag{7}$$

Therefore, A and B can be figured out via least square method based on e-p curve.

$$e = e_0 - \varepsilon(1+e_0) = e_0 - \frac{(1+e_0)p}{A+Bp} \tag{8}$$

1810 days after the end of the first filling, effective consolidation stress (p) of each soil layer can be regarded as the additional stress (σ_s) at the corresponding depth. The pore water pressure is ignored at the beginning of the second filling. According to the e-p compression test at different depth, the void ratio e0' when the effective consolidation stress is σ_s can be figured out, which is the initial void ratio of soil at the beginning of the second filling. Let the effective consolidation stress at the very beginning of the secondary preloading be $\sigma_{s1}{}^{i}$ and the effective consolidation stress after the secondary preloading be $\sigma_{s1}{}^{i} + \Delta p^{i}$. The total settlement after the secondary preloading can be figured out based on the e-p curve when Δp^{i} is the subsidiary stress of the secondary preloading at the corresponding depth. The calculation process of the e-p curve is shown in Figure 3.

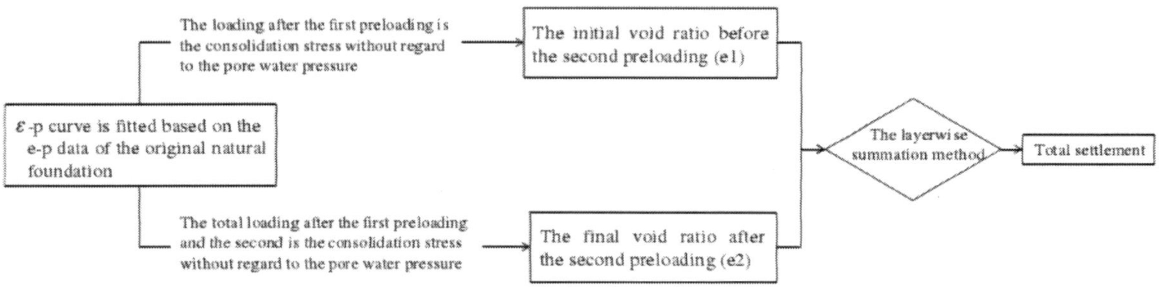

Figure 3. The calculation flow chart of the e-p curve.

According to Table 6, the total settlement after the secondary preloading is 191.3mm based on the e-p curve while the predicted value is 53.35mm. That is, the calculated value is 3.6 times as large as the predicted one because of the ignorance of the effect that the stress history makes on the settlement in calculation. Therefore, the e-p curve is unfit for the calculation of the total settlement of the secondary settlement.

Table 6. The calculation results of the secondary preloading based on the e-p curve.

Number of layer	Thickness of layer/m	Initial void ratio (e1)	Final void ratio (e2)	Layered settlement/mm
(1)	2.86	0.811	0.792	31.61
(2)	3.24	0.946	0.919	47.84
(3) -3	5.63	0.652	0.644	28.20
(3) -4	4.37	0.876	0.863	30.34
(3) -5	6.68	0.829	0.815	53.36

(2) E-logp Curve

E-logp curve shows the stress history of soil. The calculation process of the initial void ratio (e1) before the secondary preloading is the same as that of the e-p curve method. Underground water is two meters below the earth's surface. Because the first preloading makes the pre-consolidation stress of the second preloading increase, The calculation process of the e-lop curve is shown in Figure 4:

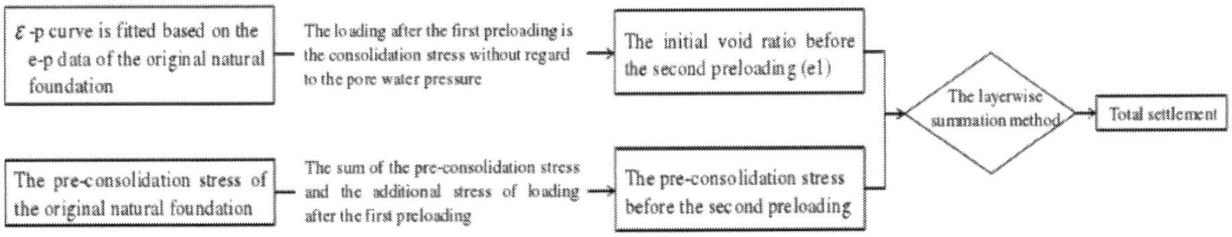

Figure 4. The calculation flow chart of the e-lop curve.

Table 7. The calculation results of the secondary preloading based on the e-lop curve.

Number of layer	Thickness of layer/m	Compression index	Swelling index	Pre-consolidation stress/kPa	Layered settlement
(1)	2.86	0.25	0.03	332.03	4.83
(2)	3.24	0.30	0.04	212.79	23.87
(3)-3	5.63	0.14	0.06	372.89	13.27
(3)-4	4.37	0.2	0.02	396.64	2.37
(3)-5	6.68	0.33	0.02	345.67	2.82

According to Table 7, the total settlement is about 47.16mm based on the e-logp curve that considers the historical influence of the stress while the measured value is 53.35mm which is slightly larger than the former. The relative error is 11.6%. Because e-logp curve method do not take the settlement caused by lateral displacement into consideration, the value of total settlement based on the e-logp curve method is slightly smaller than the measured. Therefore, e-logp curve method can meet the engineering requirements in calculating the total settlement after the secondary loading.

4. Conclusion

In this paper, the author takes the testing section of Kunshan of the Beijing-Shanghai Railway as an example and studies the calculation of the settlement of the sand pile composite foundation after the secondary preloading which has been put aside for a long time. The conclusion is shown as follows.

(1) The consolidation characteristics of the settlement of the sand pile foundation change over time. That is, drainage by consolidation turns into composite foundation. Based on the Ascend Demi-cauchy Distribution Function of the Fuzzy Mathematics Theory which describes the degree of how the sand pile foundation is attached to the vertical drains foundation and the composite foundation at some point, the time expression of the consolidation degree of the sand pile foundation is proposed. The expression is proved to be of higher accuracy by measured data.

(2) After comparison, the e-lop curve method which considers the historical influence of the soil stress is more accurate in calculating the total settlement after the secondary preloading. The post-construction settlement of the railway can be figured out with the expression of degree of consolidation mentioned in this paper.

References

1. Huang Ruijiang. The application of composite foundation in solving the problem of bridge bump. Journal of China & Foreign Highway, Vol.4, pp : 188-190, April 2007
2. Li Tao, etc. The construction technology on the foundation strengthening of the open caisson of the north anchorage of the Fourth Nanjing Yangtze Bridge. Journal of Highway,Vol.6, pp: 20-25.July 2010
3. Zheng Junjie, etc. The laboratory model experiment on the multi-element composite foundation with steel pipe pile and sand pile. Journal of Huazhong University of Science and Technology (Natural Science Edition), Vol.9, pp : 121-124.September 2008
4. Linfei, Performance evaluation of the sand compaction pile of the expressway in dealing with the liquefied soil.. Journal of Highway Traffic Science and Technology (Application Technology Edition), Vol.10, pp: 125-127. October 2010
5. Wang Xianzhi, Zheng Junjie and Li Xiangdong, Study on the laboratory model experiment of the sand pile composite foundation. Journal of Huazhong University of Science and Technology (Urban Science Edition), Vol.04, pp : 22-25. April 2007
6. Zhao Shouxian, The calculation methods of the settlement of sand pile composite foundation based on CPT. Journal of Geotechnical Engineering World, Vol.06, pp : 36-39. July 2007
7. Zhang xi, Wu Xinyi and Yin Haiqing. Reinforcement mechanism of compacting sand pile underwater and calculation methods of the settlement.. Journal of Chinese Harbour Engineering, 2010 Vol.S1, pp: 148-150.Janury 2010
8. Gu Xiangkui, Wang Xiaohui. The design of the composite foundation with compacting sand pile underwater of the gravity wharf. Port and Waterway Engineering, Vol.11, pp:227-231. November 2011

Construction Stability Analysis of Curved Continuous Rigid Frame Bridges with High Piers

Guohua Song[1], Minghui Li[2] and Delu Che[3]

[1]*School of Civil Engineering, Zhengzhou University, 450001,Zhengzhou, China*
[2]*The First Construction Co., LTD of China Construction First Group, 201100, Shanghai, China*
[3]*THenan Provincial Communications Planning and Design Institute Co., LTD, 450000, Zhengzhou, China*

Abstract. Taking the curved continuous rigid frame bridges (CRFBs) with high piers as research object, the construction stability of four different styles of main piers were analyzed by finite element (FE) method in three construction stages including the segment 0 stage, largest cantilever stage, and finished bridge stage. Rules governing the stabilities of bridges with various pier styles, height, and slenderness ratio were established. At different construction stages, pier type and pier height affect structural stability differently. Wind loads have little influence on structural stability at any construction stage, and accidental events influence the stability of bridges more than normal construction loads. There is a nonlinear power relationship of fixed functional type between the stability safety factor (SSF) and slenderness ratio in all four pier styles; this power relationship can be used to predict the stability of a new pier.

1 Introduction

Rapid economic advancement in China has created increasing demand for new road traffic construction. In Western China or remote mountainous regions, however, bridges have to span rivers, valleys, and other complex terrain. Long-span continuous rigid-frame bridges (CRFBs) are common in these areas, containing piers that are often over one hundred meters high. Long-span CRFBs have flexural structure, low cost, good adaptability, and facilitate smooth and comfortable driving. The high piers must be sufficiently flexible to ensure safely low internal forces and to minimize unwanted mechanical influence on the superstructure due to temperature, shrinkage, concrete creep, or natural disasters such as earthquakes. The piers are typically built with small sections to accommodate their tall height, generally with a height to span ratio of less than 1/10. The pier height must be even taller when a bridge crosses a mountain or deep body of water. Curved CRFBs with piers more than 50 meters high require cantilever construction technology, longer spans, and more box sections; they are also influenced by curvature, so their overall stability and local stability have considerable impact on the safety of their construction and maintenance. The overall stiffness and local stiffness of these bridges tend to be weak and must be mitigated accordingly.

Several previous researchers have studied the influence of different loads on CRFBs [1]-[4] and analyzed their structural stabilities in different construction stages [5]-[11]. Some have also discussed the mechanical principles and various methods of analyzing them [8], [12]. Others have investigated the influence of main pier design parameters on stability [3], [10], [11]. Many have conducted extensive studies on stability and continuously improved the materials and configurations of CRFBs [13]. There have been relatively few studies focused specifically on curved CRFBs [9], [10], however. C. S. Jason and his co-workers [14]-[16] have studied the stresses, behavior, and analysis method of curved steel I-girders bridges during lifting, but the construction stability of curved CRFBs is still to be investigated.

Curved CRFBs with high piers represent a combination of the advantages of curved girder bridges and CRFBs. The main span of a curved CRFB is continuous; its upper girder and lower piers are consolidated. The consolidation system allows girders and piers to bear loads simultaneously, but the type of pier substantially influences the internal forces and deformation to the bridge.

In order to investigate the construction stability of curved CRFBs, we assessed the influence of construction loads, wind loads, and other accidental loads during construction on structure stability. We used a curved CRFB in mountainous area of Western China as a practical engineering prototype, and ran the finite element (FE) program Midas/civil to build bridge models in three construction stages including the segment 0 stage, largest cantilever stage, and finished bridge stage. The construction stabilities of the models and their relationships with pier parameters were analyzed and several conclusions were drawn as discussed below.

2 Stability theories

CRFBs with high piers are a typical beam-column structure which features extrema point instability. When the structure is in an initial equilibrium state, plastic deformation occurs at high stress areas and continues to increase as the load on the structure increases. When the load reaches a certain value, the deformation rapidly spikes even as the load remains the same – at this point, the bridge is in danger of structural failure. This process can be expressed numerically as follows:

$$([K_D] + [K_G])\{\delta\} = \{F\} \quad (1)$$

where $[K_D]$ is the element geometric stiffness matrix; $[K_G]$ is the initial stress stiffness matrix; $\{\delta\}$ is the nodal displacement array; and $\{F\}$ is the internal force array on members.

Suppose $\{F\}$ increases k times. $[K_G]$ will also increase k times, so:

$$([K_D] + k\,[K_G])\{\delta\} = k\,\{F\} \quad (2)$$

If k is sufficiently large, the structure can maintain an indifferent equilibrium sate. That is, when $\{\delta\}$ develops to $\{\delta+\Delta\delta\}$, equation (2) can be rewritten as follows:

$$([K_D] + k\,[K_G])\{\delta+\Delta\delta\} = k\,\{F\} \quad (3)$$

The condition satisfying equation (2) and equation (3) simultaneously is:

$$([K_D] + k\,[K_G])\{\Delta\delta\} = 0 \quad (4)$$

Equation (4) is the characteristic equation for computing stability safety factor (SSF). If the equation has n orders, n eigenvalues will exist as $k_1, k_2, \ldots, k_n$. In a practical engineering case, only the minimum eigenvalue k_1 has practical significance, so the minimum eigenvalue k_1 is denoted as k_{cr} or as stability safety factor κ, and $k_{cr}\{F\}$ is defined as the critical load.

Critical load is expressed via Euler's Formula as:

$$F_{cr} = \pi^2 EI / (\mu L)^2 \quad (5)$$

where EI is structural stiffness; μ is a factor related to boundary conditions; and L is member length.

SSF κ is a direct assessment index reflecting structural stability. It can be expressed as the ratio of $\{F\}_{cr}$ to $\{F\}$:

$$\kappa = \{F\}_{cr} / F \quad (6)$$

where $\{F\}_{cr}$ is the total instability load array of structure; $\{F\}$ is the actual load array.

3 Model building

3.1 Bridge design

One curved CRFB model with high piers was built with spans of 61 + 104 + 61 m, curvature radius of 348 m, pier height of 63 m, and central line of piers normal to the central line of the roadway. The superstructure had a cross slope of 5 percent and was 22 m wide. It was composed of two girders, each comprising a single-box single-room section with width of 11 m and varying height by quadratic parabola. The piers were 63 m high. Other design details were as follows.

3.1.1 Design technical standards

The highway classification was an expressway with four traffic lanes, design speed of 80 km/h, traffic load of Highway grade I, and designed flood frequency of grand bridge 1/200.

3.1.2 Main construction materials

(1) *Concrete*: C50 was for superstructure, C40 was for piers and cross beams.

(2) *Prestressed tendon*: Prestressed tendons were considered only in the longitudinal direction with a nominal diameter of 15.2 mm, comprised of a steel strand of seven wires with high strength and low relaxation. The tendons had elasticity module E_p = 1.95 × 10^5 Mpa, standard strength f_{pk} = 1860 Mpa, and less than 2.5% of stress relaxation rate in 1000 hours; a large tonnage grouped anchor system and vacuum mortar suction technique were used to fabricate them.

(3) *Rebars*: The main reinforcements were HRB400.

(4) *Bearings*: Seismic steel bearings were used for end bearings; WKJ-QZ4000DX for the inner side and WKJ-QZ4000SX for the outer side.

3.1.3 Box girder size

A continuous rigid-frame girder with varying sections was used in the superstructure. The baseboard thickness d varied uniformly from 80 cm at the pier-top section to 25 cm at the midspan section, the web thickness t varied uniformly from 70 cm at the pier-top section to 50cm at the midspan section, and the girder width h varied by quadratic parabola from 6 m at the pier-top section to 2.3 m at the midspan section. The girder had a top board 11 m wide, a base board 6 m wide, and two cantilevers 2.5 m long. The detailed section size of the girder is shown in Figure 1.

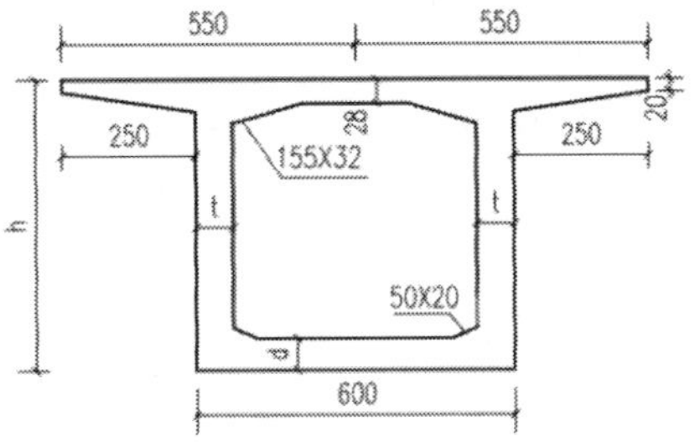

Figure 1. Cross section of box girder at pier top (unit: cm).

3.1.4 Piers

The piers of the bridge were double-limb thin-walled piers with cross beams. Each limb had a rectangular cross section of 2×6 m and the distance between two limbs was 5 m. There was a cross beam with an interval of 21 meters between any two limbs.

3.1.5 Construction process

The prestressed curved CRFB was constructed in order as the pile foundation and caps, piers, and casting segment 0. All segments but the closure segment and cast-in-site segment on the side span were cast successively via hanging basket cantilever pour symmetrically. Eleven segments were constructed on both sides simultaneously. The steel tendons were prestressed after each segment was finished. This process continued until the largest cantilever stage was completed.

3.2 FE modeling

We used the Midas/Civil program, which is applicable for analysis and design of bridge structures, underground structures, industrial buildings, airports, dams, harbors, and other engineered structures. We built the FE models of bridges including their superstructure, which was modeled with beam elements. The girder and piers were rigidly connected, the bottoms of piers were comprised of conventional bearings, and the supports on both ends were elastically connected according to support parameters. Analysis was conducted via the lumped mass method. The FE model of the bridge at the finished bridge stage is shown in Figure 2.

Figure 2. FE model at finished bridge stage.

4 Stability at segment 0 stage

4.1 Analysis under gravity and wind loads

The bridge becomes softer and increasingly susceptible to wind load as pier height increases. It is necessary to take wind loads into account to fully analyze structural stability in the segment 0 stage. The designed national wind pressure in the transverse direction of the pier was determined according to Wind-Resistant Design Specification for Highway Bridges (JTGT D60-01-2015) [17]. The wind loads across piers in our experiment were as follows.

Wind pressure on the piers across the bridge: (0.68 - 1.42) kN/m^2, with width of piers along the bridge of 6 m.

Wind load at pier's bottom across the bridge: 1.36 kN/m,

Wind load at pier's top across the bridge: 1.42 kN/m.

According to (JTGT D60-01-2015) [17], the characteristic value of wind pressure along the bridge is 70 percent of that across the bridge; the width of piers across bridge is 2 m, so the wind loads along the bridge are as follows.

Wind load at pier's bottom along the bridge: 2.86 kN/m,

Wind load at pier's top along the bridge: 5.96 kN/m.

Diagrams of these wind loads are provided in Figure 3.

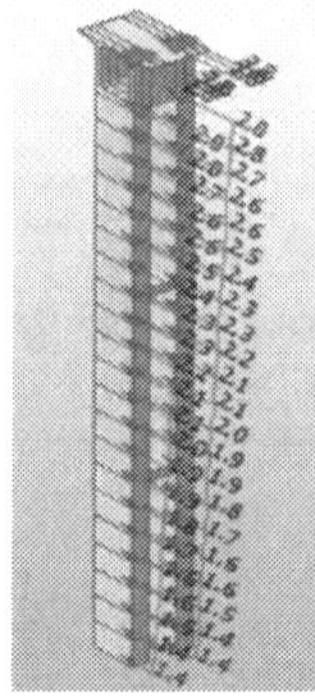

(a) Across the bridge

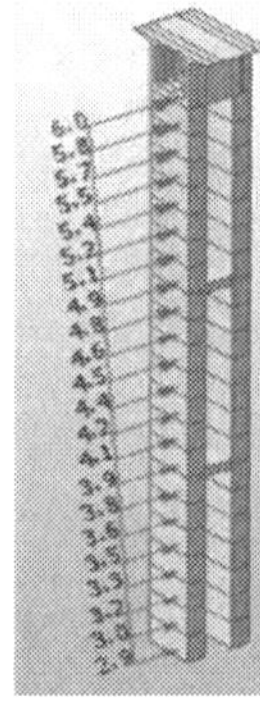

(b) Along the bridge

Figure 3. Wind load diagrams.

The load cases were classified into three types as shown in Table 1. After analysis, SSFs and first-order bucking modes were obtained as shown in Table 2.

Table 1. Load combination at segment 0 stage.

Load case No.	Load combination
LC1-1	(i)
LC1-2	(i) + (ii)
LC1-3	(i) + (iii)

Table 2. Stability at segment 0 stage under various load cases.

	LC1-1	LC1-2	LC1-3
SSF	60.27	60.26	60.26
The first order of buckling mode	Bucking along the bridge		

Table 2 shows where the overall stability of the pier is fairly uniform under wind load across the bridge or along the bridge. The first-order buckling mode of the pier extends along the bridge under all three load cases. The SSF values all well exceed 4, indicating that the structure is reasonably stable at the segment 0 stage [18], [19].

4.2 Influence of pier height on structural stability

Four models of different pier types were built: A double-limb pier with a cross beam (DLP with CB), a double-limb pier without a cross beam (DLP without CB), a single-limb hollow pier with wall thickness of 0.5 m

(SLHP with W0.5), and a single-limb hollow pier with wall thickness of 1.0 m (SLHP with W1.0). The cross sections of the piers are shown in Figure 4. For every type of pier, the model was built with varying height from 40 m to 120 m with an interval of 10 m. Stability analysis was conducted under load case LC1-3 (Table 1). The SSF curves were plotted against pier height as shown in Figure 5.

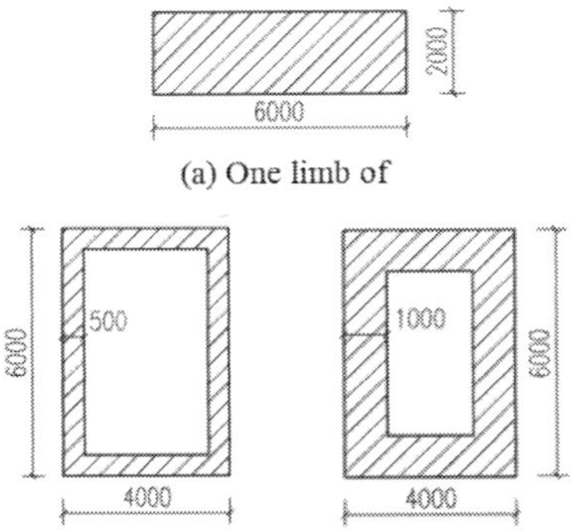

(a) One limb of

(b) SLHP with W0.5 (c) SLHP

Figure 4. Cross sections of four pier types (unit: mm).

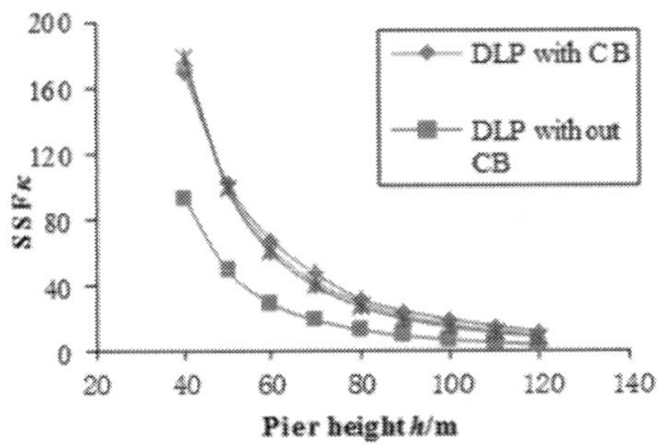

Figure 5. Curves of SSF against pier height at segment 0 stage.

4.3 Influence of slenderness ratio on structural stability

The SSF data k were nonlinearly power-fitted and polynomial-fitted against the slenderness ratio λ of the piers; the functions were generally fitted well [10]. For DLP with CB models, the functions could be expressed as equation (7) and equation (8) (Figure 6),

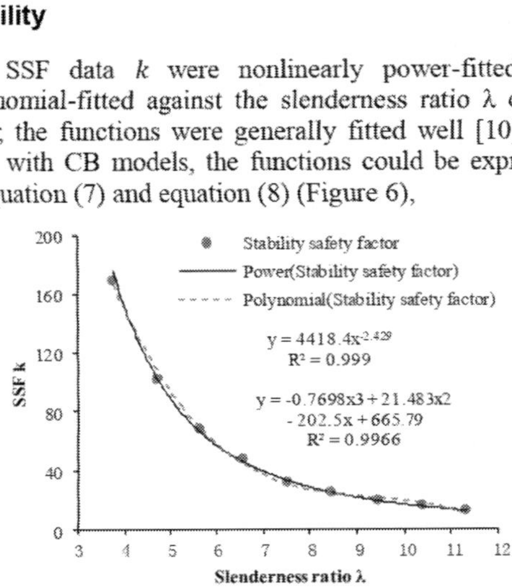

Figure 6. Different expressions of SSF vs. slenderness ratio.

$$k = 4418.4\,\lambda^{-2.429} \tag{7}$$

$$k = -0.7698\lambda^3 + 21.483\lambda^2 - 202.5\lambda + 665.79 \tag{8}$$

For a compression bar in *Material Mechanics* [20], the relationship between compressive stress σ_{cr} in the cross section and slenderness ratio λ is defined as follows,

Slenderness ratio,

$$\lambda = \mu L / i\,,\ i = (I/A)^{1/2} \tag{9}$$

Compression stress in cross section,

$$\sigma_{cr} = \pi^2 E / \lambda^2 \tag{10}$$

where A, I, i are the area, inertia moment, and gyration radius of the compressive bar, respectively; E, μ are Poisson's ratio and the length ratio of the compressive bar, respectively.

In the design of a compressive bar, denote allowable stability stress $[\sigma]_{st}$ as the product of material strength $[\sigma]$ and SSF κ which varies with the slenderness ratio λ of the bar [20]. Eqaution (7) is closer to the physical interpretation of a relationship between κ and λ in a pier; it also fitted better per our calculations than equation (8). So we chose equation (7) as the expression of the relationship between k and λ values for all subsequent analyses.

SSF k was nonlinearly fitted against slenderness ratio λ for the four types of piers. The fitted curves are shown in Figure. 7 and are expressed in equations (11) - (14).

For DLP with CB,

$$k = 4418.4\,\lambda^{-2.429} \tag{11}$$

For DLP without CB,

$$k = 99126\,\lambda^{-2.972} \tag{12}$$

For SLHP with W0.5,

$$k = 973740\,\lambda^{-2.629} \tag{13}$$

For SLHP with W1.0,

$$k = 2.0 \times 10^6 \lambda^{-2.751} \tag{14}$$

All the above functions of SSFs are expressed as power functions, so the following features of the power function $y = bx^{-a}$ (a, b are all positive numbers) merit further discussion here.

(1) The exponent a and factor b determine the curve slope. A smaller a or larger b (or vice versa) mean a steeper function curve (or a wider opening of the curve).

(2) The opening size of the power functional curve is determined by a and b together.

(3) When a and b are both large, the curvature of the function cannot be inferred.

Equation (12) yields the lowest value among equations (11) - (14), but the other three formulas cannot be effectively compared. The curves of k versus λ have almost the same shapes as the curves of the factor versus pier height (Figure 6), because the same set of pier data have the same height h.

5 Stability at largest cantilever stage

5.1 Stability during construction

We used hanging-basket cantilever construction to build the superstructure of the girder bridge. This process

https://doi.org/10.1051/matecconf/201820601017

can be summarized as: constructing main pier → segment 0 of main girder → segment 1 → stressing tendons of segment 1 → segment 2 → ...→ segment 11 → stressingtendons of segment 11. In short, the largest cantilever stage was the most unstable stage. All the worst load conditions had to be considered to properly assess the worst-possible construction girder → segment 1 → stressing tendons of segment 1 → segment 2 → ...→ segment 11 → stressing tendons of segment 11. In short, the largest cantilever stage was the most unstable stage (Figure 7). All the worst load conditions had to be scenarios in regards to structural stability. The FE model of the largest cantilever stage is shown in Figure 8. A total of 12 segments were analyzed, including pier 1 and segment 0 to segment 11. The boundary conditions of the model were rigid solidification at the pier bottom and free at the two cantilever ends.

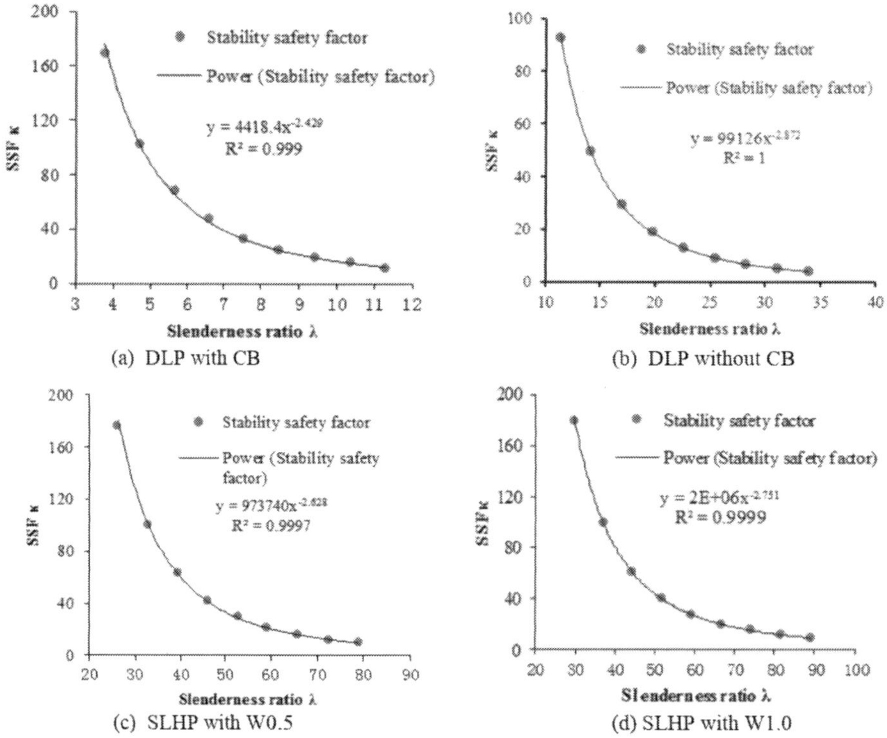

Figure 7. Nonlinearly fitted curves of SSF against slenderness ratio at segment 0 stage.

Figure 8. FE model at largest cantilever stage.

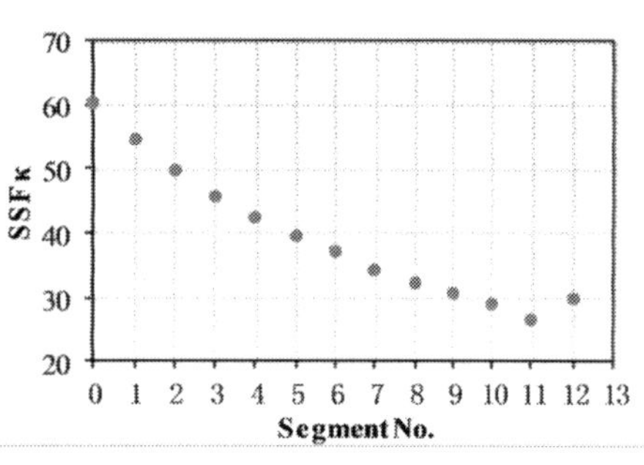

Figure 9. Stability of construction stages.

The stability throughout the construction process was analyzed from stage 0 to stage 11, i.e., the largest cantilever stage. Models at 12 construction stages from segment 0 - 11 (where the finished bridge stage corresponds stage 12) were built and analyzed per the stability under gravity alone. The resulting SSFs are as shown in Figure 9.

As shown in Figure 9, structural stability decreases as construction progresses. It reaches a minimum at the largest cantilever stage, but is resolved at the finished bridge stage. The length and the mass of the cantilever at both sides increase along the construction process while the structure becomes softer; at finished bridge stage, however, the structure has stronger whole rigidity and stability due to the support restriction at both ends.

5.2 Influence of construction loads on stability at largest cantilever stage

The template is designed so that author affiliations are not repeated each time for multiple authors of the same affiliation. Please keep your affiliations as succinct as possible (for example, do not differentiate among departments of the same organization). This template was designed for two affiliations. Construction loads at the largest cantilever stage include: (i) Gravity; (ii) longitudinal prestressing forces; (iii) loads from construction equipment and materials placed unevenly; (iv) gravity of the hanging basket with dynamic effect (1200 kN, with dynamic coefficient of 1.2 on one side and 0.8 on the other side); (v) wind load in the lateral direction (same value as at the segment 0 stage, with wind load on the girder additionally included); (vi) wind load in the longitudinal direction (same as that of the segment 0 stage); (vii) load from a falling hanging basket (impact coefficient of 2.0); and (viii) load from unbalanced concrete pouring. When pouring the concrete of the last segment, the concrete on one cantilever had been finished while that on the other side was half-finished; the half-finished concrete was converted to node concentration forces.

The above loads were combined into comprehensive load cases as shown in Table 3.

Table 3. Load combinations at largest cantilever stage.

Modeling event	Load case No.	Load combination
Casting the last girder segment	LC2-1	(i) + (ii)
	LC2-2	(i) + (ii) + (iii) + (iv)
	LC2-3	(i) + (ii) + (iii) + (iv) + (viii)
	LC2-4	(i) + (ii) + (iii) + (iv) + (v) + (viii)
	LC2-5	(i) + (ii) + (iii) + (iv) + (vi) + (viii)
Pouring the last segment when hanging basket falls	LC2-6	(i) + (ii) + (iii) + (v) + (vii) + (viii)
	LC2-7	(i) + (ii) + (iii) + (vi) + (vii) + (viii)

5.3 Influence of pier design parameters on stability at largest cantilever stage

The four types of pier models described in Section 3.2 were rebuilt and stability analysis was conducted under seven load cases to yield the SSFs shown in Table 4. We plotted SSF versus pier height curves under load case LC2-5 as shown in Figure 10.

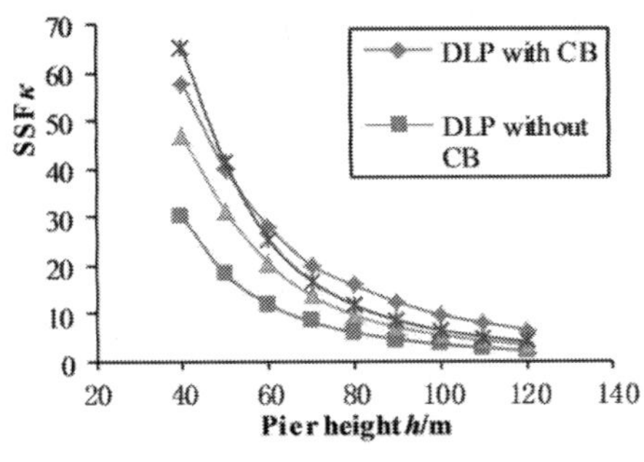

Figure 10. Curves of SSF against pier height at largest cantilever stage.

In the seven load cases shown in Table 4, the first five represent normal construction conditions; the last two represent a sudden drop of the hanging basket, i.e., accidents. The SSF is highest under dead loads such as LC2-1. Construction loads influence the structure stability under normal conditions, and SSFs become smaller under LC2-2, LC2-3, LC2-4, and LC2-5. SSFs are lowest under accident-related loads, e.g., LC2-6 and LC2-7.

Table 4. SSFs under load cases at largest cantilever stage.

Load case No.	SSF	The first order of buckling mode
LC2-1	27.57	Pier buckling in longitudinal direction
LC2-2	25.79	
LC2-3	25.38	
LC2-4	25.36	
LC2-5	25.38	
LC2-6	24.39	
LC2-7	24.41	

Figure 10 indicates that when pier height $h < 54$ m, the SSFs of SLHP with W1.0 are the highest and those of DLP with CB are second-highest. When $h > 54$ m, the SSFs of DLP with CB are the highest and those of SLHP with W1.0 are second-highest. The SSFs of SLHP with W0.5 consistently fall into third place and SSFs of DLP without CB are always the lowest.

At the largest cantilever stage, the relationships between SSFs and pier type and height differ markedly from those at the segment 0 stage. The SSFs of two single-limb piers are close to each other. Using the same method as that described in Section 4.3, we nonlinearly fitted the relationship of SSFs and slenderness ratio using the functions defined in formulas (15) - (18).

For DLP with CB,

$$k = 845.13\lambda^{-1.981} \quad (15)$$

For DLP without CB,

$$k = 9467.3\lambda^{-2.361} \quad (16)$$

For SLHP with W0.5,

$$k = 142292\lambda^{-2.422} \quad (17)$$

For SLHP with W1.0,

$$k = 376343\lambda^{-2.542} \quad (18)$$

The values of κ cannot be ranked effectively when the independent variable λ is constant, because they have the same characteristics as the power function $y = bx^{-a}$ when a and b increase or decrease simultaneously.

6 Stability at finished bridge stage

6.1 Stability analysis under gravity and wind loads

For stability analysis at the finished stage, the loads included: (i) Gravity; (ii) longitudinal prestressing forces;

(iii) secondary dead loads; (iv) wind loads in the lateral direction; and (v) wind loads in the longitudinal direction. The load cases were split into three categories as shown in Table 5.

Table 5. Load combination at finished bridge stage.

Load case No.	Load combination
LC3-1	(i) + (ii) + (iii)
LC3-2	(i) + (ii) + (iii) + (iv)
LC3-3	(i) + (ii) + (iii) + (v)

We obtained the SSFs and first-order bucking modes shown in Table 6.

Table 6. Stability at finished bridge stage under various load cases.

	LC3-1	LC3-2	LC3-3
SSF	29.424	29.418	29.413
The first order of buckling mode	Bucking along the bridge		

The overall stability of the bridge is relatively stable under wind load in the lateral or longitudinal direction. The first-order buckling mode of the bridge occurs in the longitudinal direction under all three load cases.

6.2 Influence of pier height on structural stability at finished bridge stage

Four pier types of models were again built as described in Section 4.2. Stability analysis was conducted under load case LC3-2 to plot the SSF curves against pier height as shown in Figure 11.

Figure 11 shows that pier height has varying influence on the stability of the structure. At the finished bridge stage, two types of bridges with single-limb piers have closer stability than in the other stages we assessed. The stability of the DLP with CB bridge is greater than the other three types.

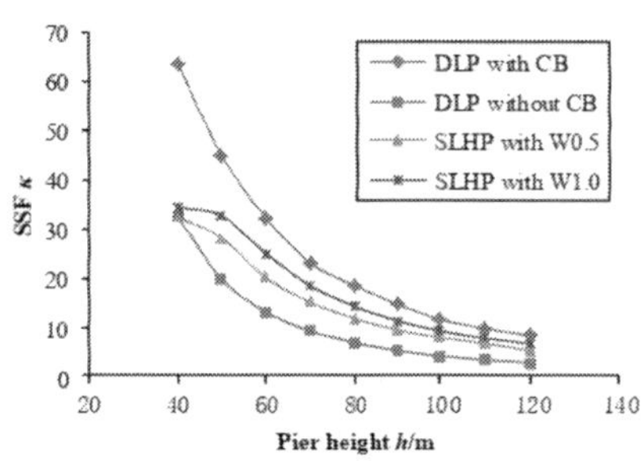

Figure 11. Curves of SSF against pier height at finished bridge stage.

We again nonlinearly fitted the SSFs and slenderness ratios using the same method as described in Section 4.3 as power functions. The functions were obtained as follows.

For DLP with CB,

$$k = 818.9\lambda^{-1.888} \tag{19}$$

For DLP without CB,

$$k = 8151.8\lambda^{-2.275} \tag{20}$$

For SLHP with W0.5,

$$k = 21378\lambda^{-1.895} \tag{21}$$

For SLHP with W1.0,

$$k = 22962\lambda^{-1.865} \tag{22}$$

The values of k cannot be determined using the above formulas when the independent variable λ is constant, because they have the same characteristics as the power function $y = bx^{-a}$ when a and b increase or decrease simultaneously.

7 Conclusions

In this study, a curved CRFB in a mountainous area of Western China was used as a practical engineering prototype to build bridge models at three different construction stages in FE software. The construction stabilities of the models and their relationships with bridge parameters were analyzed. Our conclusions can be summarized as follows.

(1) At different construction stages, pier type and pier height influence structural stability differently. The DLP with CB type has stronger stability, two types of single-limb piers have generally closer stability, and the DLP without CB type has the lowest stability. The DLP without CB type is not recommended for practical bridge engineering.

(2) Wind loads have relatively little influence on structural stability at any construction stage or the finished bridge stage. Accident events, such sudden drop of the hanging basket, do markedly influence the stability of the bridge more than normal construction loads.

(3) A nonlinear power functional relationship exists between SSFs and slenderness ratios for the four types of piers we assessed. This power functional relationship can be used to predict the stability of a new pier.

Acknowledgment

This work was supported by the National Natural Science Foundation of China (U1304515). The authors would also like to thank the anonymous reviewers for their valuable comments and suggestions.

References

1. J. Li, C. Bai, B. Yu, "Dynamic test and analysis of T-type rigid bridge damage due to excess vibration,"

Journal of Highway and Transportation Research and Development, **5**(2), 94-98, (2011)
2. Z. Zong , Z. Xia , H. Liu , Y. Li , X. Huang, "Collapse failure of prestressed concrete continuous rigid-frame bridge under strong earthquake excitation: testing and simulation," *Journal of Bridge Engineering, ASCE*, **21**(9) 04016047 (15 pages), (2016) https://doi.org/ 10.1061/(ASCE)BE.1943-5592.0000912.
3. B. Yao, J. Dong, Z. Qi, "Natural vibration properties analysis of continuous rigid frame bridge varying with consolidation damage at pier top," In: *10th Asia Pacific Transportation Development Conference*, (2014). Beijing, China. May 25-27.
4. Z. Pan, C. Fu, Y. Jiang, "Uncertainty analysis of creep and shrinkage effects in long-span continuous rigid frame of Sutong bridge," *Journal of Bridge Engineering, ASCE*, **16**(2), 248-258, (2011) https://doi.org/10.1061/(ASCE)BE.1943-5592.0000147
5. J. S. Hastings, Q. H. Zhao, E. G. Burdette, "Steel girder stability during bridge erection: AASHTO LRFD check on L/b ratios," *Journal of Bridge Engineering, ASCE*, **15**(6), 759-762, (2010) https://doi.org/10.1061/(ASCE)BE.1943-5592.0000102
6. Z. S. Harper, S. T. Edwards, G. R.Consolazio, K. R. Gurley, "Drag coefficients for construction-stage stability analysis of bridge girders under wind loading," *Journal of Bridge Engineering, ASCE*, **22**(1) , 04016110: (14 Pages), (2017) https://doi.org/10.1061/(ASCE)BE.1943-5592.0000988
7. Q. Zhao, B. Yu, E. G. Burdette, J. S. Hastings, "Monitoring steel girder stability for safer bridge erection," *Journal of Performance of Constructed Facilities, ASCE*, **23**(6), 391-398, (2009) https://doi.org/10.1061/(ASCE)CF.1943-5509.0000048
8. K.Y. Li, Z. Q. Chen, Y. Han, "Quick arithmetic of analysis of stability of double-leg thin-wall high piers during construction," *Journal of the China Railway Society*, **26**(5), 86-90, (2004) (in Chinese with English abstract)
9. H. J. Wu, Y. Q. Kang, L. Zhang, "Analysis of sensitive factors of long-span continuous rigid frame bridge's stability with thin-wall piers in the construction stage," *Advanced Materials Research*, **255-260**(2), 921-925, (2011)
10. J. L. Wang, S. H. He, "Nonlinear analysis of long-span curve rigid bridge with high piers during cantilever construction," *Journal of Highway and Transportation Research and Development*, **22**(10), 64-67, (2005) (in Chinese with English abstract)
11. J. L. Wang, S. H. He, "Nonlinear stability of long-span curve rigid bridge with high piers in whole process," *Journal of Chang'an University (Natural Science Edition)* , **28**(3), 49-52, (2008) (in Chinese with English abstract)
12. J. Zhang, Y. Tian, C. Q. Yang, B. T. Wu, Z. S. Wu, G. Wu, X. Zhang, L. M. Zhou, "Vibration and deformation monitoring of a long-span rigid-frame bridge with distributed long-gauge sensors," *Journal of Aerospace Engineering, ASCE*, **30**(2), B4016014 (9 pages), (2017) https://doi.org/10.1061/(ASCE)AS.1943-5525.0000678
13. Z. P. Bazant, L. Cedolin, *Stability of Structures: Elastic, Inelastic, Fracture and Damage Theories*, World Scientific Publishing Co. Pte. Ltd., Singapore, (2010)
14. C. S. Jason, A. H. Todd, B. W. Eric, H. F. Karl, D. E. Michael, C. S. Andrew, F. F. Jamie, J. P. Brian, "Behavior of horizontally curved I-girders during lifting," *Journal of Structural Engineering, ASCE*, **139**(4), 481-490, (2013) https://doi.org/10.1061/(ASCE)ST.1943-541X.0000674
15. C. S. Jason, A. H. Todd, B. W. Eric, H. F. Karl, D. E. Michael, C. S. Andrew, F. F. Jamie, "Comparisons of the computed and measured behavior of curved steel I-girders during lifting," *Journal of Structural Engineering, ASCE*, **138**(1) 1-10, (2012). https://doi.org/10.1061/(ASCE)ST.1943-541X.0000417
16. D. F. Jeremiah, C. S. Jason, A. H. Todd, S. Andrew, F. Jamie, D. E. Michael, B. W. Eric, H. F. Karl, "Instrumentation of a horizontally curved steel I-girder bridge during construction," *Journal of Structural Engineering, ASCE*, **141**(1), D4014006 (9 pages), (2015). https://doi.org/10.1061/(ASCE)ST.1943-541X.0001110
17. Ministry of Transport of the People's Republic of China, *Wind-Resistant Design Specification for Highway Bridges (JTGT D60-01-2015)*, China Communications Press, Beijing, China, (2015). (in Chinese)
18. J. Liu, Y. J. Liu, B. C. Yang, "A discussion on linear elastic stability safety coefficient of bridge structure," *Journal of Guangxi University (Natural Science Edition)*, **34**(6), 719-724, (2009) (in Chinese with English abstract)
19. X. L. Tong, Z. Fang, "Study on SSF of bridge structure based upon reliability index," *Journal of the China Railway Society*, **36**(10), 102-108, (2014) (in Chinese with English abstract)
20. X. F. Sun, X. S. Fang, L. T. Guan, *Mechanics of Materials*, High Education Press, Beijing, China, (2008). (in Chinese)

 https://doi.org/10.1051/matecconf/201820601018

Preliminary Evaluation of the Earthquake Hazard for Underground Electric Facility Lines in Pohang City based on ArcGIS

Young S. Kim[1], Thien Q.Tran[1], Tan M. Do[1], Inn J.Park[2], Kyeong M.Hwang[3]

[1]*Department of Civil and Environmental Engineering, Chonnam National University, Yeosu 550-749, South Korea*
[2]*Department of Civil and Environmental Engineering, Hanseo University, Sosan, South Korea*
[3]*Korea Electric Power Corporation Research Institude, Daejeon, South Korea*

Abstract. Pohang, a big city of Gyeongbuk province, South Korea, is developing rapidly in every aspect, containing many important infrastructures such as seaports, industrial factories, commercial centers, and underground electric facility systems that play an integral part in the contribution of electricity stemming from nuclear power plants. The main purpose of this research is a preliminary assessment of the earthquake hazard for underground electric facility lines (UEFL) located beneath Pohang city by using ArcGIS program. About 1000 drilling boreholes were collected as a database. All of the data were processed by filtering with respect to the depths, then the SPT-N values of unknown points were estimated by interpolation. Four interpolation methods (e.g., Inverse Distance Weighting (IDW), Kriging, Natural Neighbor, and Spline) were evaluated to find the method which gives the best representation of the SPT-N value for unknown points. The IDW method was finally selected. Afterwards, the border surface between soft- and hard-soil layers, the ground rock zonation map, the dramatic change of soil condition, and the depth of structure were determined as important scales to evaluate primarily whether the structures might be in danger or not when an earthquake occurs. In this way, proposed method can be applied to whole South Korea with over 100,000 available borehole data.

1 Introduction

Historically, the Korean peninsula has been known as a relatively safe region with respect to the earthquake hazard due to the characteristics of its location. But experts are now warning that recent earthquakes in nearby Japan have destabilized fault lines in Korea. The seismicity of Korea is mainly controlled by the active area Gongju, Gwangju, and Yangsan Fault Zone. In recent years, more and more earthquake occurred in southeastern Korea where are nearby the border of three main plates including Eurasian plate, Philippine plate, and Pacific plate [1]. Particularly, in the last two years, two big earthquakes occurred in Pohang and Gyeongju (a neighboring city) due to Yangsan Fault system's activities, measured magnitudes were around 5.4 on the moment magnitude scale. These earthquakes are the country's strongest earthquake in modern history. Furthermore, in Pohang, there are some soft-soil area, it is necessary to analyze the earthquake hazard to minimize effectively its danger. In this paper, the ground rock surface was created and SPT values of unknown points were also predicted by utilizing the IDW method based on the collected field data. The purpose of this is to evaluate primarily whether the structures might be in danger or not when an earthquake comes in Pohang area.

2 Methodology

2.1 Spatial prediction methods and criteria for comparison

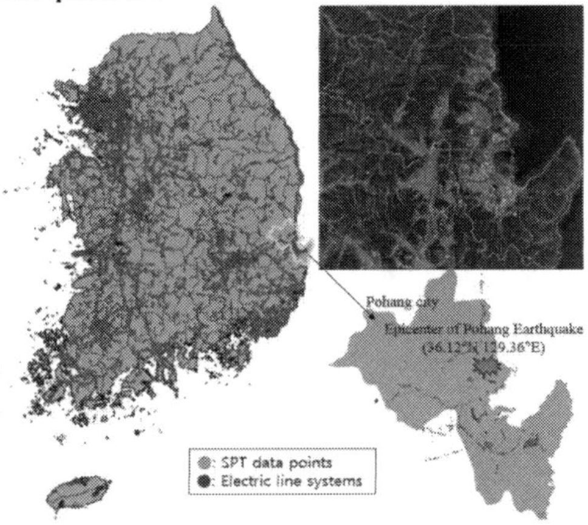

Figure 1. Data point systems, electric facilities, and map of Pohang city.

IDW estimations are based on nearby known locations. The data points (Figure 1) close to the prediction point have more effect on the predicted result than farther ones and vice versa. The formula of this interpolator is [2]:

$$Z(x) = \frac{\sum_{i=1}^{n} Z(x_i).d_i^{-p}}{\sum_{i=1}^{n} d_i^{-p}} \quad (1)$$

where: $Z(x)$ is the interpolated value for the location x; n is the number of sample data values, d_i is the distance from the interpolated point to the sample data point, $Z(x_i)$ is the observed value at location x_i, and p denotes the weighting power.

Similar to IDW method, however, in the kriging method, the weights are based not only on the distance but also on the overall spatial arrangement of the data points. Kriging estimates the unknown values and the error prediction variance value of those points, calculated as follows [3], [4]:

$$Z(x_o) = \sum_{i=1}^{n} \lambda_i Z(x_i) \quad (2)$$

$$\sigma_k^2(x_o) = \mu + \sum_{i=1}^{n} \lambda_i \gamma(x_o - x_i) \quad (3)$$

where: $Z(x_o)$ is the estimated value, λ_i are the weights, n is the number of measured values, μ is the Lagrange constant, $\gamma(x_o - x_i)$ is the value corresponding to the distance between x_o and x_i.

Spline method is similar to IDW, the Spline interpolation method fits a flexible surface as if it were stretching a rubber sheet across all the known point values. The algorithm used for the Spline tool uses the following formula for the surface interpolation:

$$S(x, y) = T(x, y) + \sum_{i=1}^{n} \lambda_i R(r_i) \quad (4)$$

where: n is the number of data points, λ_i are coefficients found by the solution of a system of linear equations, (r_i) is the distance from the interpolated point to the sample data point, $T(x, y)$ and $R(r_i)$ are defined differently, depending on the selected option (http://pro.arcgis.com).

Natural Neighbor interpolation finds the closest subset of data points to an unrecorded point and applies weights to them based on proportionate areas.

$$S(x, y) = \sum_{i=1}^{n} w_i f(x_i, y_i) \quad (5)$$

where $S(x, y)$ the estimated value, w_i are the weights, and $f(x_i, y_i)$ is the data point at (x_i, y_i) (http://pro.arcgis.com).

In this present study, in order to determine the best predictive method for the interpolation in Pohang area, a simple criterion was built. Particularly, twenty predicted points in the area were randomly selected and compared with the real SPT-N values at the corresponding depth. In this method, the most accurate method was designated 4 plus points, the second, third, and least accurate ones were 3, 2, and 1 plus points, respectively. The best method is a method getting the highest score among 4 methods and it gives predicted results closest to the real values.

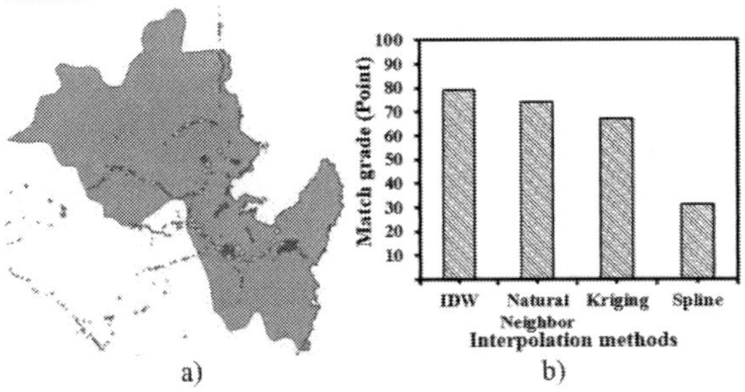

Figure 2. (a) The map of training data points, (b) Match grade of 4 interpolation methods.

Figure 2b shows that IDW is more accurate than the others three methods with 79 plus points, followed by Natural Neighbor, Kriging, and Spline methods with 74, 67, and 31 points, respectively. With the distribution condition of the input data points in Pohang city as shown in Figure 2a, the result clearly shows that IDW is the best method for predicting the unknown SPT-N values in this study. This finding is a good agreement with previous studies [2], [5].

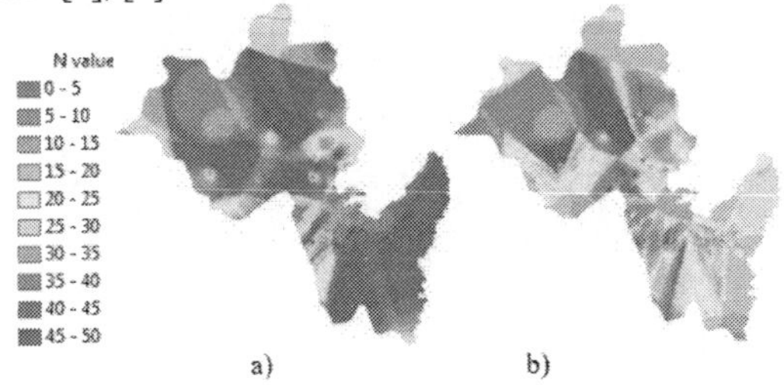

Figure 3. Predicted result map at the depth of 1.5 m (a) and 7.5 m (b) of SPT-N value.

Figure 3 shows predicted result maps of SPT-N value at the depth of 1.5 m and 7.5 m, with the dense system of the input data point that is a good condition for IDW method to optimize its advantages, the SPT-N values at unrecorded locations were effectively estimated at every depth corresponding to the nearby input data points.

2.2 Dangerous level criteria

2.2.1 Criteria 1

In this study, based on the relationship between stress ratios required to cause liquefaction and SPT- N values and relative density of soils with SPT-N value, SPT N-value of 20 is considered as a critical value for the soil to resist the earthquake hazard [6], [7]. Based on the estimated SPT N values using IDW around the electric facility lines, vulnerability about earthquake hazard was preliminarily evaluated. The depth of UEFL is thoroughly compared with the depth where is the border between the soft soil and hard soil layer ("N20 surface") at that location, and structure is considered as dangerous when its depth is shallower than "N20 surface" and vice versa. As shown below, Figure 4 presents the idea used for determining the dangerous and safe electric facility line cases.

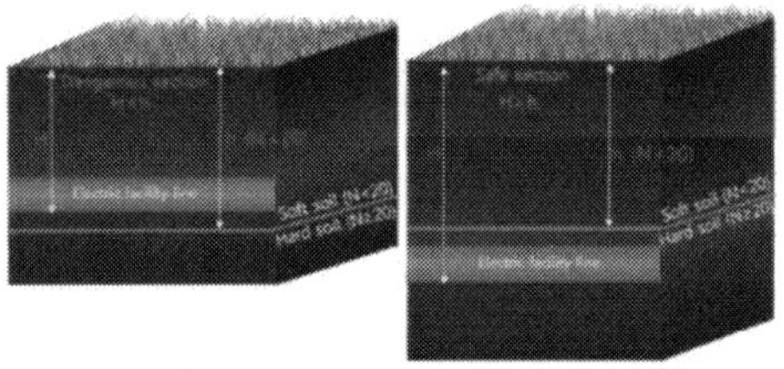

a) b)

Figure 4. (a) Simulation of the dangerous section and (b) the safe section.

For this work, the very first step is determining the border surface between soft- and hard-soil layers by filtering the SPT-N values that are smaller than 20 at the maximum depths. Afterwards, the IDW interpolation method was used to connect those values and predict the unknown values to create the complete border "N20 surface". Similar to the method determining "N20 surface", however, the "bedrock surface" was obtained by filtering the SPT-N values that are around 50 at the minimum depths. The fundamental methods defining the "N20 surface" and "bedrock surface" are illustrated in Figure 5 & 6.

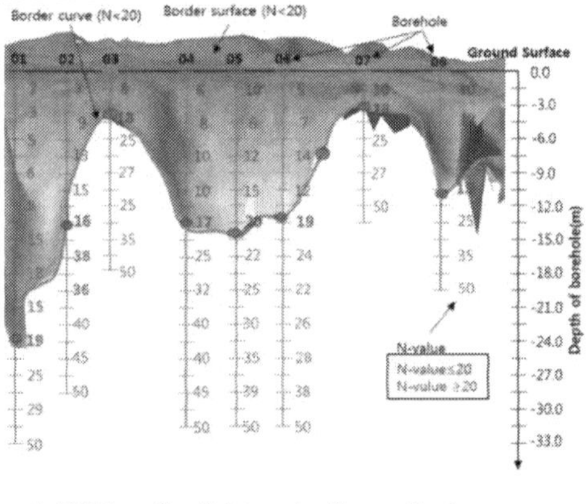

Figure 5. "N20 surface" determination method.

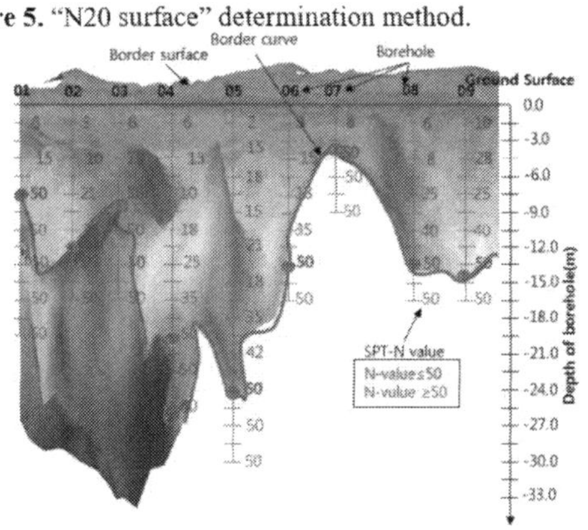

Figure 6. "Bedrock surface" determination method.

2.2.2 Criteria 2

Figure 7 shows that criteria 2 might be one of these two conditions that cause the skew deformation of UEFL when the earthquake occurs.

The first is when the electric facility line's longitudinal depth changes dramatically from the soft soil layer to the hard and vice versa. The second case is sudden changes of the soil condition surrounding the electric facility lines, particularly, the "N20 surface" changes dramatically and cut through facilities. Thus, criteria 2 is useful for evaluate danger level in 2 different cases.

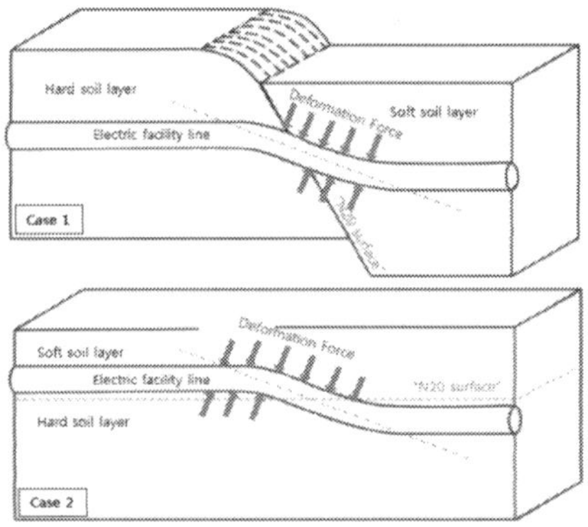

Figure 7. Explanation of two different dangerous cases.

2.2.3 Relatively horizontal deformation (RHD)

At case 2 shown in Figure 7, it is necessary to estimate the relatively horizontal deformation of UEFL when it fluctuates dramatically in soft soil areas or from the soft soil layers to the hard soil layers and conversely with.

The magnitude of relatively horizontal deformation of soil at the structural depth is estimated by using the following formulas [8]:

$$\Delta_{\text{free-field}} = h.\gamma_{\max}, \quad \gamma_{\max} = \frac{Vp}{Cs} \quad (6)$$

where: h is height of the structure, Δfree-field is free-field shear strain at the depth of the electric facility line, Vp is peak particle velocity at the electric facility line depth, Cs is effective shear wave velocity of ground surrounding the electric facility line that is determined based on an empirical correlation between shear wave velocity and SPT-N value for Korean soil deposits [9]. The model of the effect of RHD to the underground structure is presented in Figure 8.

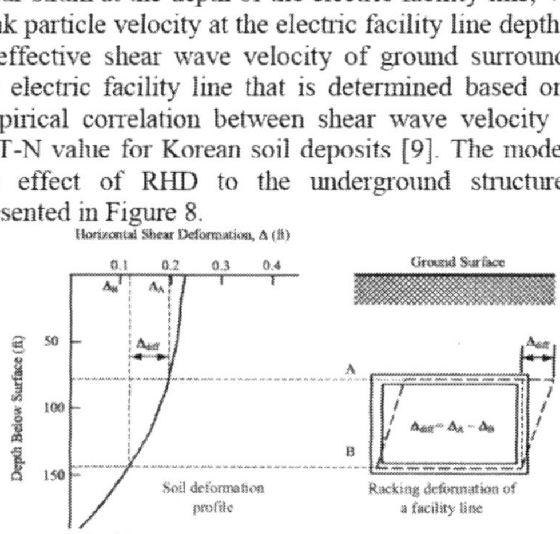

Figure 8. Typical free-field deformation affecting structures [8], [10].

The ratio of peak ground velocity (cm/s) to peak ground acceleration (g) was employed to derive the particle velocity Vp at the corresponding depths.

Table 1. Ratios of Peak ground velocity to peak ground acceleration at the surface in rock and soil [8], [10].

Moment magnitude (M)	Ratio of peak ground velocity (cm/s) to peak ground acceleration (g)		
	Source-to-size distance (km)		
	0-20	20-50	50-100
Rock[a]			
6.5	66	76	86
7.5	97	109	97
8.5	127	140	152
Stiff soil[a]			
6.5	94	102	109
7.5	140	127	155
8.5	180	188	193
Soft soil[a]			
6.5	140	132	142
7.5	208	165	201
8.5	269	244	251

[a]In this table, the sediment types represent the following shear wave velocity ranges: rock ≥ 750 m/s; stiff soil is 200-750m/s; and soft soil <200 m/s.

In order to clarify the parameters in Table 1, the design magnitude for the earthquake in Korea is 6.5 and distance from Pohang area to the Yangsan fault zone system is within 20 km. With the geographical location belongs to zone 2 (seismic zone) in Korea, the seismic numerical coefficient was designated $a_{gR} = 0.11g$ [11]

The site-specific Peak Ground Acceleration amax,s was given by the following equation [12]:

$$a_{max,s} = S.a_{gR} \quad (7)$$

where: S is ground type factor [11]

As shown in Table 2, according to the ratio of ground motion at tunnel depth to motion at ground surface, the Peak Ground Acceleration at each different depth was determined from the site-specific Peak Ground Acceleration with the following relationship:

$$a_{z,max} = C.a_{max,s} \quad (8)$$

where: C is ratio of ground motion at tunnel depth to motion at ground surface that is derived from Table 2

Table 2. Ratios of ground motion at depth to motion at the ground surface [8], [10].

Tunnel depth (m)	Ratio (C) of ground motion at tunnel depth to motion at ground surface
≤6	1.0
6-15	0.9
15-30	0.8
>30	0.7

Finally, the RHD values at corresponding depths were calculated with sufficient input parameter such as SPT-N values, shear wave velocities, the magnitude of 6.5, peak ground acceleration, the distance between sources to study locations and height of UEFL.

3 Results and discussion

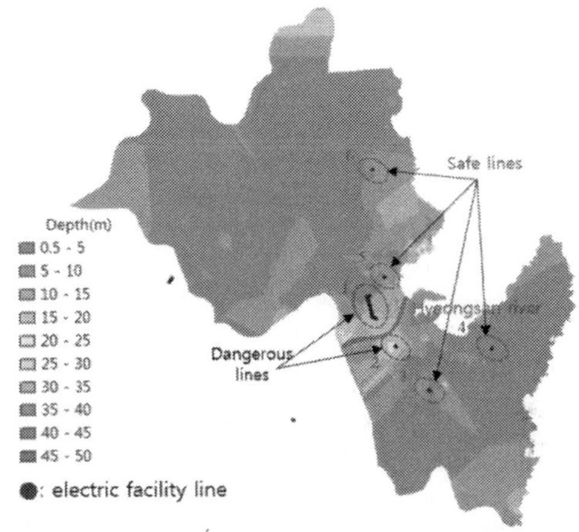

Figure 9. Zonation map of "N20 surface".

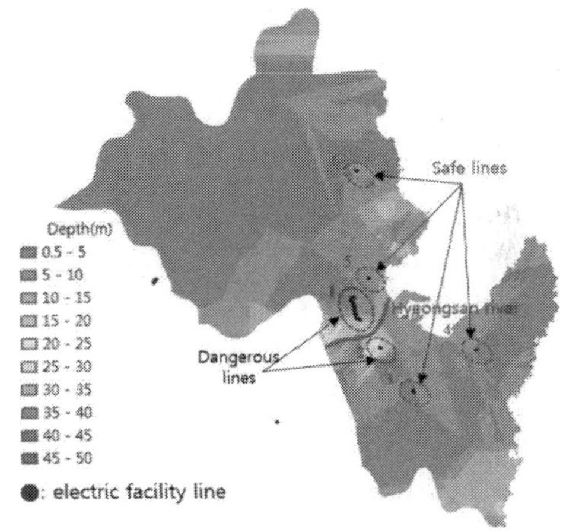

Figure 10. Zonation map of "bedrock surface".

Figure 9 shows the zonation map of "N20 surface", according to the distribution of the color spectrum results. It indicates that the "N20 surface" depth ranges from around 0.5 m to 35 m and soft soil layers are mainly located around Hyeongsan river where Line 1 and Line 2 are nearby.

Similarly, the color spectrum in Figure 10 shows the zonation map of the "bedrock surface" depth that varies in a range of 0.5-40m. It also indicates that the deep-bedrock surface is primarily concentrated around the river, one of the agents of this distribution may be the river's alluvial activities.

 https://doi.org/10.1051/matecconf/201820601018

According to the distribution characteristic of the Pohang's UEFL system, it is divided into 6 main lines consist of UEFL1, 2, 3, 4, 5 and 6. The location of lines is shown in Figure 9 & 10.

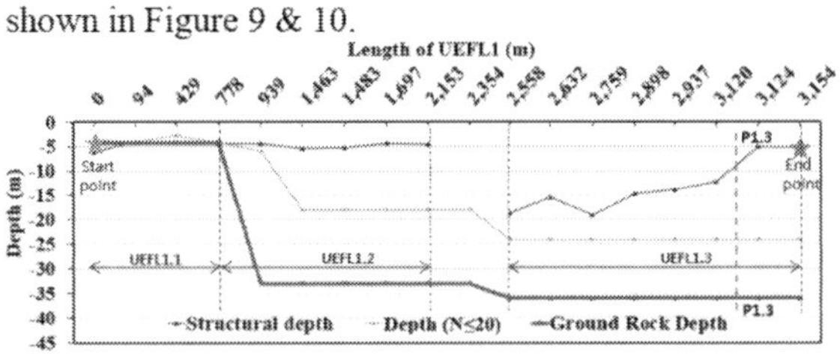

Figure 11. Longitudinal section of UEFL1 – Pohang city.

UEFL1 consists of three different types of the dangerous levels (UEFL1.1, 1.2, & 1.3). As shown in Figure 11, UEFL1.1 is located deeper than the "N20 surface" depth and the "bedrock surface" depth. It indicates that the UEFL1.1 is buried in a good soil condition, therefore this line is considered as a safe line for both criteria 1 and criteria 2. Similarly, UEFL3, 4, 5, and 6 are also considered as safe ones based on criteria 1 and 2.

Besides, Figure 11 also shows that UEFL1.2 is located at shallow depths that are much shallower than the "N20 surface" depths and "bedrock surface" depth. Consequently, it might be in danger based on criteria 1.

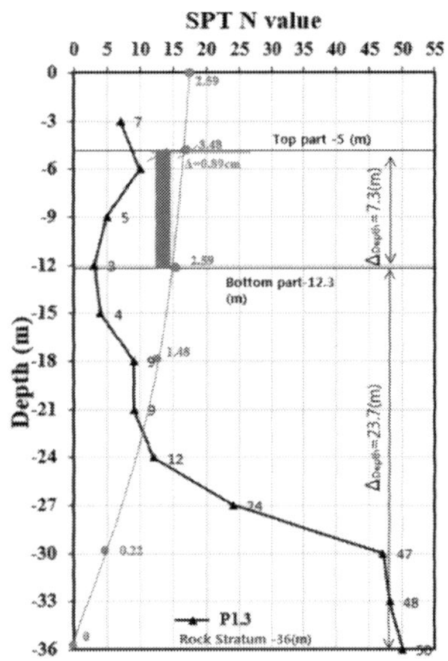

Figure 12. Vertical section P1.3 of UEFL1.3 – UEFL 1 – Pohang city.

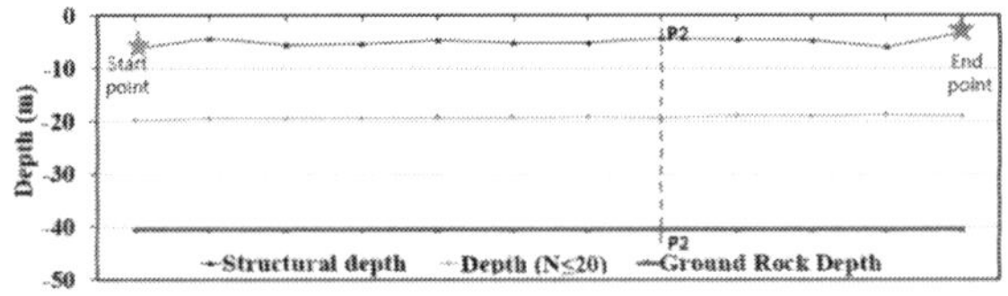

Figure 13. Longitudinal section of UEFL2– Pohang city.

In the last part of UEFL1, UEFL1.3 is placed above the "N20 surface" and "bedrock surface" at fluctuant depths. UEFL1.3 is, therefore, considered as a dangerous line for criteria 1. Furthermore, Figure 11 & 12 show that there is a dramatic change of the structure depth with a different depth between the top and bottom structure's part of 7.3 m within the horizontal distance of 4.0 m. Hence, the segment of UEFL1.3 including section P1.3 should be noted to be a dangerous part for both criteria 1 & 2, the RHD value should be estimated for checking the flexibility of this underground electric facility segment at this location.

In case of UEFL2, Figure 13 & 14 indicate that there are soft-soil layers surrounding UEFL2 which is stably located shallower than "N20 surface" and "bedrock surface". Furthermore, the maximum of soft-soil depth and the minimum of bedrock depth are much deeper than UEFL2's average depth. Thus, this line might be dangerous when an earthquake comes for criteria 1.

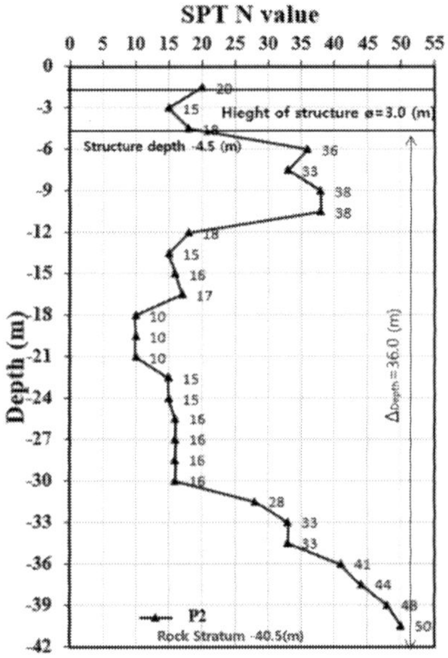

Figure 14. Vertical section P2 of UEFL2– Pohang city.

Overall, the zonation map shows that the depth of ground rock in Pohang area is generally shallow, except for UEFL1 and UEFL2 in Hyeongsan river's delta – Pohang city. Typical dangerous sections are selected based on the vertical and longitudinal variations in SPT-N value, relative horizontal deformation Δ around the structure. In addition, some sections in the line might be classified as dangerous based on the relatively horizontal deformation of the surrounding ground between the bottom and top of the structure. Based on the evaluated results from proposed method using ArcGIS, the in-situ investigation is on-going.

4 Conclusion

1. In this study, SPT N values around the underground electric facility lines in Pohang area are successfully estimated using the IDW method.

2. Based on the interpolated SPT N values, structure depth, longitudinal change in ground condition, and estimated relative horizontal deformation during the earthquake, dangerous underground electric facility lines in Pohang area can be classified and specific sections are evaluated.

Acknowledgment

We would like to express our deepest gratitude to Korea Electric Power Corporation Research Institute (KEPCO) for financial support on this study.

References

1. Myung-Soon Jun and O. Kulhánek, "Source parameters of earthquakes in and around the Korean Peninsula deduced from spectral analysis," Phys. Earth Planet. Inter., vol. 65, no. 3–5, pp. 255–266, (1991).
2. T. P. Robinson and G. Metternicht, "Testing the performance of spatial interpolation techniques for mapping soil properties," Comput. Electron. Agric., vol. 50, no. 2, pp. 97–108, (2006).
3. S. Science and S. Of, "Soil science society of america," vol. 22, no. 3, pp. 189–192, (1958).
4. G. S. Bhunia, P. K. Shit, and R. Maiti, "Comparison of GIS-based interpolation methods for spatial distribution of soil organic carbon (SOC)," J. Saudi Soc. Agric. Sci., (2016).
5. S. Naoum and I. K. Tsanis, "Ranking spatial interpolation techniques using a GIS-based DSS," Glob. Nest J., vol. 6, no. 1, pp. 1–20, (2004).
6. H. Bolton Seed, K. Tokimatsu, L. F. Harder, and R. M. Chung, "Influence of SPT Procedures in Soil Liquefaction Resistance Evaluations," J. Geotech. Eng., vol. 111, no. 12, pp. 1425–1445, (1985).
7. K. Terzaghi, R. B. Peck, and G. Mesri, "Soil Mechanics in Engineering Practice." pp. 1–529, (1996).
8. F. T. Manual and R. Tunnels, "FHWA-NHI-10-034 Errata Sheet Foreword," English, (2008).
9. C. G. Sun, C. S. Cho, M. Son, and J. S. Shin, "Correlations Between Shear Wave Velocity and In-Situ Penetration Test Results for Korean Soil Deposits," Pure Appl. Geophys., vol. 170, no. 3, pp. 271–281, (2013).
10. Y. M. A. Hashash, J. J. Hook, B. Schmidt, and J. I-Chiang Yao, "Seismic design and analysis of underground structures," Tunn. Undergr. Sp. Technol., vol. 16, no. 4, pp. 247–293, (2001).
11. K. D. Standard, "Architectural Institute of Korea. Korean Building Code; Architectural Institute of Korea," (2016).
12. JRC European Commission, C. Eurocode, and International Code Council, Eurocode 8 : Seismic Design of Buildings Worked examples, vol. 1, (2004).

Current Status of Degraded Road Bridge Slab Located in Mountainous Area

Yuji Maki[1], Tuan Minh Ha[2], Saiji Fukada[2], Kazuyuki Torii[2] and Ryohei Ono[1]

[1]Division of Environmental Design, Kanazawa University, Kanazawa, Japan
[2]School of Geosciences and Civil Engineering, Kanazawa University, Kanazawa, Japan

Abstract. In the mountainous area of the Hokuriku region, bridges are suffering early deterioration caused by salt damage, alkali-silica reaction (ASR) and frost damage. Under such a regional situation, the load carrying capacity evaluation of these bridges has become an urgent issue. In this study, a statistical method–survival analysis–was carried out by using inspection data for bridges in the Hokuriku region, and the relationship between degradation tendency of bridges and regional characteristics causing salt damage, ASR and frost damage has been studied. In addition, the degradation causes of the reinforced concrete deck of a road bridge which deteriorated early in the mountainous area of the Hokuriku region were investigated. Cylindrical core extraction was carried out to analyze the degradation causes. Polarizing microscopic observation of specimens collected from RC slabs confirmed that cracks developed from the andesite particles and cracks were confirmed to be filled with ASR gels. Therefore, ASR was a cause of the deterioration. Moreover, the degradation in the mechanical properties of concrete due to ASR was also studied and reported. The results show that the reduction in compressive strength and static elastic modulus of both dry and leakage-water parts affected the load bearing capacity of the structure.

1 Introduction

While the aging of local bridges is progressing, bridge maintenance and management expenses decrease in the fiscal situation of reduction in tax revenue due to the declining birth-rate in Japan. For bridges spanning more than 2 m, the visual inspection is conducted once every five years. Because of the current budget constraint, it is not currently possible to implement repair and reinforcement systematically. In addition, measures are repeatedly carried out without evaluating the residual strength of the bridge after inspection is completed.

In the Hokuriku region, salt damage, alkali-silica reaction (ASR), and frost damage caused severe degradation early in bridges in the mountainous area without warning [1]. Under such a regional situation, how to evaluate the load bearing capacity of these bridges has become an urgent issue. Moreover, the relationship between degradation tendency of bridges and regional characteristics causing salt damage, ASR and frost damage was also a problem that needs to be solved. As a part of the present study, this study performed a statistical analysis–survival analysis [2]–using inspection data for bridges in the Hokuriku region to solve the above problem. To be more specific, the survival time analysis was conducted for reinforced concrete slabs (hereinafter referred to as RC slab) of steel road bridges in the mountainous area constructed for the dam construction in the 1970s.

In addition, as a case study on the destruction process of the RC slab subjected to frost damage in the Hokkaido district which is a snowy region, Mitamura and colleagues [3] reports the effect of freezing and thawing on fatigue life. Besides, Ishikawa et al. [4] reported the damage situation of the RC slab which has been replaced due to fatigue and salt damage caused by the anti-freezing agent on the expressway in the Hokuriku district. Using test specimens close to full size, Maeshima et al. [5] analyzed the relationship between ASR deterioration and fatigue durability of RC slab. However, few studies evaluate the current status of RC slabs in the mountainous area of the Hokuriku district where ASR occurred.

Therefore, this study selected the RC deck slab of a steel road bridge from the group of bridges in the mountain area suspected of being deteriorated by ASR (see Figs. 1-2). Then, degradation factors were identified from various analyses by core extraction and visual observation. Then, the current status of the test bridge was evaluated by the change in the mechanical properties, the observation results of ASR gel under the polarized microscope, and the salt content analysis.

2 Survival time analysis for routes in mountainous areas

In the Hokuriku district, salt damage, ASR and frost damage are cited as reason leading to early deteriorations

https://doi.org/10.1051/matecconf/201820601019

in concrete structures, but they are closely related to site conditions. In this study, survival time analysis using bridge inspection data was carried out for the purpose of considering the deterioration tendency of the RC slab in the mountainous routes.

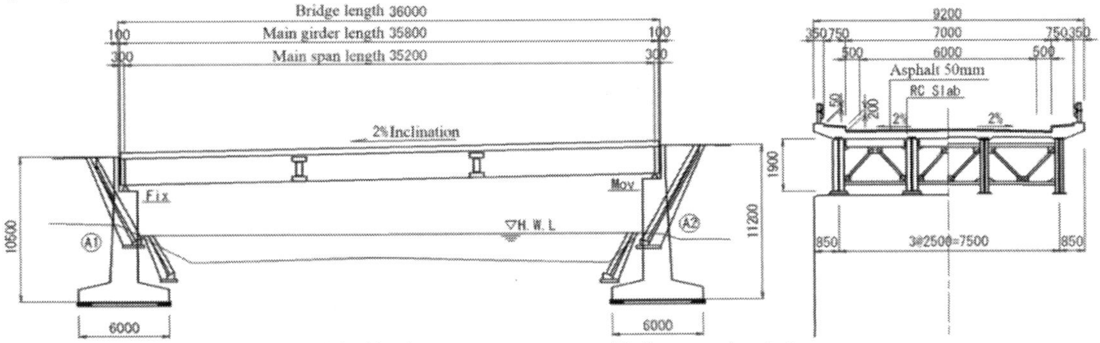

(a) Side view (b) Cross-sectional view

Figure 1. Overview of the test bridge.

(a) Upper surface

(b) Bottom surface

Figure 2. The test bridge.

2.1 Outline of survival time analysis

The survival analysis is a statistical technique focusing on the relationship between time and events from a certain point of time until an event occurs, it is often used in medical and engineering fields.

Although it is common to deal with deaths in the medical field and faults as events in the engineering field, Yamazaki et al. [6] conducted survival time analysis using inspection results in the Tohoku region as an application to the maintenance of bridges and analyzed importance and risk determination for each degradation mechanism.

The survival time T until event occurrence is set as a random variable, the probability density function $f(t)$ is expressed by the following equation:

$$f(t) = \lim_{\Delta t \to 0} \frac{\Pr(t < T < t + \Delta t)}{\Delta t} \quad (1)$$

Assuming that the cumulative probability distribution function is $F(t)$, the survival function $S(t)$ at a specific time point t is given by the following equation:

$$S(t) = \Pr(T > t) = 1 - \Pr(t \le T) = 1 - F(t) \quad (2)$$

Under the condition that the event has not occurred by the time t, the hazard function $h(t)$ indicating the instantaneous mortality rate at which the event occurs at the next moment $t + \Delta t$ is given by the following equation:

$$h(t) = \lim_{\Delta t \to 0} \frac{\Pr(t < T < t + \Delta t \mid T \ge t)}{\Delta t} = \frac{f(t)}{S(t)} \quad (3)$$

Furthermore, the relationship between the cumulative hazard function $H(t)$ and the survival function $S(t)$ and the hazard function $h(t)$ is as follows:

$$H(t) = \int_0^t h(t)dt = -\log S(t) \quad (4)$$

When dealing with the survival time, it is necessary to consider the case where no event is observed before the end of the investigation as "censored". The Kaplan-Meier method is an estimation method considering censoring, and the Kaplan-Meier estimator $S(t)$ is given by the following equation. Here, d_i is the number of deaths at time t_i, and r_i is the number of observers that may have an event just before time t_i.

$$\hat{S}(t) = \prod_{t_i < t} (1 - \frac{d_i}{r_i}) \quad (5)$$

2.2 Survival curve of target route

2.2.1 Outline of study

The bridges over a specific route in a mountainous region in the Hokuriku district were regarded as "applicable group", while other bridges were regarded as "non-applicable group". Then, the analysis was performed on both groups to investigate the relationship between the deterioration tendency of the RC slab and the location conditions. Here, the definition of event occurrence was considered as a case where damage requiring repair was found at bridge inspection. According to the inspection record of the district, the bridge soundness was set to 5 grades from grade 5 (mild) to grade 1 (severe). This study considered the healthy degree of grade 3 or less as "damaged", and it was regarded as an event occurrence.

The application of a filter such as classification class and construction year made the comparative examination on both groups possible. The number of steel road bridges having the RC slab in the target area is 276 bridges. Among them, the number of bridges of "applicable" group in the same route constructed for the dam construction in the 1970s is 17 bridges. Here, "bridge age" indicates the elapsed year from construction to inspection. Because the RC slabs of the route have not been repaired extensively so far, the restoration of bridge soundness through repair was not taken into consideration, and thereby its impact on the result was considered to be small. For the survival time analysis, the survival package of statistical software R ver.3.2.5 was employed [7].

2.2.2 Study results

Survival time obtained by the Kaplan-Meier method for the "applicable" and "non-applicable" groups is shown in Fig. 3. From this figure, while the survival curve of "non-applicable" group gradually decreases, "applicable" group shows a sharp decline after 35–45 years. Although there was a difference in the number of samples and the distribution of bridge age in both groups, it became clear that the RC slabs of "applicable" group tended to deteriorate prematurely.

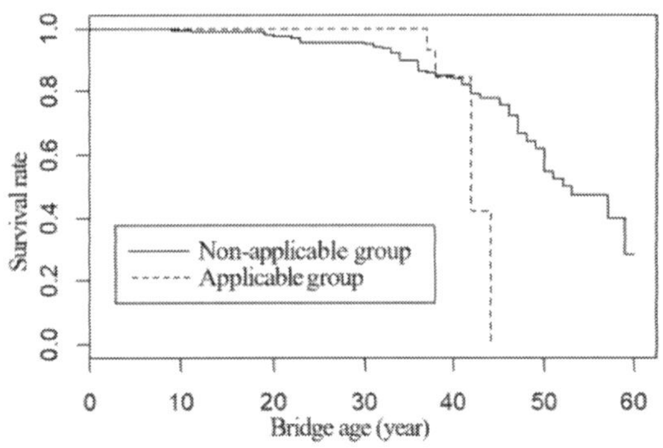

Figure 3. Survival curves of RC slab in target bridge groups.

3 Description of the test bridge

As a part of the national highway, the bridge was constructed in the mountainous area (altitude: 559 m) 45 years ago (completed in 1973). From Fig.1, the bridge is a simple composite girder bridge with a bridge length of 36.0 m, and four main girders. Moreover, because the A1 side is lower than the A2 side, the skew angle and the longitudinal gradient are 60 degrees and 2% respectively. The main girder spacing and the deck thickness are 2.5 m and 180 mm, respectively. Besides, the volume of traffic (small car + large car) traveling on this bridge and the ratio of large vehicles according to 2015 Census were 1177 vehicles per day and 20%, respectively. Because of the small traffic volume, it was difficult to consider that fatigue was a cause of deterioration in this bridge.

4 Appearance evaluation

(a) Pavement surface near the joint (b) Bottom surface near the joint

(c) Bottom surface near the span center (d) The wheel guard surface

Figure 4. The test bridge.

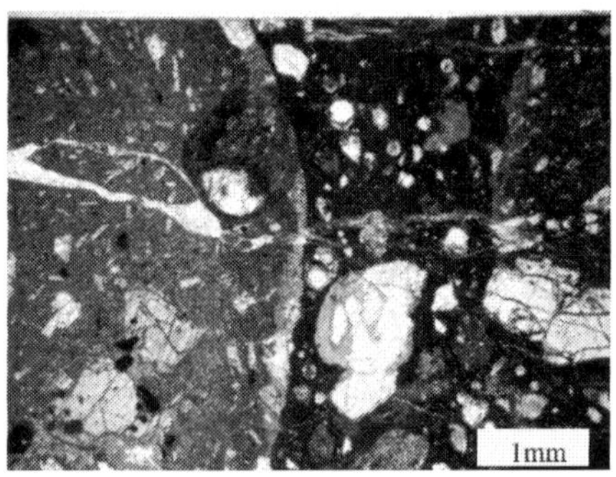

Figure 5. Andesite particles and ASR gel (observation slice 25 mm × 25 mm, thickness 20 μm) under polarizing microscope (normal nicol).

Fig. 4(a) shows the situation of patching of the pavement which seemed to have been repaired several times at the vicinity of the joint. Patches were also confirmed at some places on the pavement. In addition, efflorescence was observed at the bottom of the RC slab and near the joint (see Fig. 4(b)). Signs of water permeation also appeared under the RC slab and near the span center, as shown in Fig. 4(c). Furthermore, the wheel guard section on the bridge surface (Fig. 4(d)) had many horizontal cracks in the longitudinal direction of the bridge.

5 Diagnosis ASR-induced deteriorations using drilled cores

5.1 Overview of experiment setup

Cylindrical core samples with a diameter of 55 mm were collected from the RC slab of the test bridge so as to investigate the mechanical properties of concrete such as the neutralization depth, the compression strength, static modulus of elasticity, and the salinity.

Two sampling locations were selected with and without the efflorescence (dry part, leakage-water part) on the bottom of the slab. The slab thicknesses at the dry part and the leakage-water part (near the joint) are 180 mm and 190 mm, respectively. The pavement thickness is 50 mm for both cases.

Fig. 5 shows an observation result under the polarized microscope (normal nicol) of the concrete core collected from the water leakage part. From this figure, cracks propagated from andesite particles to cement paste and the ASR gel was filled in those cracks. Therefore, it was inferred that the main cause of deterioration in the RC slab was ASR.

5.2 Compressive strength and static elastic modulus

Compressive strength and static elastic modulus were calculated by performing compression tests using cores collected at two places, i.e., the dry section and the leakage-water section. The results are summarized in Table 1.

From this table, the compressive strength of specimens was reduced by about 30% in the leakage-water section compared to the dry part. Because the test bridge is a composite bridge and the design strength of concrete was 35 N/mm2 for the RC slab, it is inferred the reduction in compressive strength of both dry and leakage-water parts affected the load bearing capacity of the entire structure.

Regarding the static elastic modulus, the results obtained from the leakage-water part decreased to about half of those of the dry part. The previous study pointed out that the elastic modulus was significantly degraded by the ASR-induced deterioration [8], this outcome was also confirmed in results on the RC slab of the test bridge.

Table 1. Compressive strength and coefficient of static elasticity.

	Dry part	Leakage-water part
Static elastic modulus (kN/mm^2)	21.5	10.0
Compressive strength (N/mm^2)	32.3	21.2

5.3 Salt content analysis

Regarding the salt content analysis of the dry part and the leakage-water part, the sampled cores were sliced in 10 mm pitch. Then, the total chloride ion content was determined by the potentiometric titration method, and the chloride ion concentration distribution is summarized in Fig. 6. From this figure, the chloride ion concentration in the dry part is almost zero, but in the leakage-water part, it was estimated to be about 1.32 kg/m^3 at the concrete cover of the lower rebar, and the concentration on the lower surface was considerably high. In addition, the measured results did not exceed the limit value of steel corrosion occurrence which could be determined from the proposed formulas (-3.0(W/C) + 3.4; estimated W/C: 55%–60%) by Standard Specification for Concrete Structures of Japan. In addition, the corrosion of the reinforcing bars was not observed at the place where the core was collected.

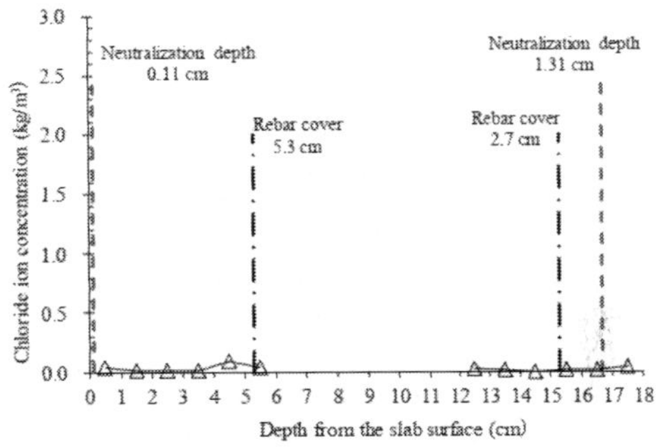

(a) Dry part

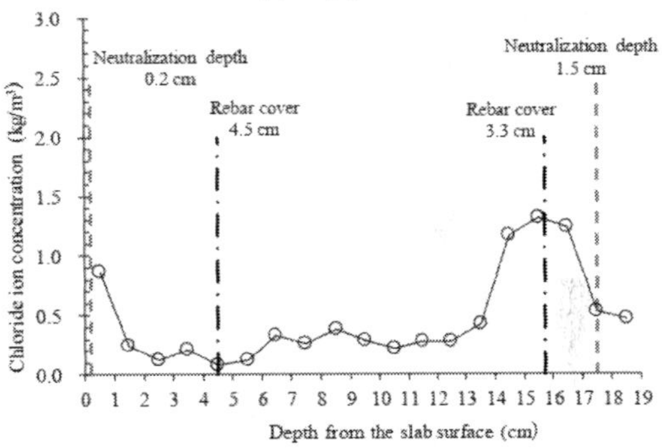

(b) Leakage-water part

Figure 6. Result of salinity analysis.

In addition, the neutralization depth is also shown in same figure. On the lower surface, it was 1.3 cm in the dry part and 1.5 cm in the water-leakage part, and neutralization was not progressed as the deck had been over 40 years old. The explanation of this outcome is that rainfall and snowfall frequently occur in the mountains in the Hokuriku district, the humidity continues to be high, and the humid environment slows the progress of neutralization. Also, because the bridge built in the 1970s does not have a waterproof layer, it is presumed that the wet environment causes ASR to progress continuously, resulting in the deterioration of the RC slab.

6 Conclusions

This study employed the survival analysis using inspection data for bridges in the Hokuriku district and analyzed the relationship between degradation tendency of bridges and regional characteristics causing salt damage, ASR and frost damage. In addition, the degradation causes and current mechanical properties of the reinforced concrete deck of a road bridge which deteriorated early in the mountainous area of the Hokuriku district were investigated by many analyses using cylindrical concrete cores. The primary results obtained from this study are summarized as follows.

Survival time analysis using inspection data was carried out for steel road bridge groups with RC slabs in the mountainous area. It was revealed that the RC slabs of the applicable group were deteriorated earlier than other bridges.

Polarizing microscopic observation of concrete cores collected from RC slabs confirmed that cracks developed from the andesite particles to the cement paste and those cracks were confirmed to be filled with ASR gels. Therefore, it was confirmed that the primary cause of the deterioration in the concrete of the RC slab was presumed to be ASR. In addition, as a result of the investigation on the distribution of chloride ion concentration, corrosion of the steel bar was not confirmed in the leakage-water part as well as the corrosion occurrence limit of the steel material was not exceeded.

As a result of the investigation on the mechanical properties of the cores, the compressive strength decreased by about 30% in the leakage-water part as compared to the dry part. Regarding the static elastic modulus, the results obtained from the leakage-water part dropped to about half of those from the dry part.

Acknowledgement

This research was promoted by the Cross-Ministerial Strategic Innovation Promotion Program [Title: Resolution of Early-aged Deterioration Mechanisms in Concrete Bridges and Development of Total Management System Based on Evaluation for Material and Structure Quality Performance, Kanazawa University] from the Japan Science and Technology Agency (JST). The authors wish to thank the concerned parties for their valuable collaboration, sub-consultants, and support.

References

1. K. Torri, T. Daidai, H. Yamato, T. Hirano, *Proceedings of the Japan Concrete Institute*, **30**, 6 (2008).
2. Y. Maki, Y. Chikata, K. Torri, S. Fukada, *Japan Society of Civil Engineers 2017 Annual Meeting*, **VI-524**, 2 (2017).
3. H. Mitamura, T. Satou, K. Honda, S. Matsui, J. Struc. Eng., JSCE, **55A**, 12 (2009).
4. Y. Ishikawa, Y. Adachi, M. Aoyama, M. Nagai, J. Struc. Eng., JSCE, **57A**, 10 (2011).
5. T. Maeshima, Y. Koda, I. Iwaki, H. Naito, R. Kishida, Y. Suzuki, K. Ohta, M. Suzuki, *Proceedings of the Japan Society of Civil Engineers E2*, **72**, 20 (2016).
6. T. Yamazaki, T. Ishida, J. JSCE, Ser. F4 (Construction and Management), **71**, 12 (2015).
7. The R Manuals, 2017, https://cran.r-project.org/manuals.html.
8. Y. Kubo, T. Ueda, T. Kuroda, N. Nomura, *Proceedings of the Japan Concrete Institute*, **28**, 6 (2006).

Parameter Analysis on the Anti-Impact Behavior of Pcfst Columns under Lateral Impact Load

W Xu , A Z Zhu and K Gao

School of Civil Engineering and Mechanics, Hubei Key Laboratory of Control Structure, Huazhong University of Science and Technology, Wuhan, P.R. China

Abstract. Concrete-filled steel tubular (CFST) structures have been widely used in civil engineering structures, due to its good behaviors under both static and dynamic loads. In this paper, numerical studies were carried out to investigate the anti-impact behavior of partially concrete-filled steel tubular (PCFST) columns under lateral impact loads. Finite element analysis (FEA) model was established using ABAQUS. To validate the FEA model, the numerical results were compared with experimental results. Moreover, parameter analysis was carried out to further study the anti-impact behaviors of the PCFST columns. The concrete filling height, the impact energy, the impact direction, and the yield strength of steel were the main parameters considered in this study. The dynamic responses under the impact load, including the impact force, the failure mode, and the displacement response, were all analyzed. The results of parameter analysis showed that the anti-impact behaviors of the PCFST columns significantly increased when the concrete filling height or the yield strength of steel increased greatly. The impact energy and direction also greatly affected the anti-impact behaviors of the PCFST columns.

1. Introduction

As a type of composite construction technique, concrete-filled steel tubular (CFST) members have been widely used in civil engineering structures, owing to their large load-carrying capacity and ease of construction [1]. With the development of urban traffic, vehicle collision became one of the main reasons causing the failure of bridge piers. Otherwise, due to the defect of anti-impact device and design, bridges may be destroyed in some serious vehicle-bridge collision. The design demand for structures to withstand impact load has increased, and using CFST members was expected to effectively improve structural anti-impact behaviors.

In the reality, the impact loads usually occurred near the bottom of columns rather than at the mid span [2]. Besides, partially filling concrete into hollow steel columns could help delay local buckling of the steel columns and was expected to improve the local anti-impact behaviors of the steel columns. Previously, the seismic behaviors of partially concrete-filled steel tubular (PCFST) members were investigated by Ge et al. [3] and Yuan et al. [4]. Usami et al. [5] performed cyclic and dynamic loads on PCFST bridge piers and found out that the PCFST columns showed good earthquake-resistance characteristics. However, investigations into the behaviors of PCFST columns under lateral impact loads were not found. The influence of parameters on the anti-impact behaviors of PCFST columns under impact loads were unknown.

This paper was attempted to study the behaviors of PCFST members under lateral impact loads. The columns were lateral impacted by pendulum hammer near the bottom. First, finite element models were established using ABAQUS to simulate the vehicle-pier collision. Second, the finite element models were validated by comparing the numerical results with experimental results. Third, to further study the anti-impact behaviors of PCFST columns, parameter analysis was carried out. The parameters including the concrete filling height (h_c), the impact energy (E_i), the impact direction (α), and the yield strength of steel (f_y) were considered. The dynamic responses under the impact load, including the impact force, the failure mode, and the displacement response, were all analyzed.

2. Finite element model

To simulate the actual condition, the material properties under impact load, the boundary conditions and the contact between the steel tube and the infilling concrete were considered. Details of the FEA models were shown in the following sections.

2.1. Element mesh

The cross-section of the rectangular steel tube was shown in figure 1. The sectional dimension of the rectangular steel tubes was 140 × 80 × 3 mm. The inner radius of the cold-formed corner was 4.5 mm. The total length of the columns was 1500 mm. The impact center was at the height of 320 mm from the bottom. The steel tube was partially filled with concrete. The steel tube and the infilling concrete were simulated using 4-node shell element with reduced integration (S4R) and 8-node brick element with reduced integration (C3D8R), respectively. The pendulum hammer was simplified to a rigid shell surface with an initial impact velocity and a mass of 350 kg, in order to simulate the vehicle impact. The rigid shell surface had a dimension of 200 × 50 mm. Furthermore, a mesh convergence study was carried out to choose an appropriate mesh density. The adopted element mesh was shown in figure 1. Moreover, stiffness-type hourglass control was used to eliminate the zero energy modes and the hourglass energy was less than 5% of total impact energy.

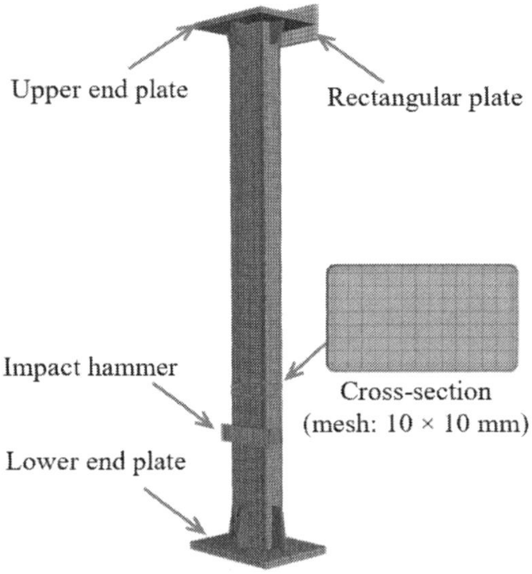

Figure 1. FEA model under lateral impact

Figure 2. Test device

2.2. Materials

A stress-strain model for cold-formed steel with a multilinear isotropic strain hardening rule [6] was used to simulate the material of the steel tube. The cross-sections of the rectangular steel tube were divided into flat zones and corner zones. The stress-strain relationship for the flat zones and the corner zones of the cold-formed sections were different. The yield strength (f_y) and ultimate strength (f_u) of the flat zones were obtained by conducting steel tensile coupon test. The yield strength of the corner zones (f_{yc}) were considered in an analysis model developed by Abdel-Rahman et al. [6], which was related to the inner radius of the cold-formed corner (r), the wall thickness of the steel tube (t) and the material properties of the flat zones. An empirical equation for f_{yc} was presented by Abdel-Rahman et al. [6]. The ultimate tensile strength of the corner zones (f_{uc}) was calculated by the equation (1) proposed by Tao et al. [7]. Cowper-Symonds power [8] was introduced to consider the effect of strain-rate ($\dot{\varepsilon}$) with a multiplier factor (D) of 40.4^{-1} and exponent (p) of 5 to achieve the dynamic yield strength .

$$f_{uc} = \begin{cases} [1.6 - 2\times10^{-3}(f_{yc} - 200)]f_{yc} & 200MPa \le f_{yc} \le 400MPa \\ [1.2 - 3.75\times10^{-4}(f_{yc} - 400)]f_{yc} & 400MPa \le f_{yc} \le 800MPa \end{cases} \quad (1)$$

The concrete damage plasticity (CDP) model in ABAQUS was adopted in this FEA model to simulate the behavior of concrete. In the CDP model, dilation angle (ψ), flow potential eccentricity (e), the ratio of the second stress invariant on the tensile meridian to that on the compressive meridian (K_c) and ratio of the compressive strength under biaxial loading to uniaxial compressive strength (f_{b0}/f_{c0}) were set as 30°, 0.1 , 2/3 and 1.16 [7], respectively. The stress-strain relationship of core concrete presented by Han et al. [9] was used to consider the increasing in the plasticity behavior of infilling concrete due to the passive confinement of the steel tube. The characteristic static compressive strength of concrete (f_{ck}) was multiplied by a dynamic increase factor of 1.25 to consider the effect of impact load [2].

2.3. Boundary conditions and contact

In order to simulate the actual boundary condition of impact, the freedoms of the lower end plate were restricted. The displacement of the upper end plate along the impact direction was restricted by the rectangular plate. A surface-based interaction with a contact pressure model in the normal direction and a Coulomb friction model in the tangential direction was used to simulate the contact between steel and infilling concrete. The detail information about the interaction could be found in Han et al. [9]. The friction factor of the Coulomb friction model was taken as 0.47 [10].

2.4. Validation of the FEA model

Impact tests were performed to study the anti-impact behaviors of PCFST column. The device of the impact test, including the column specimen, the impact hammer and the base plate, was shown in figure 2. To validate the FEA model, the numerical results including the failure mode and displacement response were compared with the experimental results, as shown in figures 3 and 4 respectively. The hollow steel tubular specimen (H0) and the PCFST specimens with concrete filling height of 700 mm (P700) and 1000 mm (P1000) were the specimens for comparisons. From figure 3, the global flexural deformation and local buckling of the test could be well predicted by the FEA model. Especially, for P700, the local buckling near the interface between the hollow steel section and the concrete-filled steel section could also be predicted. From figure 4, the numerical and experimental displacement processes were similar, both with three stages: the rising, the rebounding, and the stable stages. The displacement peak values of the FEA model were close to the test. The mean value of the ratio $\Delta_{maxFEA}/\Delta_{maxTest}$ was 1.03. Thus, a good agreement was achieved between the test and FEA model.

Figure 3. Comparison of failure modes

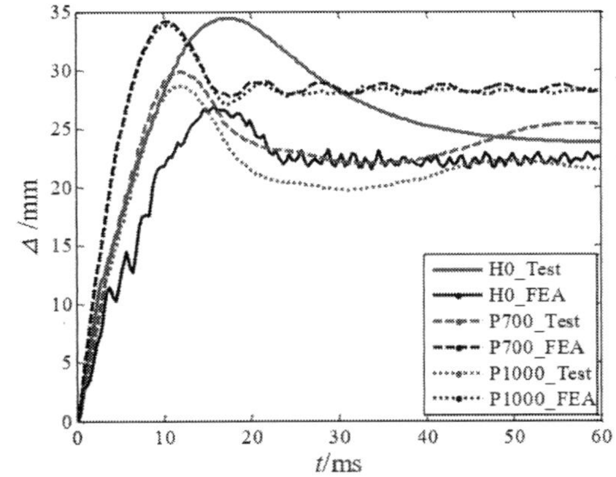

Figure 4. Comparison of displacement responses

3. Parameter analysis

To further study the anti-impact behaviors of PCFST columns, parameter analysis was carried out. As shown in table 1, the parameters including the h_c, E_i, α and f_y were considered, where α was the angle between the impact direction and the short edge of the cross-section. Different impact energies were achieved by changing the velocity of the impact hammer. The failure mode, the post-peak mean force (F_m) and the displacement peak value (Δ_{max}) were the dynamic responses for comparisons.

First, the concrete filling height significantly affected the anti-impact behaviors of PCFST columns. When h_c increased from 0 mm to 400 mm, F_m significantly increased and Δ_{max} decreased. The failure mode of M1 was similar to H0 in figure 3 and local bucklings of M2-M4 near the interface between the hollow steel section and the concrete-filled steel section were similar to P700. When h_c was higher than 1000 mm, local buckling near the interface was not observed. Furthermore, the anti-impact behavior of full concrete-filled steel tubular column (M6) was similar to PCFST columns. Comparing M1 with other PCFST models (M2-M5), it could be found that the anti-impact behaviors could be greatly improved by the infilling concrete.

Table 1. Parameter analysis on the anti-impact behaviors of PCFST columns

Model	h_c (mm)	E_i (kJ)	α (°)	f_y (MPa)	F_m (kN)	Δ_{max} (mm)
M1	**0**	8	0	540.6	104.2	41.5
M2	**400**	8	0	540.6	227.0	37.6
M3	**700**	8	0	540.6	223.5	37.0
M4	**800**	8	0	540.6	218.7	37.0
M5	**1000**	8	0	540.6	223.3	36.4
M6	**1500**	8	0	540.6	230.0	35.4
M7	700	**5**	0	540.6	202.4	25.4
M8	700	**6**	0	540.6	204.5	30.0
M9	700	**7**	0	540.6	216.3	33.8
M10	700	**8**	0	540.6	223.5	37.0
M11	700	**9**	0	540.6	228.7	40.6
M12	700	**10**	0	540.6	234.1	43.6
M13	700	8	**0**	540.6	223.5	37.0
M14	700	8	**15**	540.6	227.9	34.2
M15	700	8	**30**	540.6	212.9	31.7

M16	700	8	**45**	540.6	216.7	29.3
M17	700	8	**60**	540.6	234.3	22.6
M18	700	8	**75**	540.6	260.9	16.7
M19	700	8	0	**200**	136.1	55.6
M20	700	8	0	**300**	173.4	44.3
M21	700	8	0	**400**	212.0	37.5
M22	700	8	0	**500**	221.6	37.3
M23	700	8	0	**600**	235.1	35.4
M24	700	8	0	**700**	256.1	33.0

Second, Δ_{max}, F_m and local buckling near the interface all increased with the increase of E_i. The increase of local buckling indicated that the stress waves generated by larger impact energy propagated to the interface, which was the weak zone of columns. Thus, larger deformations occurred under larger E_i.

Third, PCFST columns under different impact directions shown different failure modes. The local buckling near the impacted corner occurred in M14-M18, as shown in figure 5. With the increase of α, global flexural deformations and Δ_{max} both decreased. Thus, the bending about the weak axis of the column was the most critical situation under lateral impact load.

Last, with the increase of f_y, for columns with h_c of 700 mm, F_m greatly increased and Δ_{max} decreased. Besides, as shown in figure 6, for columns (M19-M21) with lower f_y, larger local buckling near the interface and larger lateral displacement occurred in mid-span. Thus, increasing f_y could significantly improve the anti-impact behaviors of PCFST columns.

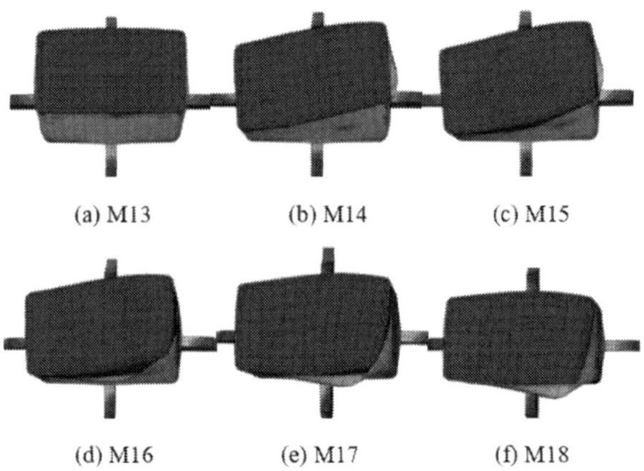

Figure 5. Local buckling near the impacted corner

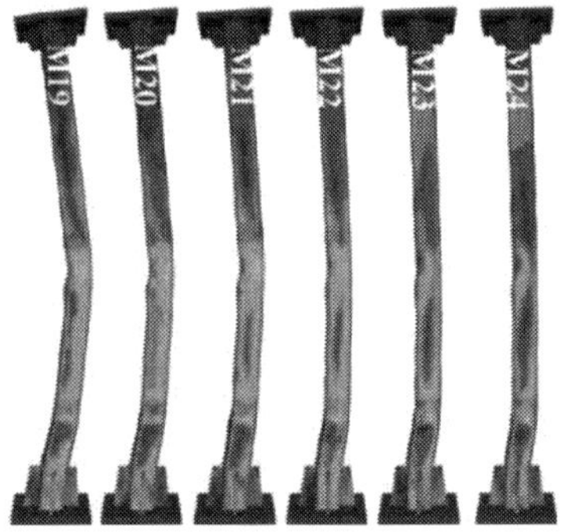

Figure 6. Failure mode of M19-M24

4. Conclusion

This paper presented a numerical investigation on the anti-impact behavior of PCFST columns under lateral impact load. The impact load was conducted at the height of 320 mm from the column bottom. A finite element model was established. The element mesh, material properties, boundary conditions, and contact were all considered in this model. Then, by comparing the numerical and experimental results, a good agreement was achieved between the test and the FEA model. Thus, the accuracy of the FEA model was validated. Parameter analysis was carried out to further study the anti-impact behavior. The results shown that the anti-impact behavior of PCFST columns increased with increasing concrete filling height and yield strength of steel. Moreover, the columns with α of 0° were the most critical situation under impact loads. And the columns under larger impact energy was prone to larger deformation.

Acknowledgement

The authors gratefully acknowledge the supports of the National Natural Science Foundation of China (No.51378233 & No.51629801) and the Fundamental Research Funds for the Central Universities (No.2016YXMS095).

Reference

1. Han L H, Li W and Bjorhovde R 2014 *J. Construct. Steel Res.* **100** 211-28
2. Yousuf M, Uy B, Tao Z, Remennikov A and Richard J Y 2014 *J. Construct. Steel Res.* **96** 54-68
3. Ge H B and Usami T 1996 *J. Struct. Eng.* **122** 1169-77
4. Yuan H H, Dang J and Aoki T 2013 *Earthquake Engng Struct. Dyn.* **42** 2197-216
5. Usami T, Ge H B and Saizuka K 1997 *Construct. Steel Res.* **41** 121-136
6. Abdel-Rahman N and Sivakumaran K S 1997 *J. Struct. Eng.* **123** 1135-43
7. Tao Z, Wang X Q and Uy Brain 2013 *J. Mater. Civ. Eng.* **25** 1306-16
8. Abramowicz W and Jones N 1984 *Int. J. Impact Engng.* **2** 179-208
9. Han L H, Yao G H and Tao Z 2007 *Thin-walled Struct.* **45** 24-36
10. Baltay P and Gjelsvil A 1990 *J. Mater. Civ. Eng.* **2** 46-49

Trial to Determine Durability and Serviceability for Swine Farm in Thailand

L Lapcharatsangroj and Chalida U-tapao

Department of civil engineering, King Mongkut's Institute of Technology Ladkrabang. 1 Soi Chalong Krung 1, Chalong Krung Road, Ladkrabang, Ladkrabang, Bangkok, Thailand 10520

Abstract. In agricultural construction, the most frequently used construction material is concrete. In particular swine farm, traditional concrete flooring system in the business is conventional concrete. However, the conventional concrete floor has several key issues on its surface. They are easily to be spalled and cracked, pig scour, rough areas, and low wear resistance of acid from animal feed. This research involves study in the ratio between Portland cement and supplementary materials (fly ash and silica fume) that is the most resistant to abrasion. Concrete samples were cured in water for 28 days, then submersed in lactic acid solution at 30 °C (pH 2-3). After exposure to the acid solution, every 7, 14, 28, and 56 days, the samples will be test for attrition on the face of concrete in accordance with ASTM C944 to simulate the acceleration reaction refer to animal behaviour and corrosion from acid of feed. From experiment, it was found that increasing the volume of fly ash and silica fume can increase concrete resistance to corrosion due to acid derived from feed.

1 Introduction

The construction of Thai swine farms is generally made of concrete. It is crucial that the swine farm floor must be smooth but non-slippery. Normally, the floor is casted by farmers or contractors who usually build it with conventional concrete or just use reinforcement in the solid floor. Concrete is one of the most frequent used as building material. It is not only for general construction but also for a specific purpose like the agricultural construction, where the floor is aggressively subjected to chemical erosion and physical force such as cleaning, acid from animal feed and animal behaviours. [1]

This document purpose is to enable an assessment of a swine building's construction and key factor of the concrete surface's strength in the building and structure is agricultural environment. [2] The surface may be exposed to aggressive composition like chemical and physical environment. Hence, concrete material for swine building floor should be high quality material that resistible against the chemical and physical attacks. [3] The information following will present a reviewing of recent performed research in durability of building materials and components in agricultural environment. Leading to the development of 'special concrete for farm floor' with a concept of increasing in resistance in erosion due to lactic acid derived from animal feed; corn, wheat, manioc, so forth. Moreover, the concrete mass is also increased the hardiness to be durable from animal behaviour such as scratching that is caused of craggy and coming apart. [4]

In fact, there are many factors that could cause damages on swine farm floors. The various sorts of cleaning equipment and methods make it more difficult to precisely and concisely identify the actual cause of the floor corrosion. It is also more complicated owing to the environmental dissimilarity in each area (urban, rural, and coastal). [1] Consequently, concrete becomes common used in every farm due to its ability that can be in any forms, durability, strength and suitability to environmental practices. Mostly, concrete is made from lime-based concrete like Portland cement concrete or other hydraulic cements. [5] It requires other processing and value-added facilities like composting and anaerobic digestion facilities as integral parts in the operations. The compressive strength depends on the proportions of compound ingredients, i.e. water-cement ratio, and the cement aggregate ratio. Furthermore, superplasticizers was used in the mixed design. [6] Superplasticizers help in removal of 15% or adding water from the mixed design or increasing concrete workability (slump).

2 Related literature

Nowadays, the demand in pork products is increasing. It causes not only the growth in numbers of swine but also the numbers of slaughter weight. In Thailand, swine farm mostly consists of concrete.[7] Eighty percent of concrete floors in swine farms were cast by farmers or contractors, usually with conventional concrete. These floors are directly subject to chemical attack by cleaners, feed acids and animal behaviours (see **Figure 1** and **Figure 2**). From **Figure 3** and **Table 1**, livestock is one of the main agricultural activities where mainly in Central region's provinces of Thailand such as Nakhornpathom, Ratchaburi, and Chachoengsao.[7] The livestock and poultry farms could be divided into two types: commercial farms (big operators) and non-commercial farms (backyard farms). This document enables an assessment of the opportunities and issues involved with constructing and operating a pig finishing building.

3.1 Methods to protect concrete structures

3.1 Concept design

Concrete samples with the materials and mix designs in Table 2 were prepared for compressive strength and abrasion resistance test and three methodologies that are used to define the results.

3.2 Testing concrete

3.2.1. Compressive Strength

Most important property of hardened concrete. Generally considered in the design of concrete mixtures. Compressive strength is affected by many factors (environmental, curing condition). Therefore, the actual strength of concrete will not be the same as the strength of specimen. Dimensions of the concrete specimens usually have the following sizes: cube specimens 15x15 cm. Empty moulds are filled with fresh concrete using a standard procedure and test compressive strength.

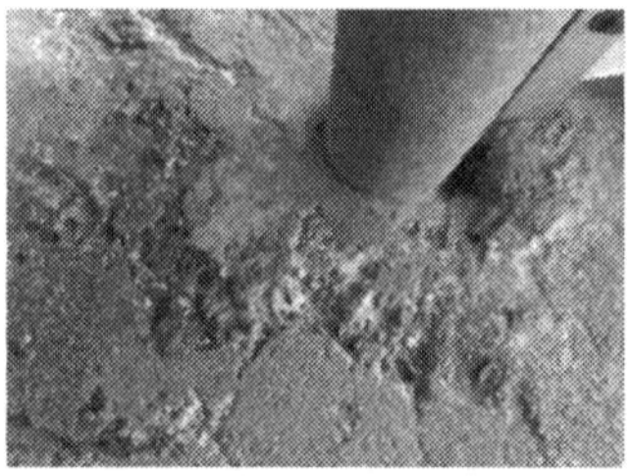

Figure 1. Spalling

Figure 2. Swine Farm

Figure 3. Map of Thailand

Table 1. Top 10 Province of Pig Farm

Province	Region	Number of Pig (Heads)	Area (sq.m.)	%
Ratchaburi	Central	1,468,839	19,094,907	19
Nakhonpathom	Central	914,718	11,891,338	12
Chachoengsao	Central	456,419	5,933,451	6
Chonburi	Central	310,758	4,039,854	4
Chiangmai	North	297,217	3,863,821	4
Nakhonratchasima	North-east	246,482	3,204,262	3
Saraburi	Central	205,330	2,669,286	3
Nakhonsithammarat	South	1,468,839	19,094,907	2
Surathani	South	914,718	11,891,338	2
Lopburi	Central	139,402	1,812,222	2

Table 2. Description of mix design

Detail/ Mix design	Conventional concrete	10FA	25FA	8SA22FA
Cement type (ASTM C150)	I	I	I	I
Cement (kg/m^3 of concrete)	320	325	270	220
Water (kg/m^3 of concrete)	180	160	165	160
w/c ratio	0.55	0.44	0.44	0.44
Fine aggregate (kg/m^3 of concrete)	840	800	800	800
Coarse aggregate (kg/m^3 of concrete)	1,060	1,110	1,110	1,110
Admixture – Type A&D – Type F	1,770 cc.	3,600 cc.	3,600 cc.	3,950 cc.
Supplementary cementing material (% of cementitious material content)	-	35 kg (10%) fly ash	90 kg (30%) fly ash	105 (22%) fly ash + 25 kg(8%) silica fume

3.2.2. Abrasion Resistance Test (ASTM C944)

This test could cover a process of determining the resistance of either concrete to mortar to the abrasion. This method's specimens could be cured lectic acid on the surface of samples cube specimens in lectic acid. The tested surface should be formed or finished and positioned at the contact of cutter on the plane. For the highly resistant concrete, there will be required an additional abrasion test. The test shall involve minimum two minutes and conducted on three separated areas of each representative concrete's surface. the test result will be able to describe more comprehensive as the following chart shown;

3.2.3. Moh's scale Test

The material's resistance of being scratched is defined as 'hardness'. The test is proceeded by placing a sharp point of one specimen on another unmarked surface specimen then attempting to create a scratch on it. The test will compare the resistance of a mineral being scratched with ten mineral references. It is known as Moh's Harness Scale as illustrated on **Figure 4**. Mohs Scale, from 1 (softest) to 10 (hardest) based on mineral type 1 = Talc 2 = gypsum 3 = Calcite 4 = Fluorite 5 = Apatite 6 = Feldspar 7 = quartz 8 = Topaz 9 = Corundum

Figure 4. Moh's scale (Test Kit)

4 Results and discussions

The use of Pozzolans is not only help in strengthen and sealing the concrete, but also have more benefit in other features when this raw material is added or mixed with other materials. These are the benefit when applying with fly ash (FA) and silica fume (SF) in order to increase density and long-term pozzolanic action. This process can be ties up free lime, show fewer bleed channels and decreases permeability in the result. Moreover, Pozzolans can be combined with free lime to increase structural strength over time. [7] However, Actual water demand of Conventional concrete and 8FA22FA was highest content. While the actual water content of 30FA samples was equal. 30FA and 8SF22FA samples gained the similar initial slump at 22-21 cm. **Figure 5 and 6.** Shows all cement samples had no adverse effects on slump loss properties and setting times of 8SF22FA were shortest. While the setting times of other samples were similar. (5-6 hrs.). **Figure 7.** shows all age, focusing at the 28-day compressive strength (the critical value), 8SF22FA gains the highest compressive strength when compare with 30FA, while Conventional concrete gains developed the lowest compressive strength at all ages.

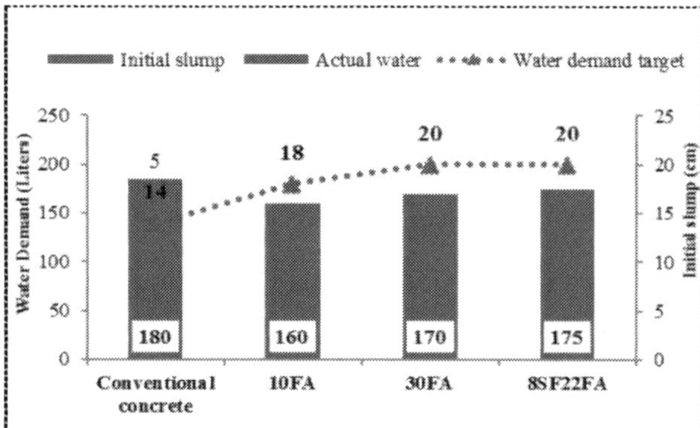

Figure 5. Actual water demand

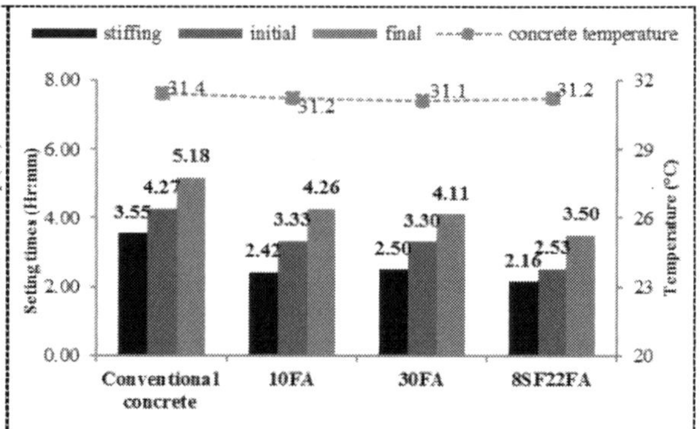

Figure 6. Setting times

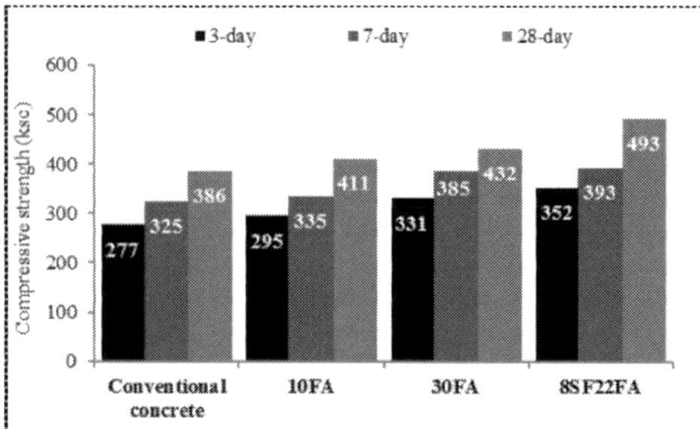

Figure 7. Compressive strength

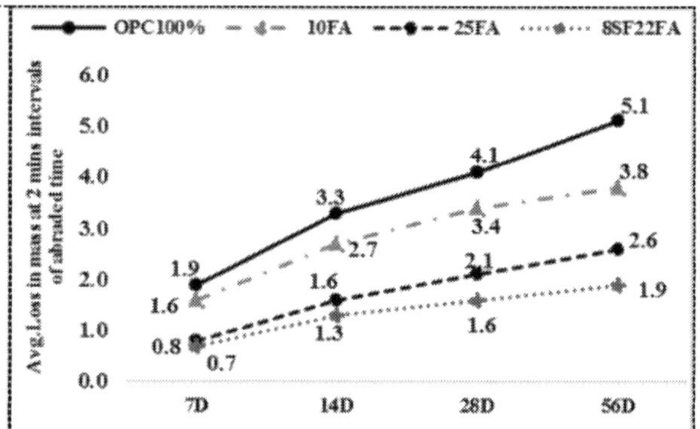

Figure 8. ASTM C944 abrasion testing

The results for the ASTM C944 abrasion testing (single load) are presented in **Figure 8** the average weight loss (grams) per abrading of 8SF22FA samples has the best result (average 1.9 g in 56 days) while the samples of Conventional concrete and with various % replacement of Pozzolan material are worth 56-Day weight loss of all cement. **Figure 9-12.** The results for the Moh's scale testing are presented at 28 days. 8SF22FA gains the highest hardness around 8-9 when compare with other samples. No.8 (Topaz) is silicate mineral created from combination of aluminium and fluorine. No.9 (Corundum) is crystalline from of aluminium oxide and one of the basic rock-forming mineral.

Figure 9. OPC 100%

Figure 10. 10FA

Figure 11. 25FA

Figure 12. 8SF22FA

5 Conclusions and recommendations

Portland cement is the most useful ingredient and popular of concrete. Eventually, Portland cement concrete usually does not have good properties to acids. However, some weak acids can be tolerated as in **Figure 5.** 8SF22FA is a solution developed for highly aggressive agricultural environments, especially piglet farms compared to standard concrete and fly ash- concrete (10FA and 25FA). The 8SF22FA samples can offers a higher acid and abrasion resistance, better workability denser concrete structure and higher final strength. Lactic acid derived from feed that can react with free lime on the concrete and produce exceedingly soluble calcium salts may cause an extremely aggressive impact on the floor. [4] When the salt is leached, the concrete porosity will be raised in contrast to pH in the pores, which will be dropped. Therefore, the durable concrete mix provided the better solution for swine farm surface spalling. Increase volume of fly ash and silica fume can increase resistance of concrete from corrosion due to acid, significantly increase compressive strength at all ages, and increase abrasion resistance (ASTM C944). It is important and required to balance the strength and workability on the concrete work and slippery resistance. In the future, the investigation of the chemical and physical environment will be more significant and intensive in the most exposed areas of the farm buildings. [1] This knowledge can be applied and useful as a tool for material and components selecting with sustained durability.

Acknowledgments

These articles, Part I Feasibilities to Develop Concrete Floor of Swine Farm in Thailand, Part II Trial to Determine Durability and Serviceability for swine farm in Thailand. The author would like to express her very great appreciation to Asst.Prof.Dr. Chalida U-tapao for valuable and constructive suggestions also like to thank my parents and friends for providing unfailing support and continuous encouragement and through the process of researching and writing this thesis. This accomplishment would not have been possible without them. Thank you.

Reference lists

1. N. De Belie, M. Debruyckere, D. Van Nieuwenburgs and B. De Blaeres 1997 Attack of concrete floors in pig houses by feed acids: Influence of fly ash addition and cement-bound surface layers. J. Agric. Engng Res. 68 pp 101-08
2. Shelton, David P. and Harper and James M 1982 G82-623 An overview of concrete as a building material. Historical Materials from University of Nebraska-Lincoln Extension. 603., Issued December 1982 (12,000 printed)
3. Makanjuola Oki, Paul A L Anawe. 2015 A Review of Corrosion in Agricultural Industries. *J. Phys Sci Inter* 5(4): 216-222, Article no.PSIJ.2015.022 ISSN: 2348-0130
4. B. Svennerstedt Durability of building materials and components in agricultural environment. *J. Agric. Engng Res.* **75**, Issue 3, March 2000, p. 225-241
5. Kramer, W. A. & Braam and C. R. 1997 Deterioration of concrete exposed to lactic and acetic acid. *Proceedings of the International Symposium on Concrete for a Susttiable Agriculture.* 2 1-24 May 1997. S tavanger
6. Tikalsky PJ, Carrasquillo PM, Carrasquillo RL 1988 Strength and durability considerations affecting mix proportioning of concrete containing fly ash. *ACI Mater J.* **85**(6) pp 505–11
7. Livestock statistics Agricultural Statistics of Thailand., 2015. Department of Livestock Development (DLD)., http://www.oae.go.th/main.php?filename=index___E N

Effects of Shrinkage Reducing Agent and Expansive Additive on Plastic Shrinkage of Mortar at Different Temperatures

Sarapon Treesuwan[1], Komsan Maleesee[1] and Shigeyuki Date[2]

[1]Department of Civil Engineering, Faculty of Engineering, King Mongkut's Institute of Technology Ladkrabang, Bangkok 10520, Thailand
[2]Department of Civil Engineering, Faculty of Engineering, Tokai University Kanagawa, Japan

Abstract. In the construction, it is inevitable to perform plaster work in hot weather which causes the dehydration and rapid shrinkage on the paste during the early age. This research shows the studies of reducing the plastic shrinkage of mortar during the early age with such additives as the Shrinkage Reducing Agent (SRA), the Expansive Additive (EX), and the Fly Ash (FA) in controlled temperatures at 30 °C and 40 °C, with relative humidity between 60% and 70% according to the ASTM C1579-06 standard, with the strain gauge installed at 0.5 cm.from the surface. The shrinkage rate was measured starting from the Initial Setting Time and every 10 minutes afterwards for 24 hours. The results show that high temperature effects the cracking and how to use different formulas of additive under different circumstances is considerably important. To use only one additive is not sufficient in high temperature. To use the SRA in addition to the EX enhances better expansion than to use only the EX. Moreover, it is recommended to pay close attention in adding large amount of the FA into mortar with the EX and SRA added which extremely enhances the expansion and potential cracking.

1 Introduction

There are several methods in reducing the shrinkage of mortar. One of the methods is to use such inorganic materials and additives as the Shrinkage Reducing Agent (SRA) and the Expansive Additive (EX) which can be added during the mixing process[1-4]. Different types of additive to be added with the cement give different results depending on such factors as the property and quantity of the cement used, the mixing ratio, and the curing method. Therefore, thorough study on each type of additives and on the manufacturer's instructions is highly recommended because these additives are beneficial when used appropriately. However, the recommended amount may not always suitable for all operations under different circumstances. Hence, tests is needed to be conducted and the amount used should be controlled. The plaster work is commonly done during the day when the temperature is ordinarily high especially for a tropical country where the average temperature is 30°C and, in summer, possibly 40°C while most tests are conducted at 25°C, which hardly is in accordance with real construction. The high temperature is one of several important factors that affects the plastic shrinkage which usually occurs during the early age, i.e., within the first 24 hours [5-6], by effecting the cracking on the surface which not only is difficult to repair but also acts upon the long-term durability of the building structure system. According to the standards from the American Concrete Institute (ACI), the Expansive Additives are divided into 3 types: (1) K-type, with anhydrous hauyne (CaO $*Al2O_3*CaSO_4$) as its basis; (2) M-type, with gypsum ($CaSO_4$) as its basis; and (3) S-type, with tricalcium aluminate (C_3A)and gypsum ($CaSO_4$) as its bases, while, according to the Japanese Standards, the Expansive Additives are divided into 2 types: (1) C-S-A type, which approximates to the K-type according to the ACI standards, producing the ettringite; and (2) CaO type, with Calcium Oxide (CaO)as its basis, producing Calcium Hydroxide($Ca(OH)_2$). In this experiment, the EX with Calcium Oxide (CaO) was used as the main additive. The EX helps reduce the shrinkage due to such factors as the expansion from water absorption and the forming of pores and crystals during the hydration reaction process[7]. The Shrinkage Reducing Agent (SRA) is a chemical substance that lowers the surface tension by decreasing the capillary stress on the surface of the paste and reserves the inner moisture which helps to reduce the evaporation on the surface, thus, reducing the shrinkage of the paste on the chemical and environmental aspects[-8 9]. The fly ash is the residual material from coal combustion during the electricity generation. It is round, smaller than 1 micrometre to 150 micrometres in size, generally in finer powder form than the cement, with specific gravity between 2.00 and 2.60. It possesses cementitious property that is called pozzolanic material while other pozzolanic materials generally possesses little or no cementitious valuebut will, in finely divided form and in the presence of sufficient moisture, chemically react with calcium hydroxide at ordinary temperatures to form compounds having cementitious properties. The fly ash is therefore considered beneficial as an additive to be added to the cement mixture[10]. Experiments on the plastic shrinkage during the early age and studies on the 3 types of additive are rare. This research thus focuses on the influence of the plastic shrinkage reduction during the early age, i.e., within the first 24 hours, of mortar added with the fly ash, the shrinkage reducing agent (SRA), and the expansive additive (EX) at 30°C and 40°C. The analysis was conducted conforming to the ASTM C1579-06 standard, using the strain gauge to measure the shrinkage near the surface.

2 Materials and experiments

2.1 Materials

Table 1. shows the materials used in the experiment and Table 2. shows the chemical composition. The ordinary Portland cement (C), the fly ash (FA) Class C – according to the ASTM C618 standard, the expansive additive (EX) CaO type of chemical used for shrinkage reduction and the shrinkage reducing agent (SRA) Polyoxyakylene alkyl ether type– Tetraguard®AS20 were used in this experiment along with fine sand prepared in surface-dry, saturated condition with fineness modulus (F.M.) of 3.05, 0.54% of absorption, and bulk specific gravity of 2.60.

2.2 Experiments

Table 3. shows the mortar mixing ratio. The water to binder(W/B) ratios were 0.7, the shrinkage reducing agent was used to replace the water at 0 and 2% for binder. To replace the cement with the EX and FA, the EX was determined to stay unchanged at 15% which is the recommended maximum amount, while the amount of FA was adjusted in accordance with the amount of the binder left in the composition.The Hobart mixer was used to mix the mortar by mixing the water and admixtures at low speed for 30 seconds before adding the binder using medium speed and adding the sand in 30 seconds before using high speed. The entire mixing duration was determined for 3 minutes.The setting time was determined following the ASTM C807 standard at 30°C and the plastic shrinkage was measured following the ASTM C1579-06 standard. The relative humidity was controlled in the range of 60-70%. The shrinkage wasmeasured at 30°C and 40°C by controlling the heat from the heat lamps with the waterproof strain gauge for cement work applicable for the range of temperatures between -10 °C and 70 °C installed in the middle of the mold above the crack riser at 0.5 centimetre from the surface. The plastic shrinkage was measured and recorded once every 10 miutes for 24 hours starting from the initial setting time. Figure 1 and Figure 2 show the plastic shrinkage measurement.

Table 1. Materials and Physical Properties

	Kind	Symbol	Specific gravity	Blaine (cm2/g)
Cement	Ordinary Portland Cement	C	3.15	3,400
Inorganic additive	Fly ash	FA	2.23	3,800
	Expansive additive	EX	3.16	3,450
admixture	Shrinkage reducing agent	SRA	0.99	-

Table 2. Chemical Compositionof Binders

Kind	SiO_2(%)	$Al2O_3$(%)	$Fe2O_3$(%)	CaO(%)	MgO(%)	Na_2O(%)	K_2O(%)	SO_3(%)
C	21.45	4.64	3.05	64.99	0.88	0.09	0.59	2.47
FA	24	10	21.2	32.6	2.08	0.1	3.57	1.1
EX	4.2	1.1	1	74	0.51	-	-	16.5

Table 3. Mortar Mixing Ratio

W/B ratio	FA Replacement ratio(Bx%)	EX Replacement ratio(Bx%)	SRA Dosage (Bx%)
0.7	0,20,40	0,15	0,2

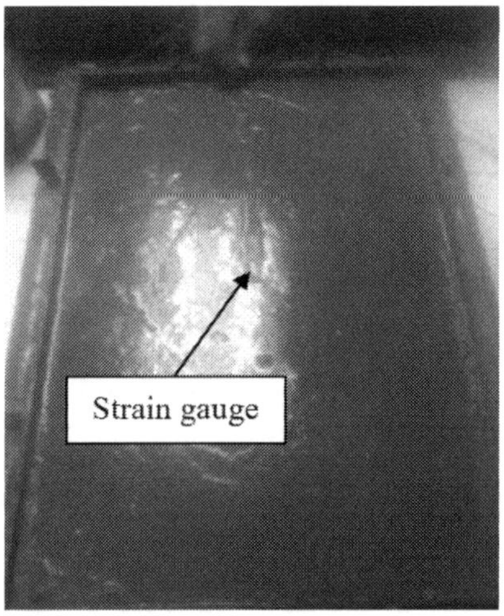

Figure 1. The installation of strain gauge in the middle of the mold at 0.5 cm. from the surface

Figure 2. The temperature and relative humidity control for environmental stability according to the experiment plan

3 Results and discussions

Figure 3, 4 and 5 show the results of plastic shrinkage during the early age of mortar added with the EX, SRA and FA at 30°C and 40°C in comparison with that of the control mortar (C 100%) 24 hours after molding in accordance with the ASTM C1579-06 standard. The result shows that the shrinkage and expansion rate highly occurs since the initial setting time and the rate decreases after approximately 10 hours after molding. It is obviously seen that, as the temperature rises, the shrinkage rate tends to increase for all formulas of mixture, especially for the control mortar (C=100%) at 40°C, the risk of cracking is highest in this regard. Mortar with 15% of EX added expands extremely during the early age at ordinary temperatures. The mortar added with 2% of SRA extremely reduced the plastic shrinkage both at 30°C and at 40°C by 24.4%and 57.6% respectively after being compared to the control mortar at the same temperature. However, at 40°C, though the SRA extremely reduced the shrinkage, the measurable plastic shrinkage result was higher than 1000 με. Therefore, there is still a high risk of the cracking on the surface to use only the SRA at high temperature. To add the FA helps to reduce the shrinkage during the early age. The mortar's shrinkage rate decreases in accordance with the increasing ratio of the FA added. This decreasing shrinkage may be the result of the roundness of FA which lessens the water requirement of mortar, causing the excess water on the surface[11-12], resulting the extension of the evaporation period. Moreover, the cement replacement reduces the hydration reaction between the cement and the water which is a cause of the shrinkage[13]. The expansion rate shall increase when the EX+SRA are added. This increasing expansion may be the result of the CaO crystal forming of the EX which normally creates the calcium hydroxide in the form of Hexagonal plate but to add the SRA changes the crystal forming to be in the form of Prismatic needle which is longer in shape[14]. To replace the cement with 15% of EX +2% of SRA and 20% of the FA does not clearly result in the expansion compared to adding 15% of EX + 2% of SRA but, when 40% of the FA was replaced, the expansion rate increases for more than 50% compared to adding 15% of EX + 2% of SRA at 25 °C and at 40 °C respectively. Nevertheless, care should be taken in terms of compressive strength because the large amount of FA added can reduce the strength during the early age.

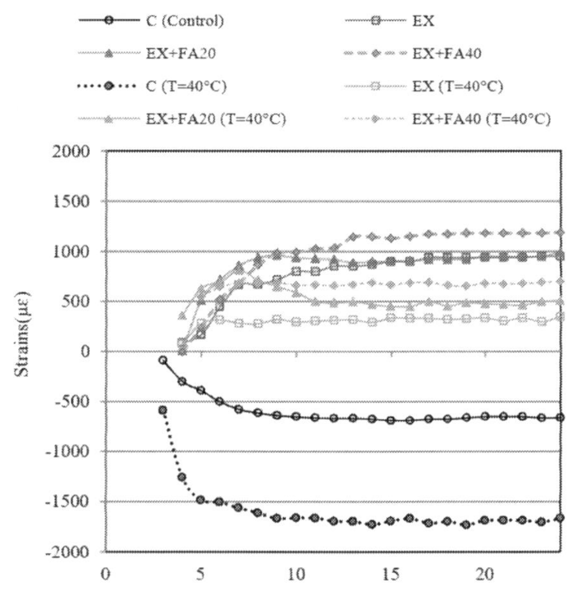

Figure 3. Early age plastic shrinkage results for EX and EX+ FA at 30°C and 40°C

https://doi.org/10.1051/matecconf/201820602002

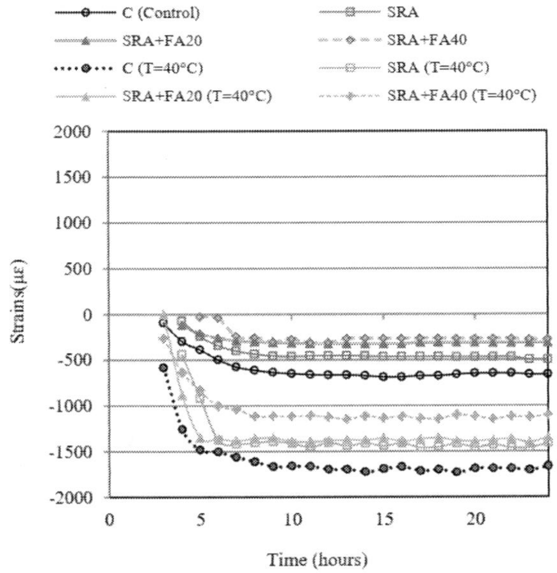

Figure 4. Early age plastic shrinkage results for SRA and SRA + FA at 30°C and 40°C

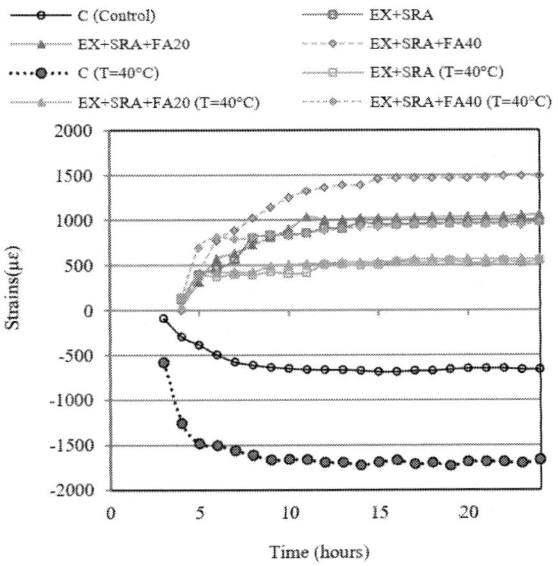

Figure 5. Early age plastic shrinkage results for EX+SRA and EX+SRA + FA at 30°C and 40°C

Figure 6 shows the comparing of plastic shrinkage during the early age of mortar mixed with the FA, EX and SRA at 25°C and 40°C 24 hours after casting according to the ASTM C1579-06. It was obvious that, at higher temperature, the shrinkage amount tended to increase with every mixture, especially with the controlled mortar (100% cement) at 40°C, there was a high risk of cracking. As such, to reduce the shrinkage by using the EX, SRA and FA must be considered carefully and used suitably depending on different environments. The shrinkage evaluation during the early age is important especially in high temperature where there is more risk of plastic shrinkage and cracking

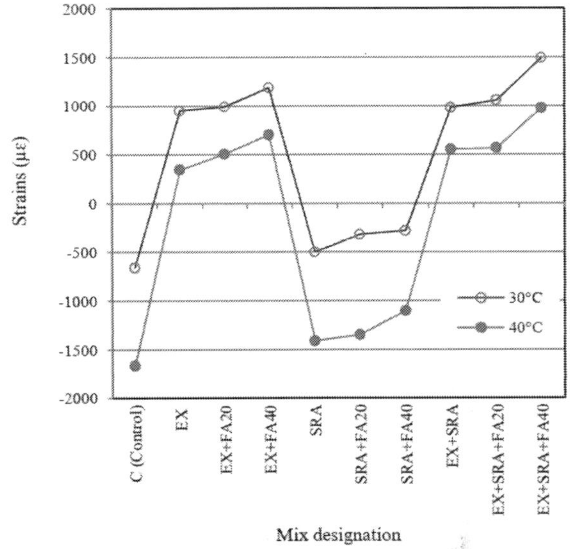

Figure 6. To compare Early-age plastic shrinkage of mortar mixed with FA, EX and SRA at 30°C and 40°C at 24 hrs. after casting

4 Conclusion

From the experiment, summaries can be stated as follow.

1) To replace the water with 2% of SRA helps to reduce the plastic shrinkage in all ranges of temperature but may not be sufficient when the temperature is up to 40 °C and it helps reducing the shrinkage to add the SRA+FA in accordance with higher amount of the FA added.

2) To add 15% of the EX increases the plastic expansion rate of mortar during the early age, i.e., the expansion rate increases when the ratio of replacing the cement with FA is raised.

3) To add 15% of EX + 2% of SRA increases the plastic expansion rate of mortar during the early age more than to add only the EX or SRA. The expansion rate extremely rises when the ratio of replacing the cement with FA especially for 40% is raised.

4) In all formulas of mixture, the plastic shrinkage rate during the early age rises as the temperature changes from 25 °C to 40 °C.

5) It is recommended to adjust the amount of EX, SRA and FA used in different circumstances to meet the appropriateness as each environmental circumstance differently affects the plastic shrinkage during the early age.

More research is required in different conditions such as steam curing and other properties such as strength and crystalline structure.

References

1. Palacios M and Puertas F, Y 2005, Cement and Concrete Research 37 pp 691-702
2. Saliba J, Rozière E, Grondin F and Loukili A, Y 2011, Cement and Concrete Composite 33 pp 209-217
3. Carballosa P, García JL, Revuelta D, Sánchez JJ and Gutiérrez JP, Y 2015, Construction and Building Materials 93 pp 223-229
4. Choi H, Lim M, Kitagaki R, Noguchi T and Kim G, Y 2015, Construction and Building Materials 84 pp 468-476
5. Uno P J, Y 1998 ACI Materials Journal, July-August pp 365-375
6. Holt E and Leivo M, Y 2004 Cement & Concrete Composites 26 pp 521-530
7. Nagataki S and Gomi H, Y 1998 Cement and Concrete Composite 20 pp 163-170
8. Wyrzykowski M, Trtik P, Müncha B, Weiss J, Vontobel P and Lura P, Y 2015 Cement and Concrete Research 73 pp 238-245
9. Bentz D P, Geiker M R and Hansen K K, Y 2001, Cement and Concrete Research 31 pp 1075-1085
10. Chindaprasirt P,JaturapitakkulC and Sinsiri T,Y 2005 Cement & Concrete Composite 27 pp 425-428
11. Termkhajornkit P, Nawa T, Nakai M, and Saito T, Y 2005, Cement and Concrete Research 36 pp 473-482
12. Nawaz A, Julnipitawong P, Krammart P, Tangtermsirikul S, Y 2016, Construction and Building Materials 102 pp 515-530
13. Treesuwan S and Maleesee K, Y 2017, Advances in Materials Science and Engineering vol 2017 pp 1-11
14. Maltese C, Pistolesi C, Lolli A, Bravo A, Cerulli T and Salvioni D, Y 2005, Cement and Concrete Research 35 pp 2244-2251

The Damage of Fly Ash -Slag Concrete under Sustained Loading in Marine Environments

J Gong[1], J Cao[2] and Y F Wang[1]

[1] *School of Civil Engineering, Beijing Jiaotong University, Shangyuancun 3, Haidian District, Beijing 100044, PR China.*
[2] School of Civil and Architectural Engineering, Nanchang Institute of Technology, Nanchang 330099, PR China.

Abstract. The durability damage of concrete structures in the actual engineering environment is often the result of the interaction of the load and the environment, climate and other multiple factors. In this paper, the test researches were carried out on the concrete mass and dynamic elastic modulus under the effects of sustained loading, sulfate attack and dry-wet circulation,and the test phenomenon and results were discussed and analyzed. Then this paper proposed the durability damage model of the concrete under the effects of sustained loading, sulfate attack and dry-wet circulation considering the compressive stress level, the sulfate solution concentration and the age of corrosion as test factors, and the rationality of the model was discussed.

1 Introduction

Neville [1] considered that it needs to recognize the importance of the interaction of multiple factors in the concrete durability theoretical research and practical engineering application. There are mainly two methods for the damage analysis of concrete structures under the coupled actions of multi-factors.

One method is that the damage models were established considering chemical ion corrosion, and the required parameters were determined through the analysis of test results. Tixier et al.[2,3] established the model to calculate the expansion of concrete which was under sulfate attack applying the physical and chemical mechanics. Saetta et al.[4-6] firstly calculated the mechanical damage and chemical damage separately and then coupled them, established the chemical mechanical model to simulate the sulfate salt corrosion of cement-based materials. Kuhl et al.[7,8] considered the precipitation of calcium in the concrete as chemical damage variable and combined with the concrete mechanical damage, and then established the chemical mechanical damage model to calculate the deterioration of mechanical properties.But a large number of unknown parameters needed to be identified in these models, and these key material parameters were determined empirically in the past.

Now the mathematical statistical method about the evolution of concrete durability damage based on the test results is widely applied.Schneider and Chen [9] conducted the compressive strength tests of the concrete members in different bending stress levels which were in the ammonium sulfate and ammonium nitrate solution with different solubility, through the regression analysis of the test data, they supposed the model of concrete time varying compressive strength considering the load and sulfate attacks. Maes et al.[10] conducted theoretical researches on the fatigue properties of the prestressed bridge structures which are under stress corrosion, and provided the model of cracking leading from stress corrosion. Xi et al.[11] studied the degeneration of concrete damage and durability caused by the diffusion process of the chemical ions, and proposed the united differential equation to solve the interaction problem .

This paper conducted experimental study on the concrete damage under different experimental conditions, providing the basis for proposing a concrete damage calculation model under the effects of sustained loading, sulfate attack and dry-wet circulation.

2 Experimental program

2.1 Material properties

A Portland cement with a specific density of 3.1 g/cm^3 was used.The coarse aggregates were natural crushed limestone, with size of 5-25 mm. The used FA is classified as low calcium Class II. The design of mixture for the concrete specimen is shown in Table 1.

2.2 Experimental parameters and test apparatus

The test factors considered in this study include load stress level(0,15% and 30%), sulfate solution concentration(1%, 5% and 10%), immersion ways:long-term immersion(LM) and wet-dry cycle(DW).A new test apparatus was developed based on the guidelines provided by GB/T 50082-2009 [12],which is shown in figure1.

Table 1. Design of the concrete specimen mixtures

Cement (kg/m^3)	Sand (kg/m^3)	Aggregate (kg/m^3)	Water (kg/m^3)	Fly ash (kg/m^3)	Slag (kg/m^3)	Water reducer (kg/m^3)
296	729	1050	172	81	63	8.80

The test factors considered in this study include load stress level(0,15% and 30%), sulfate solution concentration(1%, 5% and 10%), immersion ways:long-term immersion(LM) and wet-dry cycle(DW).A new test apparatus was developed based on the guidelines provided by GB/T 50082-2009 [12],which is shown in figure1.

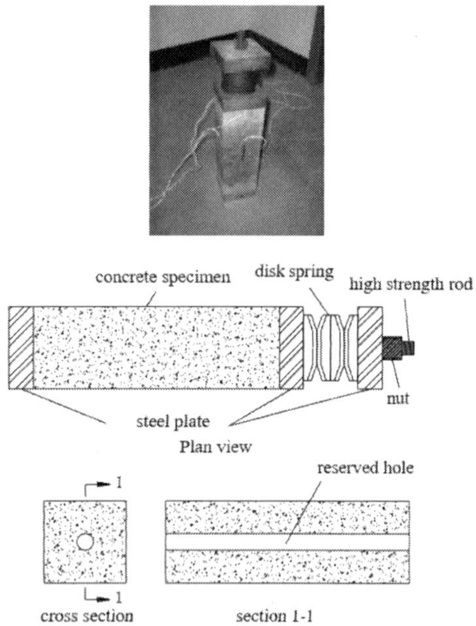

Figure 1. Loading equipment

3 Analysis of test results

3.1 Mass

The mass changes of the concrete which is in the long-term sulfate corrosion environment are shown in figure 2.

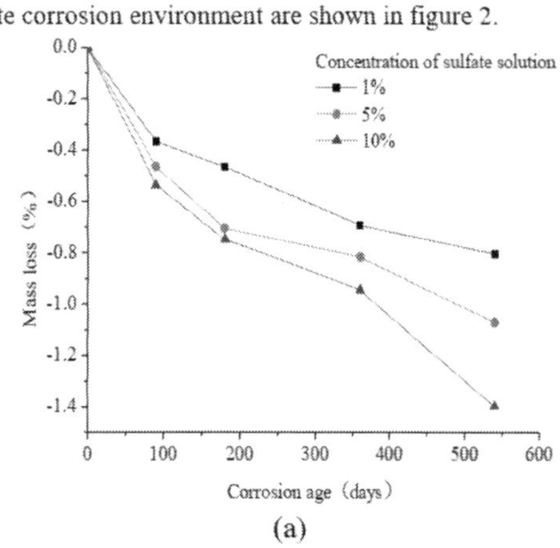

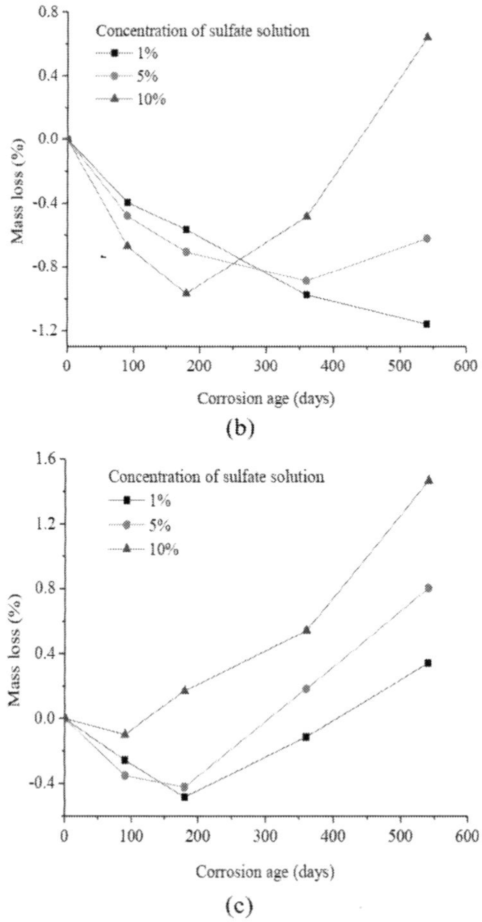

Figure 2. Mass change of concrete under long-term immersion of sulfate attack (a)the stress level is 0(b) the stress level is 15%(c) the stress level is 30%

From figure 2, when the stress level is 0, mass loss rates of the concrete in solution with different concentrations are all negative. When the stress level is 15%, masses of concrete in solution with different concentrations have different changes with the corrosion age's increasing. For the solution with concentration of 1%, mass increases along with the corrosion age's development. For the concentration of 5%, concrete's mass first increases and then decreases. When the stress level is 30%, masses of concrete in solution with different concentrations all first increase and then decrease with corrosion age's development.

The mass changes of the concrete which is in the wet and dry sulfate corrosion environment are shown in figure 3.

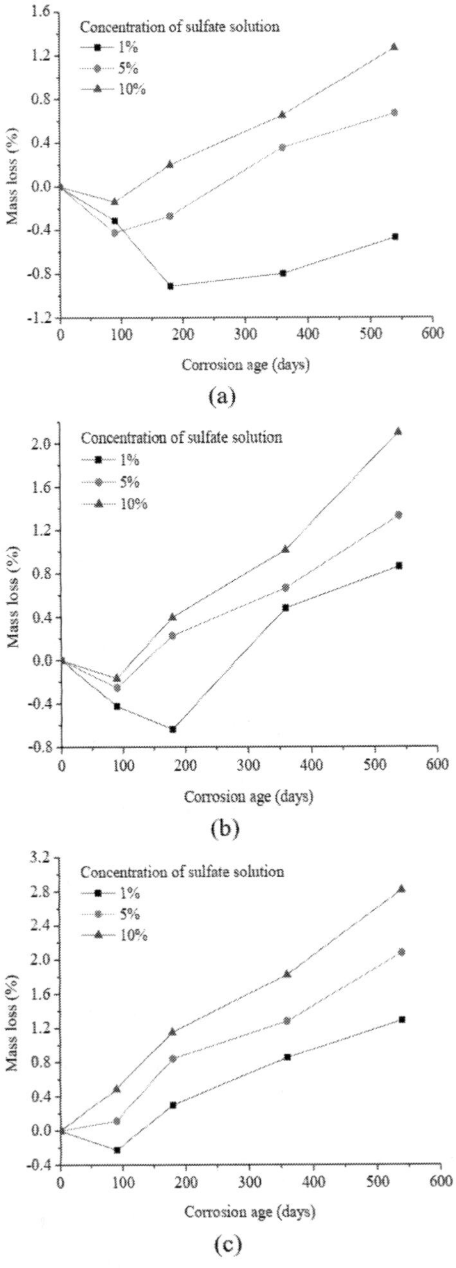

Figure 3. Mass change of concrete under dry-wet cycle and sulfate attack effects (a)the stress level is 0(b) the stress level is 15%(c) the stress level is 30%

From figure 3, when the stress level is 0%, masses of concrete in solution with different concentrations all first increase and then decrease with corrosion age's development. When the stress level is 15%, masses of concrete in solution with different concentrations all first increase and then decrease with corrosion age's developing, it is similar to the trend without load. When the stress level is 30%, mass of concrete in solution with different concentration decreases along with the corrosion age's developing, and the loss rate is proportional to the concentration value. For the specimens DW-S30-C5-D540 and DW-S30-C10-D540, the first two layers of concrete both peeled off, their mass loss are more serious.

3.2 Dynamic elastic modulus

The changing curves of relative dynamic elastic modulus of concrete within different stress level, different dipping and different concentrations of solution with the corrosion age's developing are shown in figure 4 and figure 5.

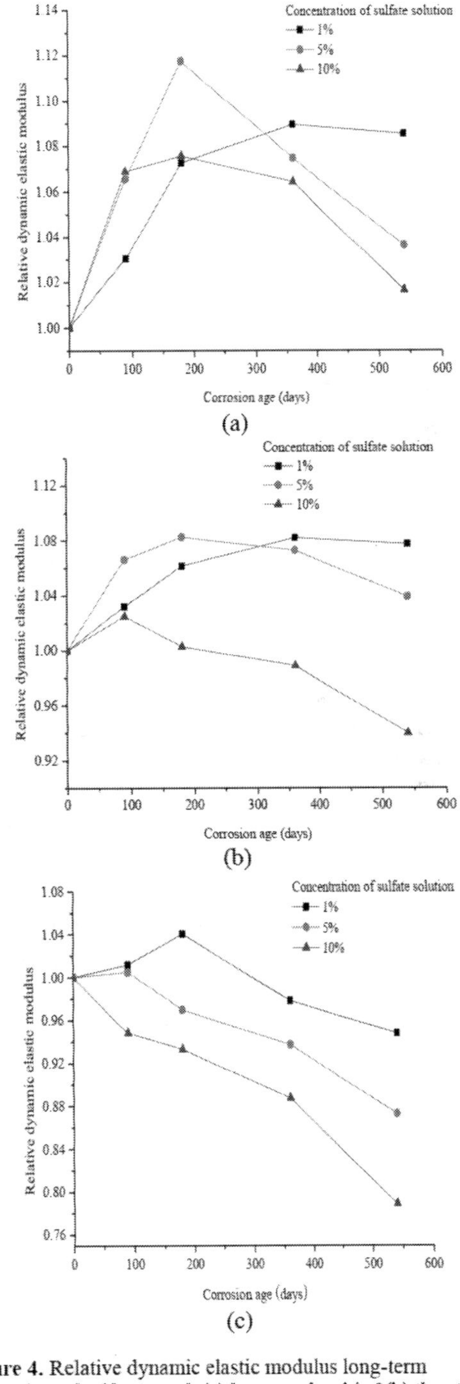

Figure 4. Relative dynamic elastic modulus long-term immersion of sulfate attack (a)the stress level is 0(b) the stress level is 15%(c) the stress level is 30%(Unit:104MPa)

As shown in figure 4, in the long-term immersion environment, the relative dynamic elastic modulus of concrete in the sulfate solution with concentration of 1% increases with the corrosion age's developing when the stress level is 0 and 15%. For the concrete in solution with concentration of 5% and 10%, their relative dynamic elastic modulus first increase and then decrease with the corrosion age's increasing when the stress level is 0 and 15%.

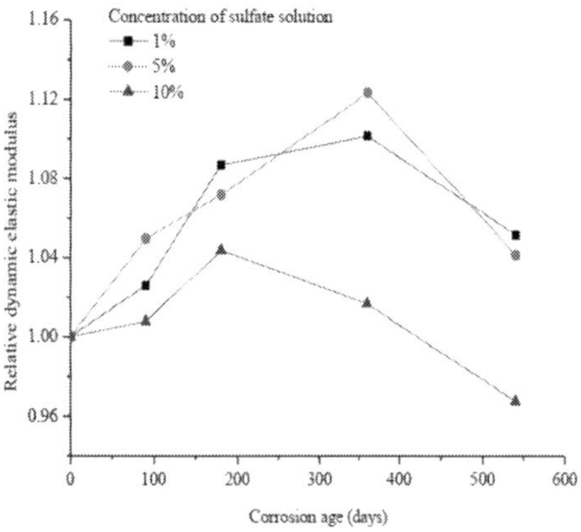

(a)

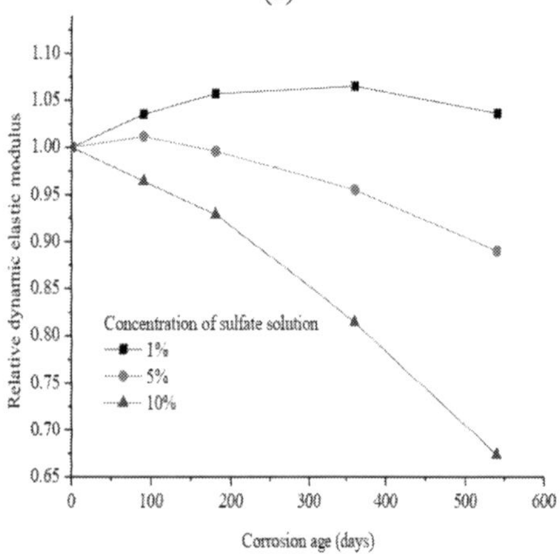

(b)

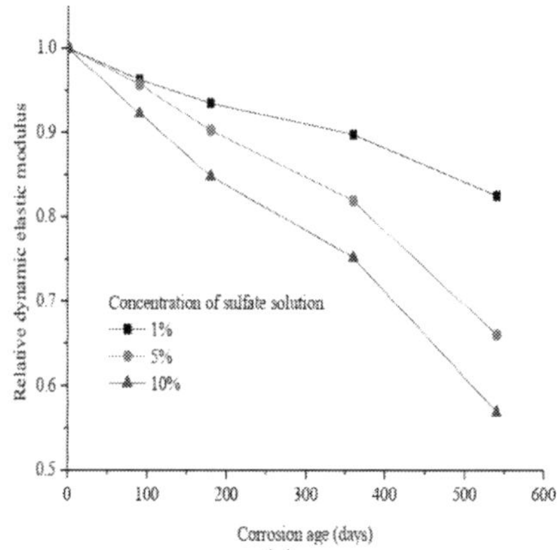

(c)

Figure 5. Relative dynamic elastic modulus under dry-wet cycle and sulfate attack effects (a)the stress level is 0(b) the stress level is 15%(c) the stress level is 30%(Unit:104MPa)

From the figure 5, when the stress level is 0, relative dynamic elastic modulus of the concrete in sulfate solution with various concentrations all first increase and then decrease with the corrosion age's developing. When the stress level is 15%, relative dynamic elastic modulus of concrete in the 1% and 5% concentrations of sulfate solution first increase and then decrease with the corrosion age's developing. For the situation that stress level is 30%, relative dynamic elastic modulus of the concrete in various concentrations of sulfate solution all decrease with the corrosion age's developing.

4 Damage model of concrete

This paper regards the dynamic elastic modulus as the evaluation indicator of the durability damage of concrete ; the prediction model of the evolution with the corrosion age of dynamic elastic modulus of concrete under the situation of different stress level, different immersion way and different concentrations of sulfate solution is presented. Specifically as shown in equation (1):

$$D = a + b \cdot t + c \cdot t^2 \quad (1)$$

Table 2. Regression coefficients of formula (1)

Immersion way	Stress level	concentration	a	b	c	R2
Long-term immersion	0	1%	1.220E-03	-4.719E-04	5.808E-07	0.981
		5%	-9.969E-03	-6.687E-04	1.211E-06	0.822
		10%	-1.112E-02	-5.346E-04	9.871E-07	0.876
	15%	1%	-1.785E-04	-4.155E-04	5.071E-07	0.997
		5%	-1.042E-02	-5.423E-04	9.230E-07	0.879
		10%	-5.544E-03	-1.072E-04	4.255E-07	0.932
	30%	1%	-4.679E-03	-1.728E-04	5.359E-07	0.808
		5%	-3.967E-03	8.925E-05	2.835E-07	0.981

		10%	1.157E-02	2.459E-04	2.094E-07	0.974
		1%	8.104E-03	-6.646E-04	1.021E-06	0.936
	0	5%	6.187E-03	-7.388E-04	1.185E-06	0.921
		10%	3.381E-03	-3.187E-04	6.935E-07	0.892
		1%	-1.220E-03	-4.227E-04	6.665E-07	0.996
Wet-dry cycle	15%	5%	-4.093E-03	-4.637E-05	4.813E-07	0.992
		10%	3.517E-04	3.177E-04	5.314E-07	0.999
		1%	5.891E-03	2.833E-04	4.234E-08	0.987
	30%	5%	5.360E-03	3.587E-04	4.675E-07	0.994
		10%	1.078E-02	6.404E-04	2.301E-07	0.990

In the equation (1), t is the sulfate corrosion age; a, b, c represents the trial-related coefficient respectively.Based on the test data, researchers conducted regression fitting and finally got the fitting coefficients in the equation (1), as shown in table 2.

From figure 6 to figure 9, the evolutionary relationship of the damage rate with corrosion age of concrete is shown.

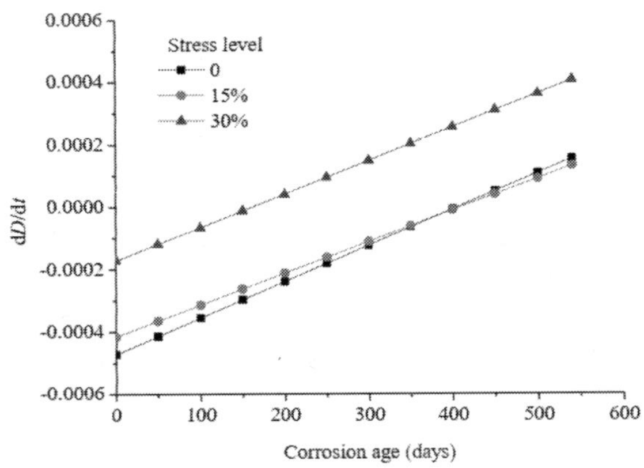

Figure 6. Evolution of concrete damage rate under different stress levels (LM-C1)

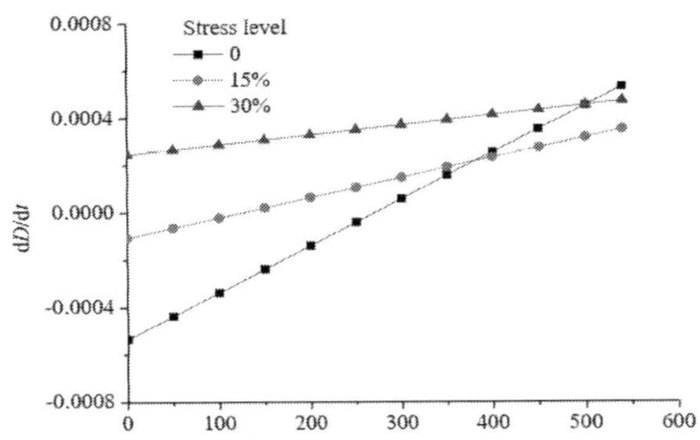

Figure 7. Evolution of concrete damage rate under different stress levels (LM-C10)

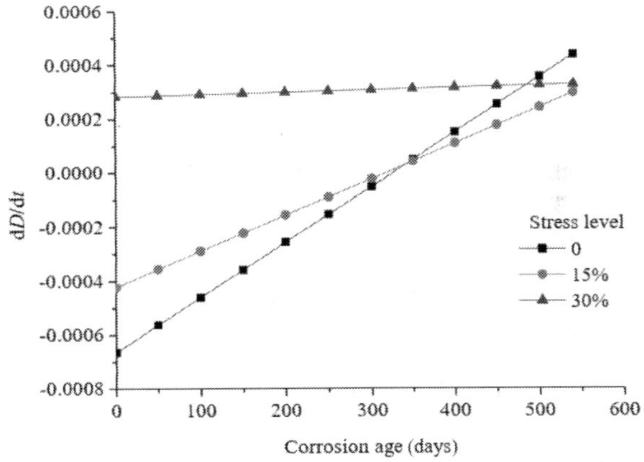

Figure 8. Evolution of concrete damage rate under different stress levels (DW-C1)

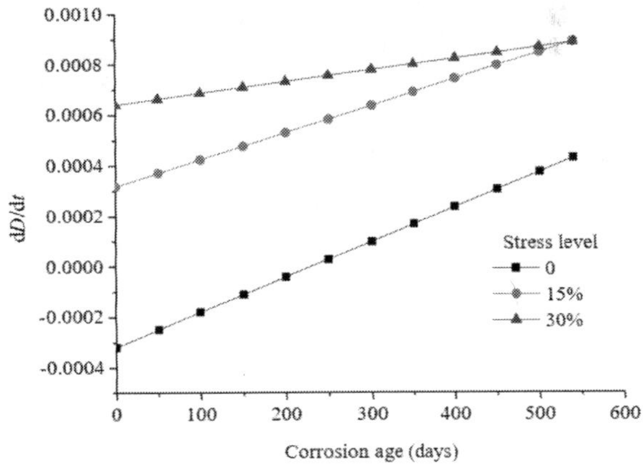

Figure 9. Evolution of concrete damage rate under different stress levels (DW-C10)

From figure 6 and figure 7, we can know that in the environment of long-term immersion, when the concentration of corrosion solution concentration is 1%, the concrete damage rate increases with the stress level's increasing but the incremental damage rate has no significant change with the improvement of stress level. The increment of concrete damage rate increases more and more slowly with the stress level when the concentration of corrosion solution is 10%. From figure 8 and figure 9, in the environment of wet and dry cycles and when the corrosion solution concentration is 1%, the concrete damage rate increment increases more and more slowly with the stress level. Before corrosion age reaches 350 days, the concrete damage rate increases with corrosion age and after 350 days, the concrete damage rate has the trend of first increasing and then decreasing with corrosion age. When the concentration of corrosion solution is 10%, the concrete damage rate increases with corrosion age but the incremental damage rate decreases with the improvement of stress level.

5 Conclusions

In this paper, the test researches were carried out on the concrete mass and dynamic elastic modulus under the effects of sustained loading, sulfate attack and dry-wet circulation,and the test phenomenon and results were discussed and analyzed,and the durability damage model of the concrete under the effects of sustained loading, sulfate attack and dry-wet circulation was proposed, then the rationality of the model was discussed.

6 Acknowledgments

This study was funded by the Jiangxi Province Natural Sciences Fund Projects of China [grant number 20161BAB206120].

References

1. Neville A. Consideration of durability of concrete structures: past, present, and future [J]. *Mater. Struct.*, 2001, **34(2)**: 114-118.
2. Tixier R, and Mobasher. Modeling of damage in cement-based materials subjected to external sulfate attack. I: formulation [J]. *J. Mater. Civ. Eng.*, 2003, **15(4)**: 305-313.
3. Tixier R, and Mobasher. Modeling of damage in cement-based materials subjected to external sulfate attack. II: comparison [J]. *J. Mater. Civ. Eng.*, 2003, **15(4)**: 314-322.
4. Saetta A. Mechanical behavior of concrete under physical-chemical attacks [J]. *J. Eng .Mech-ASCE*, 1998, **124(10)**: 1100-1109.
5. Saetta A V, Scotta R, and Vitaliani R. Coupled environmental-mechanical damage model of RC structures [J]. *J. Eng .Mech-ASCE*, 1999, **125(8)**: 930-940.
6. Saetta A. Deterioration of reinforced concrete structures due to chemical-physical phenomena model-based simulation [J]. *J. Mater. Civ. Eng.*, 2005, **17(3)**: 313-319.
7. Kuhl D, Bangert F, and Meschke G. Coupled chemo-mechanical deterioration of cementitious materials. Part I: Modeling [J]. *Int. J. Solids Struct.*, 2004, **41(1)**: 15-40.
8. Kuhl D, Bangert F, and Meschke G. Coupled chemo-mechanical deterioration of cementitious materials. Part II: Numerical methods and simulations [J]. *Int. J. Solids Struct.*, 2004, **41(1)**: 41-67.
9. Schneider U, and Chen S W. Modeling and empirical formulas for chemical corrosion and stress corrosion of cementitious materials [J]. *Mater. Struct.*, 1998, **31(10)**: 662-668.
10. Maes M A, Wei X, and Dilger W H. Fatigue reliability of deteriorating prestressed concrete bridges due to stress corrosion cracking [J]. *Can.J. Civi. Eng.*, 2001, **28(4)**: 673-683.
11. Xi Y P, and Ababneh A. The coupling effects of environmental and mechanical loadings on durability of concrete [C]. *Proc. Int. Conf. Adv. Concr. Struct.*, Xuzhou, China, 2003: 354-365.
12. Minstry of Housing and Urban-Rural Development of the People's Republic of China. *Standard for test methods of long-term performance and durability of ordinary concrete* [S]. GB/T 50082-2009. Beijing: China Architecture & Building Press, 2009.

Impact of the Content Variation of the Different Recycled Wastes to the Properties of Concrete Paving Block

X Wang[1], C S Chin[1,2] and J Xia[1]

[1] *Civil engineering department, Xi'an Jiaotong-Liverpool University, Suzhou, Jiangsu, 215000, China*
[2] *Dean of Learning and Teaching & Former Head of Department, Xi'an Jiaotong-Liverpool University, Suzhou, Jiangsu, 215000, China*

Abstract. Recycled concrete aggregates have been widely studied and used in concrete products nowadays. However, other recycled wastes, such as glass, have not been involved too much in recycled aggregate concrete studies. This paper aims to study the impact of the content variation of the different recycled wastes to the properties of the concrete paving block. In this paper, not only recycled coarse concrete aggregates, crushed glass are also used as the recycled aggregate in the concrete paving block in different replacement levels. According to test the properties of blocks mixed with different recycled wastes, the experimental results indicate that: (1) adding recycled concrete coarse aggregate (RCCA) in the blocks can decrease the blocks' strength, and increase the water absorption. The suggested replacement levels for RCCA is 60%; (2) mixing crushed glass (CG) in the concrete paving blocks as a type of coarse aggregates can obviously improve the blocks' strength and decrease the blocks' water absorption.

1. Introduction

In China, the construction industry has developed rapidly since the late of the 1990s [1]. High-rise buildings are established, and the old buildings and constructions are demolished at the same time. Due to the irreplaceability of concrete in the construction industry, large quantities of concrete needs to be widely used in engineering, and this situation leads to the generation of large volumes of construction waste. Thus, using recycled demolished concrete as the aggregate in the new construction products could reduce the utilization of the limit-resource in nature and reduce the environmental burden at the same time.

However, construction wastes do not only represent recycled concrete aggregates which contain crushed concrete, crushed bricks, and granites. Glass, woods, plastic, rubbers and so on are also counted as construction wastes [2]. Nowadays, concrete paving blocks cast with recycled aggregates have been studies and manufactured in many countries [3–6]. Thus, the concrete paving block is selected as the type of concrete product to be manufactured and investigate in this paper.

In this paper, not only recycled coarse concrete aggregates, the crushed glass are also used to investigate the impact of the content variation of the different recycled wastes to the properties of concrete paving blocks. Plus, for all the experiments in this paper, all the parameters of the production machines, as well as the mode of production, are in accordance with the actual industrial production standards of the company named Suzhou Construction Material Recycling Application Co., Ltd.

2. Experimental program

2.1. Materials

The entire materials used in this series of experiments are listed as follow:

- Natural coarse aggregate;
- Natural fine aggregate;
- Ordinary Portland cement (42.5MPa compressive strength at 28 days);
- Recycled concrete coarse aggregate (contain crushed concrete, crushed stone, and crushed bricks);
- Crushed glass.

In this experiment, there are two types of construction wastes are used: recycled concrete coarse aggregate (RCCA) and crushed glass (CG). RCCA was collected from a local recycled aggregate manufacture factory. Crushed glass was sourced from a local glass recovery plant and crushed into 5mm~2.5mm in the laboratory. Figure 1 shows the graphics of recycled wastes used in this experiment.

(a) RCCA

(b) Crushed glass

Figure 1. Characteristics of recycled wastes.

2.2. Mix proportion and replacement levels

In order to investigate the impact of the content variation of the different recycled wastes to the properties of concrete paving blocks, two different types of waste materials were used to replace the coarse aggregates in the concrete mixture with 28-days target strength of 30MPa. The original mix proportion was designed according to the "mass method" in Chinese standard JGJ55-2011 [7]. The concrete paving blocks which were cast by replacing the natural coarse aggregate by weight with different replacement levels of recycled materials.

According to the previous articles and experiments results, the replacement levels of RCCA in the experiment were determined as 20%, 40%, 60% 80% and 100%. As the same as RCCA, crushed glass was also used to replace the coarse aggregates in the blocks. According to the previous articles and trial test results [8], the final replacement levels of crushed glass were decided as 10%, 20%, 30% and 40%,

2.3. Specimens preparation

The specimens are designed as typical concrete paving blocks. The whole procedures, including preparing the concrete mixture, casting concrete paving blocks and curing, were in accordance with the actual industrial production standards of the company in Suzhou. Plus, the parameters of the production machines (a small hydraulic brick making machine) were set as same as the company's requirements. Figure 2 shows the image of the brick making machine. All the concrete mixtures were sent into the small hydraulic brick making machine to be compacted under 80kN force to be cast into 200mm*100mm*60mm concrete paving blocks [9].

2.4. Test and requirements

Both British standard BS EN -1338: 2003 [10] and Chinese standard GB 28635-2012 [11] stipulate the requirements for concrete paving blocks. In this series of experiments, four properties of the concrete paving block were tested: Dimensions, compressive strength, tensile splitting strength, and water absorption.

Figure 3 shows the universal test machine (UTM) for testing block's strength

3. Test results

Figure 2. Hydraulic brick making machine.

Figure 3. Universal testing machine.

3.1. Dimensions

All the concrete paving blocks in this series of experiments were manufactured by a small hydraulic brick making machine. The size of molds in the machine is 200mm*100mm*60mm. In this case, all the blocks' dimensions are 200mm*100mm*60mm, and the error of each side was less than 2mm. Figure 4 and figure 5 show the front and side faces of one concrete paving block.

Figure 4. Front face of concrete paving block.

Figure 5. Side face of concrete paving block.

3.2. Compressive strength

Figure 6 shows the relationship between the different replacement levels of materials and the compressive strength of specimens.

Specimens mixed with RCCA maintained a stationary compressive strength until the replacement level reached 40%. The compressive strength fluctuated as 41.07MPa, 38.04MPa and 40.38MPa when replacement levels were 0%, 20%, and 40% respectively. After the replacement levels were higher than 40%, the compressive strength almost linear decreased from 40.38MPa at 40% to 21.04 MPa. According to the BS EN 1338 [10], the compressive strength of concrete paving blocks should equal or higher than 30MPa, which indicated that the specimens with replacement level higher than 60% were not meet the specification [11].

As another material replacing the coarse aggregates in this series of experiments, the crushed glass had a positive impact on the concrete paving blocks' compressive strength. According to the figure 6 results, the specimens had a higher compressive strength when the replacement levels of crushed glass were 10%, 20%, 40% and 60%, and the strength peak the highest value, 43.45MPa, at the replacement level equals 20%.

3.3. Tensile splitting strength

Figure 7 demonstrates the relationships between the different research materials' replacement levels and the tensile splitting strength of concrete paving blocks.

Specimens mixed with RCCA had an even higher tensile splitting strength 11.73MPa at replacement level was 20%. However, the tensile splitting strength almost linear decreased from 11.73MPa to 8.80MPa when the replacement level increased from 20% to 80% and bottomed at 6.37MPa with 100% RCCA in the specimen.

According to the test results, using crushed glass as the coarse aggregates in the concrete paving blocks had a positive effect on the blocks' tensile splitting strength. The overall tendency was that the tensile splitting strength kept increasing with the replacement level changed from 10% to 40%, and reached the highest value 12.79MPa at 40%. When the replacement level was 30%, the tensile splitting strength was only 11.38MPa. However, this value was still a little bit higher than the blocks mixed with natural aggregates.

3.4. Water absorption

According to figure 8, it is clear to observe that for the blocks mixed with RCCA, the increase of the replacement levels of the material increases the specimens' water absorption. Furthermore, for blocks mixed with the RCCA, the water absorption value could maintain 3.91% at the replacement levels 20%, and then peaked 7.60% while the replacement levels were 100%.

Crushed glass is proved to be a material that can reduce the water absorption when it mixed into blocks. The values of water absorption ranged from 3.50%, 3.24% 3.65 and 3.72% while the replacement levels were 10%, 20%, 30% and 40%.

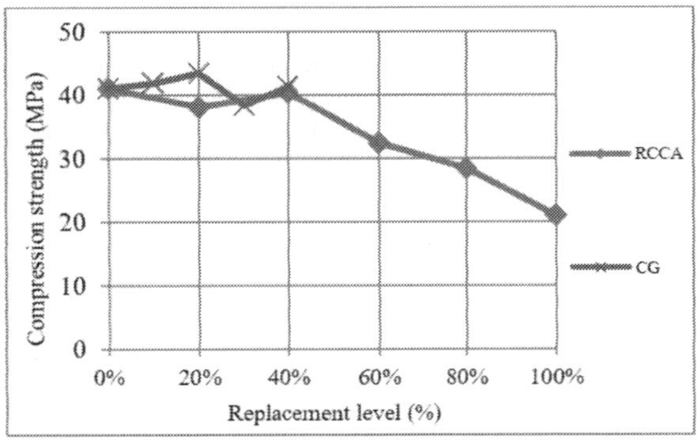

Figure 6. Relationship between the different replacement levels of materials and the compressive strength of specimens

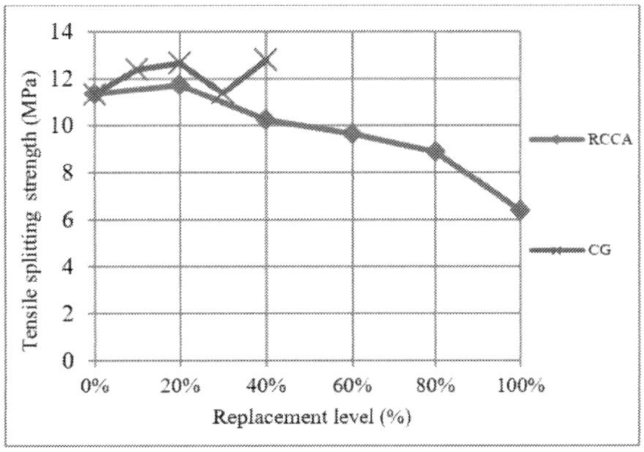

Figure 7. Relationship between the different replacement levels of materials and the tensile splitting strength of specimens.

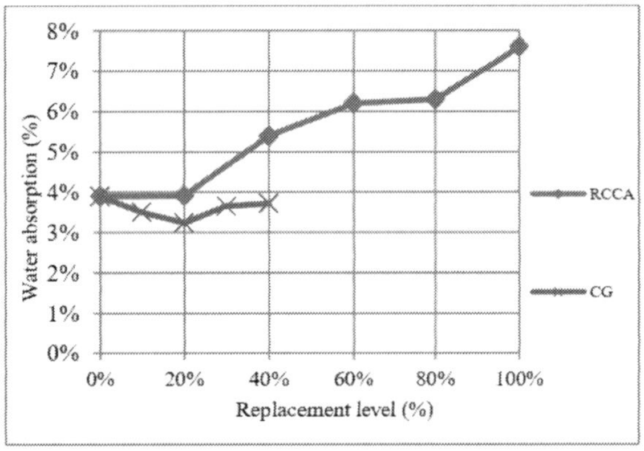

Figure 8. Relationship between the different replacement levels of materials and the water absorption of specimens

4. Conclusions

According to the experiment results in the properties of concrete paving blocks mixed with different types of recycled wastes, the conclusions can be drawn as follow:

- Using the RCCA as the coarse aggregate of the concrete paving blocks can lead a significant decrease of both compressive strength and tensile splitting strength of blocks, and also increase the water absorption of the concrete paving blocks. In order to meet the standard requirements, the replacement level of RCCA should be less than 60%.
- Replacing the coarse aggregate with crushed glass could enhance the strength of concrete paving blocks. With the replacement level ranging from 10% to 40%, all concrete paving blocks mixed with crushed glass had higher strength and lower water absorption than the blocks mixed with natural aggregates. When the replacement level of crushed glass at 20%, the concrete paving blocks had a highest compressive strength and lowest water absorption. The tensile splitting strength of concrete blocks reached the maximum value at replacement level of crushed glass equalling 40%.

Acknowledgement

The authors gratefully acknowledge Xi'an Jiaotong-Liverpool University for the finical support. We would also like to thank Suzhou Construction Material Recycling Application Co., Ltd for providing recycled concrete aggregates and machine parameters for the experiments.

Reference

1. Xing Z, Liu Y and Cai N 2011 Research and application benefit for reycled concrete *Low Temp. Archit. Technol.* 1 6
2. BS EN 933-11 2012 Tests for Geometrical Properties of Aggregates *Austrian Stand. Inst.*
3. Zaidi A M A 2009 Assessment of recycled aggregate concrete *Mod. Appl. Sci.* **3** 47
4. da Silva F M, Gachet Barbosa L A, Lintz R C C and Jacintho A E P G A 2015
5. Ferreira I K, Gachet-Barbosa L A, Cecche Lintz R C, Russo Seydell M R and Jaquiê Ribeiro L C L 2013 Evaluation of the Behaviour of Mortar with the Addition of Rubber *Adv. Mater. Res.* **742** 456–60
6. Knapton J 1988 A review of concrete block pavements at ports in the UK *Proceedings, 3rd. International Conference on Concrete Block Paving, Rome* pp 380–92
7. Ding W and Leng F G 2011 Specification for mix proportion design of ordinary concrete (JGJ55-2011)
8. Jin W, Meyer C and Baxter S 2000 " Glascrete"-Concrete with Glass Aggregate *ACI Mater. J.* **97** 208–13
9. Soutsos M N, Tang K and Millard S G 2011 Concrete building blocks made with recycled demolition aggregate *Constr. Build. Mater.* **25** 726–35
10. BSI 1338:2003 2003 Concrete Paving Blocks–Requirements and test methods *Br. Stand. 76s*
11. SAC 2012 Precast concrete paving units *GB 28635-2012.*

Engineered Bamboo's Further Application: An Empirical Study in China

WQ Xing, JL Hao, I Galobardes, SB Wei, ZT Chen, KS Sikora

Xi'an Jiaotong-Liverpool University, Suzhou, P. R. China

Abstract. The trend of sustainable development has been followed more globally nowadays. There is an urgent need for ecological friendly construction materials in China since the use of concrete produces pollution and leads to inefficient use of energy. Bamboo can effectively address the sustainability trends due to: lightweight construction, little adverse impact on the environment, and anti-seismic properties. However, bamboo products utilization has been very limited. Therefore, the purpose of this study was to investigate the feasibility to further utilize bamboo products in the construction industry in China. A review of engineered bamboo products utilization in China, and oversees was conducted. Based on this, a survey focusing on the current situation and suitability of bamboo products was carried out to study their feasibility, and preliminary testing on bonding shear strength for different configurations are presented. The results are promising, in particular the relatively low clamping pressure of 0.6 MPa seems to be sufficient for bonding, and it is clear that engineered bamboo products have a great potential in China. However, further more comprehensive studies on the performance of the laminated bamboo are needed.

1. Introduction

Bamboo is a traditional material used for construction in ancient China. In general bamboo's mechanical and durability properties are comparable or excessing values of the materials used traditionally in construction, like concrete or timber [1]. China is the main origin of bamboo in the world, considering the species, distribution area and quantity [2]. Bamboo is distributed in a wide range in China, see Figure 1; in four provinces: Fujian, Zhejiang, Jiangxi and Hunan, it occupies about 60.8% of China bamboo forest area [3]. In the world, there are bamboos of total 150 genera, 1,253 species and 2,200 million hectares. Bamboos in China have 43 genera, 707 species and the distribution area of bamboo in China covers about 700 million hectares [2]. The major bamboo species in China is Moso bamboo.

Figure 1. China's bamboo growing province map [3]

2. Current utilization of bamboo

2.1 China

Bamboo is mainly applied to construction industry without further processing. After economic reform policy (the program of economic reforms termed 'Socialism with Chinese characteristics' in P. R. China that was started in December 1978) the development process of bamboo-related industry made a considerable progress. New bamboo-based products such as bamboo ply and laminated bamboo strips appear to deal with the problem of poor forestry resources to some extent [4]. Nevertheless, these bamboo products are rarely used in construction filed; they are often used for furniture, floors and walls. Based on The Third National Industrial Census [5], China has 1,120 companies involved in bamboo products and only 9.2% are related to construction industry.

2.2 World

Potential for bamboo construction is mainly in Latin America and Southeast Asia because both regions abound in bamboo [6]. Columbia has the advanced technology of raw bamboo building construction and more than 700 low-cost bamboo residences has been built in Costa Rica. Southeast Asian countries such as Vietnam, India, Indonesia and Thailand have been researching the usage of raw bamboo in construction.

Although bamboo in Europe is not abundant or widely distributed, authorities of few European countries have made some achievements about technology of bamboo construction to follow the trend of sustainable development [6]. For instance, in Germany there have been established the complete standard to guide the construction of bamboo-made building. In general, glued laminated bamboo and bamboo based panel are two types that can be used in modern bamboo structures [7]. Their strength and stiffness promote the bamboo to be the material for multi-storey and large-span construction.

3. Physical and mechanical properties of bamboo

Bamboo is one of the fastest-growing plant on earth, whose growth rate of bamboo can be up to 91 cm per day [8]. Its rate of productivity and cycle of annual harvest outstrips other naturally growing resource; three years is enough for bamboo to be mature for construction usage. Besides, bamboo as a plant can absorb a large amount of CO_2 [9]. Generally, bamboo could be divided into two kinds, clumping and running [3]. The clumping types are typically of tropical or subtropical origin, with high growing rate. On the other hand, running bamboo can withstand occasional low winter temperatures, but it needs several years to be mature. In theory, the mechanical properties of cut bamboo elements depend on the botanical species, the age at which the bamboo has been cut, the moisture content, the position along the culm, and the position of the nodes. Bamboo is functionally graded composite and orthotropic material, its load bearing capacity is related to the distribution of the fibers, and having high strength along and low strength transversal to its fibers respectively. Bamboo has a low density but high stiffness, its shrinkage is smaller than timber, and the elasticity and toughness are high enough. Moreover, due to the unique shape of bamboo, bending resistances of bamboo are same for each internode [10]. As for nodes, they are weaker in compression and bending than internodes [9].

With regard to the disadvantages of the bamboo, raw cut bamboo is a hollow tube, sometimes with thin walls, and therefore it is hard to join pieces of bamboo. Secondly, the outer skin of a culm cannot be glued without treatment, because of its high silica contents. Furthermore, bamboo also can have some defects due to an action of insects. Low natural durability of bamboo indicates that it needs preservative treatment, which is difficult in many cases [11].

4. Challenges and potential with new developments

4.1 Engineered bamboo products

Engineered bamboo possesses the advantages of raw bamboo while overcomes some of its disadvantages. In recent years, several typical engineered bamboo products such as bamboo ply, strand woven bamboo and glued laminated bamboo have been researched and produced, and they can be composited with other structural materials such as timber and reinforced concrete. These products are made from veneers, strands and sliced piece of bamboo, which are combined into a specific structure and then bonded together with adhesives by high-pressure and/or high-temperature. This process improves significantly parameters when comparing to natural product allowing utilization of species of lower properties that were not considered before for high performance usage.

4.2 Preliminary testing: Materials & Methodology

To assess the shear strength of adhesive bonds that is a critical property for bamboo utilization in big scale construction engineered bamboo panels an experimental program has been initiated at Xi'an Jiaotong-Liverpool University. Commercially available, one-ply vertically laminated bamboo panel made from Chinese Moso bamboo with age of 3 to 5 years is used for this testing to prepare the specimens. The panel cut to specimens with bonding area of 30 mm × 50 mm glued using a hybrid polymer adhesive (HPA). Specimens are manufactured following the grain directions of bamboo that parallel and perpendicular to the grain, and cross laminated directions that cross laminated vertically and horizontally, to the bamboo material. Different clamping pressures, 0.6 MPa, 0.8 MPa and 1.0 MPa are applied in order to find out the optimal production pressures. Testing method follows the methodology developed for laminated timber by Sikora et al. [12], with the uniform rate of 3 mm/min. Block shear tests of specimens in four different configurations, end-grain, perpendicular to grain, cross laminated vertically and cross laminated horizontally were carried out.

4.3 Results and discussion

Table 1 presents results for 4 different loading configurations and for 3 different clamping pressures, and each result represents average value for the tests of 10 specimens.

As the percentages of bamboo grain failure were 0.0% for all specimens, it can be claimed that the failure is attributed to the failure of adhesive. In addition, there is no direct tendency for all the results since they perform similarly under different loading configurations. Therefore, it can be generally concluded that different clamping pressures have not significant effects on shear strength of HPA glued bamboo specimens. Moreover, almost all the bonding shear strengths of these specimens have exceeded 1.5 MPa, which is the standard strength data provided by manufacture industry of HPA adhesive. The experiment results give an evidence that although HPA is used for furniture and sealing normally, it provides the proper interface bond and penetration to the natural bamboo products as well. This confirms that bamboo panels have reasonable high strengths in both directions, higher than hybrid polymer adhesive. Furthermore, the lowest clamping pressure used (0.6 MPa) is sufficient for bonding the bamboo elements, and the highest pressure (1 MPa) seems to be more suitable for cross laminated bamboo production, presenting the highest strength results.

https://doi.org/10.1051/matecconf/201820602005

Table 1. The bonding shear strength of glued laminated bamboo (coefficients of variation in brackets)

CHARACTERISTIC	BONDING SHEAR STRENGTH (MPa)		
CLAMPING PRESSURE	0.6 MPa	0.8 MPa	1.0 MPa
End-grain	1.80 (58.9%)	1.61 (17.4%)	1.73 (23.7%)
Perpendicular to grain	1.81 (20.4%)	1.67 (20.4%)	1.74 (28.2%)
Cross laminated vertically	1.57 (14.0%)	1.38 (24.0%)	2.00 (27.0%)
Cross laminated horizontally	1.43 (27.3%)	1.58 (24.7%)	2.12 42.0%)

4.4 Challenges from Chinese perspective

Bamboo utilization in construction in China facing several possible challenges. First of all, there is no national standard for bamboo construction in China. Until now, the application of bamboo in constructions still follows the Code for Design of Timber Structures [7], but the characteristics of bamboo are different from timber. Lack of standard, as a result, limits the development of bamboo products utilization in construction. Secondly, the high cost of bamboo construction is also an important issue affecting its improvement. The production scale of bamboo frame is still at an initial stage and the production efficiency is low; machinery cost is high as well. With regard to consumers, they normally consider initial cost of material purchase the most. Although the cost of bamboo frame construction is lower than other materials considering maintenances cost, most consumers just focus on the short-term benefits, which also obstructs the use of bamboo [13]. Moreover, technology limits the development of bamboo construction. In China, timber construction has been neglected for the past few decades, accordingly, the knowledge on bamboo technology is also outdated. There is only one university, Hunan University that delivers bamboo construction courses at present in China, which limits bamboo technology development in the near future [7].

4. 5 Potential for China

Firstly, environment in China has been polluted severely, partially due to the use of concrete in construction industry, which productions requires high energy consumption. Bamboo as a material providing renewable, durable, energy-efficient and cost-efficient solutions, and it can act as substitute of concrete to some extent to mitigate the problem of vast energy consumption and severe environmental impact. There is much space for bamboo to develop since 99% buildings in China are constructed with concrete [14]. Additionally, since China has the richest bamboo resources in the world, bamboo is abundant enough for Chinese construction industry to satisfy its huge requirements in terms of quantities. Moreover, China and other countries have researched and developed new technologies related to engineered bamboo products [7], therefore engineered bamboo products are more likely to be competitive in cost, strength and sustainability, than other traditional materials. Furthermore, a series of standards related to bamboo buildings are under development and review in China. The potential of introduction regulations for bamboo building construction is being studied so that to promote the range of bamboo building usage [6]. With these specific construction regulations, bamboo is more suitable for the particular areas of China: urban areas where low-rise house is allowed only and suburb areas with lower land price and beautiful scenery. What is more, there is an increasing demand for various types of structure houses in metropolitans such as Shanghai, Beijing and Shenzhen. Another possible potential for bamboo is related to active seismic zones in China. Light weight and ductility of bamboo allows to absorb seismic energy without collapse.

5. Survey of the industry

5.1 Introduction and methodology

To investigate the current situation and suitability of bamboo for the manufacture of laminated panel a survey of the construction industry was made. A questionnaire was designed (in Chinese and English language) consisting of over 20 questions related to current usage of different materials for construction projects, and predictions for future, based on the current trends. Feedback was obtained from carefully selected 9 industry members representing consulting and architecture companies who are the main decision-makers on the type of the construction material used for the projects.

5.2 Results and discussion

From feedback, participants were mainly from private companies (7 participants) with less than 300 employees and the other two participants represented public company and public institution. Their field of business covers consulting (5 of 9), architecture (3 of 9) and research (1 of 9). Although area served are national (77.8%) or worldwide (22.2%), their companies are concentrated on Shanghai and Suzhou.

The three most important characteristics that participants take into consideration when selecting materials are strength (66.7%), durability (55.6%) and fire resistance (55.6%), see Figure 2. However, material cost (44.4%) and construction cost (44.4%) are also important for these companies. As for engineered bamboo, two participants put forward that it is suitable for residence as well as low-rise building, and retail building and leisure facility, respectively. From the perspective of companies, availability in the market is the biggest barrier to limit the use of engineered bamboo. Other factors include material cost, technology and fire performance are also the obstacle the development of these materials. And, more remarkable, several people mentioned lack of knowledge about material will limit its usage, both literature review and this questionnaire reflect this problem a lot.

The participants of the survey were asked to rank the materials: Cross-laminated timber (CLT), engineered bamboo, steel, concrete, masonry and timber-frame from the different aspects. According to overall tendency of the feedback, the marks of characteristics for engineered bamboo can be concluded generally in a way as shown in Table 2.

Engineered bamboo products were considered as the best performing in the aspects of aesthetics and environmental friendliness. It was found that engineered bamboo can be cost-efficient. In addition, its strength performance be can as good as concrete and steel, or even better with the special manufacturing.

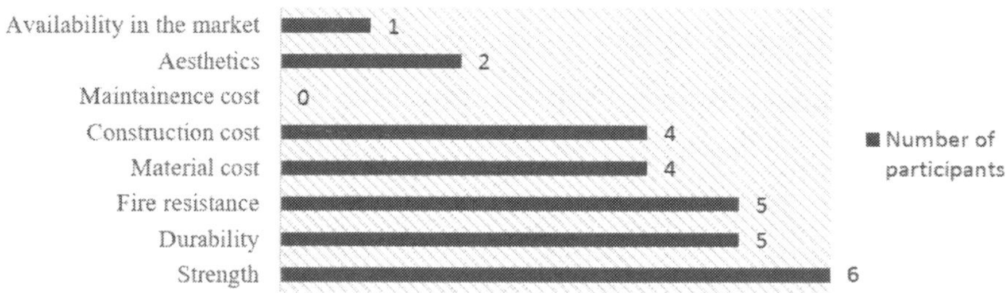

Figure 2. Participants' response on: 'The most important characteristics of material in construction'

Table 2. The characteristics marking of construction materials (Worst: 0 to Best: 5)

CHARACTERISTIC	MARK	CHARACTERISTIC	MARK
Acoustic performance	2.67	Aesthetics	4.44
Availability in the market	2.11	Material cost (low to high)	3.33
Construction cost (low to high)	2.89	Maintenance cost (low to high)	2.89
Durability	1.89	Strength performance	3.22
Fire performance	2.11	Environmental friendliness	3.67

6. Conclusions

This study gives indication that engineering bamboo products have potential for utilization in construction industry in China as there are so many advantages of using bamboo. These are good strength to weight ratio, anti-seismicity, environmental friendliness, and economy when prefabricated. However, there are still many obstacles for engineered bamboo technology development, including: lack of relevant legislations, and design and construction codes. To improve the situation, it is suggested that more research effort from the main stakeholders, such as government and universities in China should be done to address different aspects of bamboo's further utilization in China. Moreover, the industry survey reviled that bamboo can have great potential in China due to the availability and properties. This paper presents preliminary results for bonding performance, the crucial property for engineered bamboo. The preliminary tests confirming this claim that when parameters are optimized, glued bamboo has the relatively high bonding shear strength, sufficient for construction. The relatively low clamping pressure of 0.6 MPa seems to be sufficient. However, it is recommended that further comprehensive studies on mechanical performance of engineered bamboo panels are carried out.

Acknowledgments

The authors would like to thank Xi'an Jiaotong-Liverpool University support for funding the Summer Undergraduate Research Fellowship (SURF) to undertake the project.

References

1. Kumar G and Ashish DK. Review on Feasibility of Bamboo in Modern Construction. *SSRG International Journal of Civil Engineering*. 2015: 4: 66-70.
2. Tan GY, Yang L. *Tectonics of Bamboo*. Nanjing: Southeast University Press, 2014.
3. Mera FAT, Xu CY. Plantation management and resource economics of bamboo in China. *Ciencia y Tecnología*. 2014: 7(1): 1-12.
4. Feng MZ, Zhao RL, Gao L, Zhang SB. Application and status of structure bamboo plywood. *Wood Processing Machinery*. 2012: 2: 45-51.
5. Yang XS. The current situation and future tendency of bamboo industry in China. *Innovation and Industrialization*. 2003: 2: 40-41.
6. Ju TW. Research and application of modern bamboo construction system [Master thesis]. Hunan University, 2013.
7. Xiao Y, Li J. The state of the art of modern bamboo structures. *Industrial Construction*. 2015: 45(4): 1-6.

8. David F. The book of bamboo. Sierra Club Books, 1984.
9. Janssen JJA. Building with Bamboo: A handbook. 2rd ed. Practical Action Publishing, 1995.
10. Sun N, Zhang DB. Mechanical properties of bamboo. *China Academic Journal Electronic Publishing House* 1997: 19(3): 77-79.
11. Wang G, Jiang ZH, Liu XS, Fei BH, Zhang Y, Qiu YX. Status and future development of demoestic and foreign bamboo standards. *China Wood Industry* 2012: 26(4): 43-45.
12. Sikora K. S., McPolin D., Harte A. M., Shear strength and durability testing of adhesive bonds in cross-laminated timber, The Journal of Adhesion, Volume 92, Issue 7-9, 2016
13. Zhang S. Research on the Application of bamboo-based composites material in the building [Master thesis]. Chongqing Jiaotong University, 2015.
14. Fei BH. The Analysis of Timber- framed Building Development in China. Paper presented at: *International timber construction industry exchange conference, 2010, China.*

 https://doi.org/10.1051/matecconf/201820602006

Evaluation of Experts' Preferences for Selection of Suitable Industrialized Building Systems

Yidnekachew Daget[1] and Hong Zhang[2]

[1]*PhD candidate, School of Architecture, Southeast University, China, email: tesmammay@yahoo.com*
[2]*Professor, School of Architecture, Southeast University, China. email:zhangh555@aliyun.com*

Abstract. Experts' preferences for building material selection are a largely under-explored part of a material research. This study evaluates the experts' preferences to identify the suitable industrialized building system for application in Addis Ababa. Ten industrialized building systems were considered as alternatives for comparison. These systems have been used in a housing development in different parts of the world. A relevant literature review and contextual analysis were conducted. An analytical hierarchy process and an Expert Choice Comparion platform were employed as a research technique and tool to evaluate the professionals' level of preferences with regard to the housing systems. The findings revealed the most important preferential attributes and the suitable industrialized building systems for an application in housing development. The decision criteria and the experts' preferential method used in this study can help decision-makers and housing developers in developing countries make effective evaluations and decisions.

1 Introduction

The material research regarding the adoption of industrialized building system (IBS) has increased driven by the project performance opportunities it offers particularly in terms of sustainability, 'lean' efficiency in design and rapid construction process [1, 2]. Despite the significant potential, the adoption of IBS in the African context is slow. Improving the technology requires a better understanding of preferential factors influencing the selection of IBS in the context of unique construction-industry characteristics [3]. Experts' preference study has the potential to enrich material evaluation research resulting in materials that meet the necessary attributes. However, it is largely under-explored parts of material research particularly its application for housing development. Consequently, research gaps have emerged relating to the systematic decision-making of IBS selection based on experts' preference. This paper contributes to this body of knowledge by conducting an experts' preference evaluation of proposed IBS for adoption to the city of Addis Ababa. As shown in Figure 1, ten IBS were chosen for evaluation based on their application in a housing development.

The city of Addis Ababa has a high rate of population growth and urbanization resulted in massive pressure for rapid affordable housing development [4]. The housing market can not sufficiently provide the existing housing demand. One of the main challenges is the lack of systematic decision-making for the selection of suitable housing system [5]. This study can help to address the challenges in the given context.

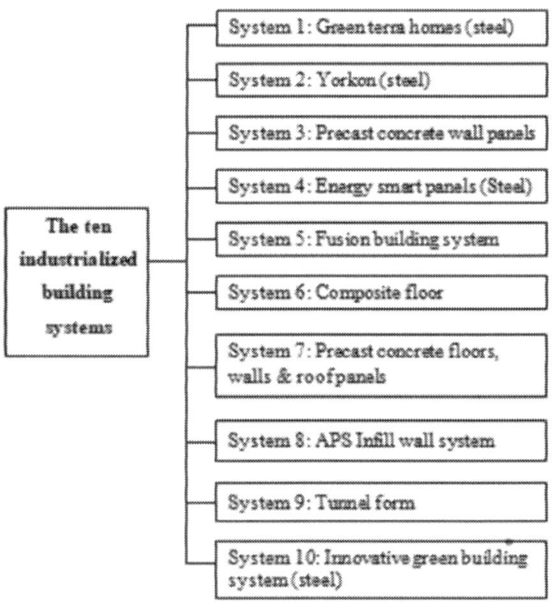

Figure 1. The ten industrialized housing systems.

2 Decision-factors for building system selection

Many studies have discussed that systematic decision-making is important for identifying the appropriate building system [1, 6]. The decision-making process involves all stakeholders in the construction industry [7]. But, the processes of evaluating building systems are not well-studied regarding African context. Holistic decision criteria are important to consider in order to select the appropriate building system for application [8]. Kaitilla (1991) has shown that when choosing building materials, most people strive to fulfil tangible, intangible, and environmental variables [9]. A broad range of decision criteria other than the traditional technical and economic factors help to make effective decisions [10]. The environmental issue is one of the factors which broaden the scope of selection criteria [11]. It is becoming a growing concern in the construction industry.

Ilozor, Okoroh, Egbu, and Archicentre (2004) have presented that the success of a housing development is directly related to the right selection and adoption of suitable building systems [12]. This is well-known among off-site construction professionals where an incorrect decision resulted in failure due to the expensive cost of the imported material, lack of expertise and incompatibility to the local context [13]. Abdullah and Egbu (2010) have suggested that applications of new IBS are based on its recognized benefits in the construction industry [14]. Despite the pressing housing market's need in developing countries, companies that tried to introduce IBS met with failures. Therefore, it is significant to consider meaningfully all decision criteria in order to make right decision according to the local context.

3 Methodology

3.1. Research design

With the purpose of identifying the most appropriate among the ten building systems, the research employed survey-based study approach to comprehend a general opinion of practice on decision factors and to provide a thorough explanation [15, 16]. Prior to the survey and the overall examination, an investigation was carried out with designated industrialized building experts from the local construction industry. The experts were interviewed to categorize the relevant factors, the encounters, and the resolutions for the implementation of IBS in Addis Ababa. Furthermore, they were invited to assess the importance, the reliability and the rationality of the questionnaire and made amendment prior to dissemination. Analytical hierarchy process (AHP) and Expert Choice Comparion (ECC) were used to rank the weighting of the factors and the alternatives to identify the suitable building systems. The outline of the research design is illustrated in Figure 2.

3.2 Decision factors and evaluation

An overall survey was carried out to select professionals with prominent qualification to involve into the AHP examination. As the practice in the implementation of IBS is inadequate, there are limited numbers of experts with the essential proficiency in various groups in the Addis Ababa's building industry. Experts' examined the relevance of the proposed factors based on the holistic criteria listed by Pan, Wong, and Hui to make sure that comprehensive criteria were set for the building system assessment [2]. As listed in Table 1, the modified framework for various selection criteria was categorized under six level-one factors and thirty level-two factors to prioritize the weights of each factor.

This study employed the analytical hierarchy process (AHP) to assist the assessments of experts. AHP's consistency evaluation helps to provide trustworthy importance scales based on a relative rank of the selection criteria and the building systems [17]. The pairwise judgment of AHP assessment and examination was carried out using the platform of Expert Choice Comparion (ECC). ECC is an online-based decision support tool which delivers an easy access for decision-makers to carry out the pairwise comparisons. Since the consistency ratio for individual expert judgment and the overall judgment was less than 0.1, it is assumed that the evaluation was consistent and valid.

Table 1: The relevant decision factors

Level one	**Building industry**	**Supply chain**	**Infrastructure**	**Sustainability**	**Users needs**	**Socio-cultural**
Level two	Labour	Material source	Transportation	environment	Number of stories	Public acceptance
	Expertise and experience	Components	Energy	Pollution	Flexibility	Regulatory laws
	Construction proficieny	Machinery	Water	Exposure	Quality	Vested interests
	Procurement	Production capacity	Sewerage	Reusability	Affordability	Communication
	time	Innovation	waste disposal	efficiency	Site location	Culture

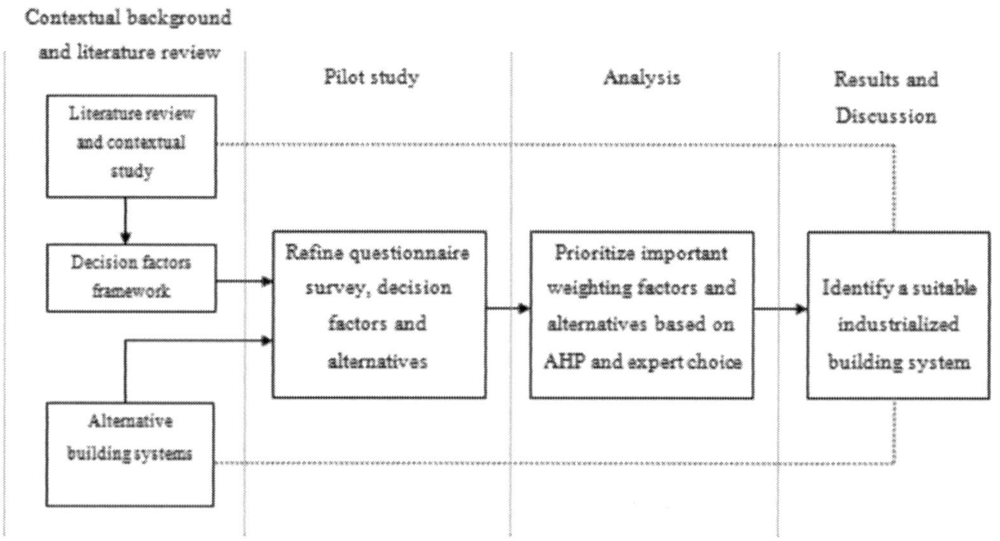

Figure 2. Research design diagram

3.3 Data collection analysis

Qualitative and quantitative data were collected. The qualitative data were gathered based on interviews and literature. The data were analyzed by the content analysis method, following structuring of the decision hierarchy. The quantitative data were collected and analyzed using Expert Choice Comparion (ECC). The general decision-making was examined through the following steps. As shown in Figure 3, primarily, the AHP decision model was structured into objectives, level-one factors, level-two factors, and alternatives to weigh the relevant factors and the alternatives. The pairwise comparison was carried out with ten selected experts using ECC to define priority vectors at level-one and level-two factors. Then, a numerical evaluation and normalization were done according to the assessment of the level of importance to identify the finalization of priorities. Lastly, a pairwise comparison was carried out by thirty identified experts and practitioners within the Addis Ababa construction industry who are key players in the precast manufacturers, suppliers, contractors, academician, engineers, and developers with an average of fifteen years of experience. The assessment indicated that the appropriateness of the ten housing systems was based on a pairwise judgment scale of 1-9 conducted to every level-two factors. Finally, the analyzed results of the alternatives were consequently weighed; then, the most appropriate IBS for Addis Ababa's housing development were selected.

4 Results and discussion

4.1. Priority of decision factors

As listed in Table 2, the finalized score for each level-one factor reveals that users needs, supply chain, and building industry are the most decisive factors for the selection of IBS in Addis Ababa. These factors are the key component part of a housing construction industry. Previous studies also revealed that the top three factors have prime importance in many developing countries [2, 18]. These factors are related to the challenges of the provision of affordable housing. One of the main challenges is lack of appropriate materials with a required quality and affordable price. This is a common challenge in most developing countries as the ability of the population to pay for a decent housing is very low. This resulted in the situation where the majority of city dwellers live in poor housing conditions. Therefore, the application of cost-effective IBS not only contributes to the high rate of affordable housing development but also improves the housing conditions. In addition, the success of a large-scale housing development is also dependable on the availability of sufficient production capacity and supply of building components.

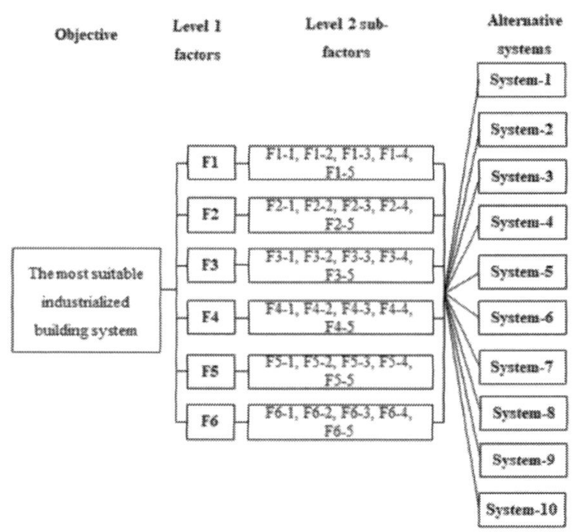

Figure 3. The Analytical Hierarchy Process (AHP structure).

In Addis Ababa, the manufacturing capacity of developing new cost-effective materials and sufficient amount building components is not well developed [19]. Innovation is the best way to push forward the boundary of the performance of building components and the method of construction in order to solve the problem related to a wide application of IBS for affordable housing development. The problem is aggravated by the disorganized nature of the existing supply chain management. These challenges have become a bottleneck for an extensive long-term application of IBS [5]. However, the issue of sustainability was the least priority. This implies that the concern given for sustainability in housing development by construction professionals is less due to the pressing demand for housing with respect to the low capacity of the population to own a decent and quality housing.

The significance of the results of this study lies in experts' preference study. The level of importance of the decision-factors depended on the rich experience of the experts. As Wilson (1971) meant it is the reality of the view that people hold that ultimately determines choices. According to the central point of choice theory, any choice is intended to produce and advance pleasure, happiness, and benefit [20]. It is argued that choices produce satisfaction, especially when one chooses the item of highest utility value while discriminating against those with low utility value [21]. The priorities of the major decision criteria showed that the need and satisfaction of the customer got prime importance over the other. This is related to the utility value of the choice theory. In this sense, choices can be used to predict individual or group knowledge, attitudes, and values. In summary, the results showed that there is a wide range of factors influencing choices of building materials.

Table 2. Level-one and level-two factors analyzed ranking.

Level 1	Priority Vector %	Rank
Building industry	14.36	3
Supply chain	25.06	2
Infrastructure	9.60	5
Sustainability	4.59	6
Users needs	39.28	1
Socio-cultural	7.11	4

4.2 Priority of the ten IBS

The priority rank of the ten IBS shows that the top three systems are the most suitable for Addis Ababa. As it is shown in Table 3, the three systems are all precast concrete systems. They are precast concrete slab and stairs (System 7), precast concrete column, beam, cross-walls, floors and stairs (System 6), and precast concrete wall panels (system 8). The findings from this exercise were very interesting indeed. As it is shown in Figure 4, after a close scrutiny of the materials that were chosen, two large, distinct groups emerged. On the one hand, those that scored highly on the preference list included all precast concrete systems. On the other hand, those that scored very low are all steel systems.

Table 3. Ranking of the alternative housing systems

The ten industrialized housing systems	(Priority Vector %)	Priority ranking
S1: Green terra homes (steel)	5.96	10
S2: Yorkon (steel)	6.26	9
S3: Precast concrete wall panels	15.38	3
S4: Energy smart panels (Steel)	8.24	5
S5: Fusion building system (steel)	7.79	6
S6: Precast concrete column, beam, cross-walls	16.24	2
S7: Precast concrete slab and stairs	17.22	1
S8: APS Infill wall system	9.40	4
S9: Tunnel form	6.62	8
S10: Innovative green building system (steel)	6.87	7

As it is listed in Table 4, the common characteristics of the suitable IBS are load-bearing, easily erectable, sound and fireproof, good for application in medium to high rise residential buildings, and economical in large scale production. The systems are about 25% more productive than the conventional method of construction [22]. They are open systems that are compatible with both conventional and industrialized components. The three systems are semi-industrialized since they only provide parts of the building, and exclude the finishing and mechanical-electrical elements. One type of semi-industrialized process involves "pre-cutting" of housing components and partial assembly to reduce onsite labour and erection time resulting in panelized houses. Another type of semi-industrialization process involves 'frame and infill" techniques, which permit smaller prefabricated panels to be installed in a structural frame [23]. The top ranking systems use both the panelized, and frame and infill technique and perceived as the favoured housing system in Addis Ababa.

Table 4: Characteristics of the suitable housing system

Structural	Architectural	Industrialized process	Erection	Economic and social
Durability	Flexibility in plan	Concrete components	Easy handling	Lower cost for high quality
Load bearing components	Open systm	Middle degree of mechanization	Compatible with conventional system	Adaptability to the needs of various socio-economic groups
Fire, sound and water proof	Medium to high rise residential	Panelised, and frame and infill techniques	Rapid construction	Need for semi-skilled/ unskilled labour

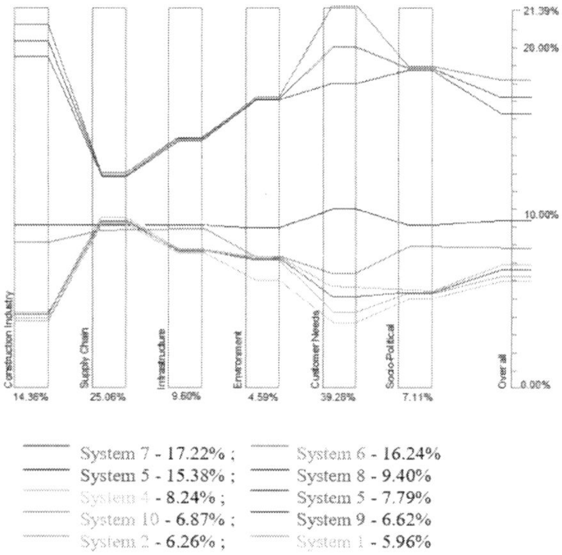

Figure 4: Perfomance analysis of the ten IBS with respect to the level-one decision-factors

As it is shown in Figure 5a, 5b, and 5c, the top three systems has priority percentage in terms of the three most important decision factors. These factors are some of the main reasons for introducing IBS in housing development as well as to solve the housing supply shortage by improving construction productivity. These imply that the three systems positively contribute to improving the productivity rate of affordable housing construction as well as enhancing the capacity of the local construction industry for high-quality products and workmanship. The products are manufactured based on the overall plans, systematic designs, uniform standards, mass production and standardized shapes. The modular component system can be adjusted to varying financial capabilities of beneficiaries. Local input materials can be used for production with a request for semi-skill specialization. The components can also be better utilized and wastage kept to a minimum. In addition, it significantly reduces the use of formwork compare to the conventional systems. In Addis Ababa, wooden formwork is commonly used for construction and has contributed to significant loss of the natural forest [24]. Therefore, housing systems which reduce the use of formwork are appropriate for Addis Ababa. This made the housing systems viable option to apply them for cost-effective mass housing development.

However, these systems have limitations, such as lack of sufficient supply chain, jointing problems; heavyweight, capital-intensive production process, costly transportation and lack of flexibility to use. Continuous improvement is required to prevent cracking around the joints. In general, the prevailing incomplete supply chain of building materials and components in the local construction industry and the poor manufacture capacity is another challenge for the wide application of the housing systems.

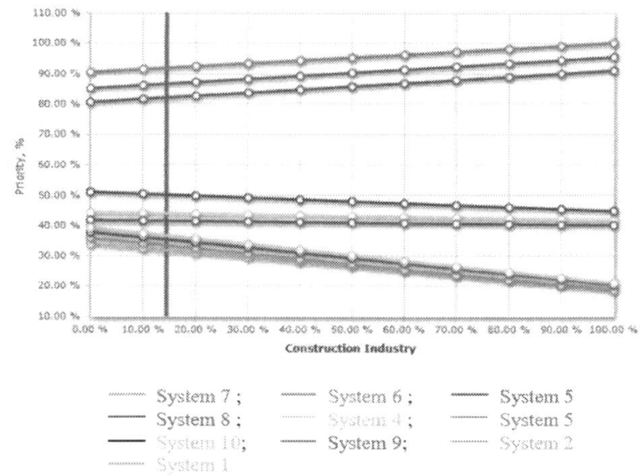

Figure 5a: IBS performance based on the construction industry

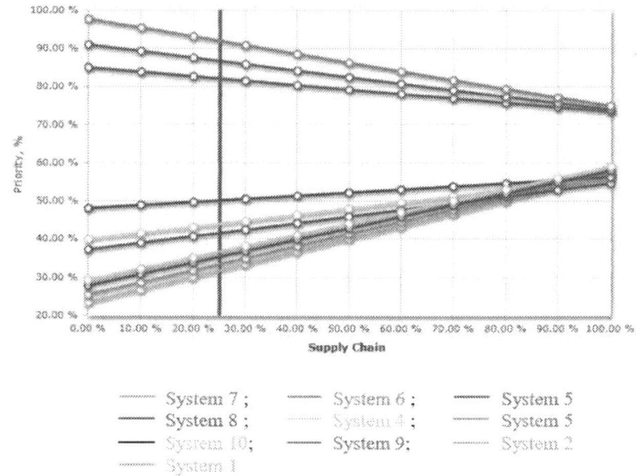

Figure 5b. IBS performance based on the Supply chain.

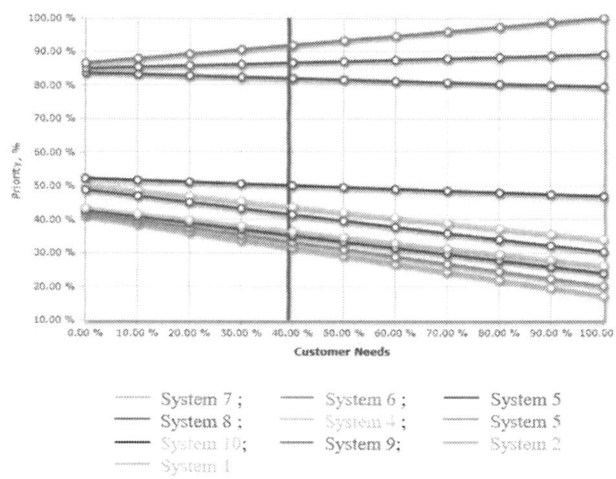

Figure 5c. IBS performance based on the customer needs.

5 Conclusions

Material research has suffered from an over-emphasis on the engineering performance of new materials and the concomitant under-emphasis of experts' perceptual and holistic decision-criteria. In this study, experts' preferences for materials were explored. The study demonstrates the utility of research on the influential attributes of material and the systematic decision-making for the adoption of suitable IBS. The findings revealed that three types of precast concrete systems are suitable for application in Addis Ababa. These IBS systems are selected due to their characteristics of being economically competitive, use of the local material resource, less use of energy, and ease of constructability, and compatibility with conventional components. The AHP rankings of selected IBS are in line with the current construction status in Addis Ababa. However, they have limitations such lack of sufficient supply chain, and jointing problems.

The Addis Ababa market is still underdeveloped, owing to lack of advanced technologies and enough investment capital. To succeed dealing with these challenges, the Addis Ababa city government has to take actions in order to develop the capacity of quality products and to support developers to improve these precast concrete systems according to the local context. Hence, deciding the most suitable IBS for Addis Ababa needs to take into account the affordability, social benefits and environmental aspects of the local context. In addition, the important decision factors should be considered as part of holistic criteria for better design and organization to apply IBS. The Experts' preferences research method used can help to contribute to the body of knowledge of material science and used as a tool for housing developers in order to make effective decisions.

References

1. A.S. Zakaria, T. Gajendran, T. Rose, G. Brewer, AEDM. 1-24 (2017)
2. Y. Pan, F.K.W. Wong, E.C.M. Hui, *Modeling risk management in sustainable construction*. 161-8 (2011)
3. F.W. Kabo, *Low cost housing design and provision : a case study in Kenya*. University of Michigan. (2006)
4. UN-Habitat, *The state of Addis Ababa*. (2017)
5. H. Teka, *National report on housing and sustainable urban development*. (2014)
6. A.G.F. Gibb, *Off-site Fabrication*. Whittles Publishing. (1999)
7. W. Pan, A.G.F. Gibb, A.R.J. Danity. INT J HOUS SCI APPL. **32 (1)**, 61-79 (2008)
8. S. Conards, A. Othman, *Industrialized building systems for low-cost housing projects*. Lamber (2011)
9. S. Kaitilla, ARCHIT BEHAV. **7(3)**, 205-22. (1991)
10. K. Nassar, W. Thabet, Y. Beliveau, AUTOMAT CONSTR. **12 (5)**, 543-60 (2003)
11. W. Pan, A.G.F. Gibb, A.R.J. Dainty, CONSTR MANAGE ECON. **25(2)**, 183-94 (2007)
12. B.D. Ilozor, M.I. Okoroh, C.E Egbu, BUILD ENVIRON PLANN. **39(3)**, 327-37 (2004)
13. D. Richards, *Which offsite system is right for you*. (2004)
14. M.R. Abdullah, C.O. Egbu, *CIB World conference*. (2010)
15. A. Bryman. *Research methods and organization studies*. Unwin hyman. (1989)
16. S.L. Jackson. *Research Methods and statistics. Cengage learning*. (2015)
17. T.L. Saaty, INT J SERV SCI. 1(1). (2008).
18. C. Atkinson, A. Yates, M. Wyatt, *Sustainability in the built environment*. Watford. (2009)
19. HDAP, *Ten years plan for housing development and administration*. (to be published)
20. M. Wilson, *Religions and the transformation of society*. Cambridge university press. (1971)
21. C. Norberg-Schulz, *Intentions in architecture*. Allen Unwin ltd. (1963)
22. Y.H. Pan, Application of industrialized housing system in major cities in China. Hongkong polytehnique university. (2007)
23. I. Majzub. *System building hardware*. Wiley. (1980)
24. S. Kebede, *The impact of alternative construction tehnology on condominium housing project*. Ethiopian civil service university. (2013)

Impact Resistance of High-Performance Steel Fiber Concrete under Long Term Exposure to Environmental Conditions

Abdulhameed Umar Abubakar , Tulin Akçaoğlu and Khaled Marar

Department of Civil Engineering, Eastern Mediterranean University, Gazimagusa, North Cyprus, Mersin 10 Turkey

Abstract. An investigation is presented here that studied the influence of long-term environmental conditions exposure on cylindrical specimens after initial 28 day moist curing. The specimens were tested after six months subjected to Mediterranean weather elements in North Cyprus. Specimens with 150 mm diameter by 60 mm long cylinders were tested under impact load using a modified version of the testing equipment described in ACI 544.2R. Numbers of blows to create first and damage cracks were determined, and subsequently the impact energies calculated. It was seen that there is as much as twenty times improvement in impact energy of fibered concretes at fiber volume of 2 % when compared with plain concrete. This will go a long way to improve the impact resistance of industrial floors and slabs on grade.

1. Introduction

1.1 Background

It is known that the addition of discrete discontinuous fibers to concrete helps in increasing the ductility as a result of the resistance to pull-out offers by the fibers. Over the years, various types of fibers of different forms and geometries have been successfully applied to concrete making process with measured success. These include steel fibers, natural fibers, polymer fibers, glass and synthetic etc. In respect of steel fibers, they come in different geometries such as straight, deformed, machined chip, hooked-end etc. and are regulated by ASTM A820 which establishes the tolerance limits for aspect ratio (length-to-diameter), and ability of the fiber to bond properly is dependent on the aspect ratio which ranges from 20 – 100.

In the application of steel fiber concrete, there are some important terms to ponder upon as defined in ACI 544.1R [1] such as *first crack strength* which is the maximum stress that a material is capable of developing without deviation from linearity. *Toughness* which is the area under the load – deformation curve obtained from flexural strength test up to a known deflection. The energy required to crack a standard specimen of known dimension under impact conditions is known as the *impact strength*.

Some studies reported in ACI 544.1R [1] stated that the peak loads in high-strength steel fiber-reinforced concrete were as much as 40 % higher than that of the plain concrete, and the fracture energy under impact is increased by a factor of as much as 3.5 times. In [2] it was reported that an increase in as much as 80 times is observed when 5 % silica fume is used with 2 % fiber volume having aspect ratio 83. The work of [3] investigated the relationship between compression toughness and impact energy for high-strength fiber-reinforced concrete, and reported a logarithmic relationship existing between them. The work of [4] conducted a statistical analysis on impact strength for first crack, failure strength and reliability strength on a hybrid steel-polypropylene fiber-reinforced concrete with a drop-weight hammer. They reported a slight improvement in hybrid fiber-reinforced concrete compared with the conventional steel fiber concrete. The use of mixed aspect ratio fibers on steel fiber concrete has been explored on 100 x 100 x 500 mm beams by impact loading. It was seen that longer fibers gave a better performance especially at 2.0 % fiber volume. Some researchers [5] reported on the use of hybrid oil palm shell fiber concrete utilizing steel and polypropylene at different fiber volume to test the impact strength using a 10 kg drop weight hammer over a height of 0.6 m with a panel specimen.

This study is devoted to the investigation of impact str ength of a cylindrical specimen of known dimension that has undergo different cycles of curing beyond the age of 28 days subjected to impact load of 13.5 kg hammer.

1.2 Significance of the study

In this study, the influence of environmental conditions mimicking the actual condition that the concrete would be subjected to while in service. Concrete specimens were exposed to weather elements after initial moist curing for 28 days. Previous studies have focused mostly on the 28 days resistance, and we believe that a study that investigate the effect of long term curing under different

https://doi.org/10.1051/matecconf/201820602007

conditions is relevant and increase our knowledge in the subject area.

2 Experimental procedure

2.1 Materials

Blast-furnace Slag Cement CEM II/B-S 42.5 N that conforms with ASTM C 595-17 having a specific gravity of 3.15 and silica fume with 95 % SiO_2 content was utilized at 10 % of the cement content. Tap water was utilized which conform to BS EN 1008-02, and a high range water reducer GLENIUM 27 conforming with ASTM C 494 brown in colour was applied. Aggregates used were crushed limestone rock conforming to the specification of ASTM C 33 and sieve analysis was conducted in accordance with ASTM C 136 and the result presented in Table 1. Relative density (SSD) performed for both fine and coarse aggregates were based on ASTM C 127 and ASTM C 128 and were 2.68 and 2.85 respectively, also absorption (%) determined based on the same standards were 3.0 % and 0.7 %. Bulk density was based on ASTM C 29 were 2083 kg/m^3 and 1203 kg/m^3. Voids in aggregates were 25 % and 50 % respectively for fine and coarse aggregates. The percentage of materials finer than 75μm to ASTM C117 was 13%. Overall, the mix design used in this study is presented in Table 2 with constituents given in kg/m^3.

Hooked end steel fiber based on the specification of A STM A820 with a length of 30 mm, diameter of 0.5 mm a nd tensile strength of 1250 MPa was used.

2.2 Methods

Mixing operation was done as prescribed for steel fiber in ACI 544.1R [1] with fiber additions at 0, 0.50, 0.75, 1.00 , 1.25, 1.50, 1.75, and 2.00 % by volume. Specimens prep ared for impact test were produced from 150 mm by 300 mm cylinders subjected to the weather elements from late June 2017 to early January 2018 approximately 6 months . However, the specimens were first moist cured in the cu ring tank at the initial 28 days before placing them outsid e the laboratory at a good spot subjected to environmental conditions. The specimens were then cut into 150 mm by 60 mm cylinders using a 3 mm thick power saw, five spe cimens were produced from each single cylinder and the t wo ends of the specimen were thrown away to have simil ar end conditions in all the specimens utilized as can be s een in Fig. 1. Apparatus utilized for the impact test was a modified version of a drop weight hammer (Fig. 1a) deve loped in the department based on the recommendations of ACI 544.2R [6] and described in detail in [2 – 3]. It consists of a 13.5 kg hammer weight falling over a height of 0.3 m. The impact energy (E) is calculated from the following formula:

$$\text{Impact energy, E (Nm)} = mgh\,.N \qquad (1)$$

Where
m is the mass of the hammer (kg), g is the acceleration due to gravity (9.82 m/s^2), h is the height of drop hammer (m), and N is the number of blows either to reach first visible crack or cause substantial damage to the specimen called damage crack.

The number of impact to produce either the first or da mage crack is recorded, and the impact energy calculated. The mean of the impact energies were calculated as a res ult of the significant variations observed in the first crack number of blows.

Figure 1a. Modified drop weight impact testing machine

Figure 1b. Damage crack in control specimen (0% fiber)

Table 1. Sieve analysis of aggregates.

Sieve sizes (mm)	12.5	9.5	4.75	2.36	1.18	0.6	0.3	0.15	0.075
% passing coarse	100	100	9.4	0.2	0.2	-	-	-	-
% passing fine	-	-	100	82	48	29	17	7	3

Table 2. Mix design utilized for HPSFC.

Material	Cement	Water	Coarse (0-10mm)	Fine (0-5mm)	Silica Fume	HRWR*
Quantity (kg/m^3)	470	165	1050	700	47	14

*HRWR– high range water reducer

Table 3. Average number of blows and the relationship with impact ductile crack

Fiber Volume (%)	Sample No.	First Crack No. of Blows	Average No of First Cracks Blows (N_f)	Damage Crack No. of Blows	Average No of Damage Crack Blows (N_c)	Impact Ductile crack $\mu = N_f/N_c$
0	A1	2		2		
	A2	2	2.33	2	2.33	1
	A3	3		3		
0.5	B1	2		4		
	B2	2	2	3	3.67	1.84
	B3	2		4		
0.75	C1	2		6		
	C2	1	2.33	6	6.67	2.86
	C3	4		8		
1	D1	3		19		
	D2	5	4.67	11	14.67	3.14
	D3	6		14		
1.25	E1	3		20		
	E2	7	7	17	21	3
	E3	11		26		
1.5	F1	4		12		
	F2	6	5.33	17	13	2.44
	F3	6		10		
1.75	G1	7		20		
	G2	5	5.33	14	16	3
	G3	4		14		
2	H1	19		41		
	H2	24	21.33	63	50	2.34
	H3	21		46		

Table 4. First crack impact energies

Fiber Volume (%)	Sample No.	No. of Blows	Impact Energy (Nm)	Mean	SE Mean	St Dev	Impact Effectiveness
0	A1	2	79.54				
	A2	2	79.54	92.8	13.3	23.0	-
	A3	3	119.30				
0.5	B1	2	79.54				
	B2	2	79.54	79.54	0.00	0.00	-0.14
	B3	2	79.54				
0.75	C1	2	79.54				
	C2	1	39.77	92.8	35.1	60.7	0
	C3	4	159.08				
1.00	D1	3	119.3				
	D2	5	198.86	185.6	35.1	60.8	1
	D3	6	238.63				
1.25	E1	3	119.3				
	E2	7	278.40	278.4	91.9	159.1	1.99
	E3	11	437.48				
1.50	F1	4	159.08				
	F2	6	238.63	212.1	26.5	45.9	1.29

	F3	6	238.63				
1.75	G1	7	278.40				
	G2	5	198.86	212.1	35.1	60.7	1.29
	G3	4	159.08				
2.00	H1	19	755.65				
	H2	24	954.50	848.4	57.8	100.1	8.14
	H3	21	835.19				

Table 5. Damage crack impact energies

Fiber Volume (%)	Sample No.	No. of Blows	Impact Energy (Nm)	Mean	SE Mean	St Dev	*Impact Effectiveness
0	A1	2	79.54				
	A2	2	79.54	92.8	13.3	23.0	-
	A3	3	119.3				
0.5	B1	4	159.08				
	B2	3	119.3	145.8	13.3	23.0	0.57
	B3	4	159.08				
0.75	C1	6	238.63				
	C2	6	238.63	265.1	26.5	45.9	1.86
	C3	8	318.17				
1.00	D1	19	755.65				
	D2	11	437.48	583.3	92.8	160.7	5.29
	D3	14	556.79				
1.25	E1	20	795.42				
	E2	17	676.11	835	105	182	8
	E3	26	1034.05				
1.50	F1	12	477.25				
	F2	17	676.11	517	82.8	143.4	4.57
	F3	10	397.71				
1.75	G1	20	795.42				
	G2	14	556.79	636.3	79.5	137.8	5.86
	G3	14	556.79				
2.00	H1	41	1630.61				
	H2	63	2505.57	1989	265	459	20.43
	H3	46	1829.47				

*Impact effectiveness = [(HPSFC – HPC)/(HPC)

3 Results and discussion

Results presented in Fig. 2 showed that the number of blows required to produce the first crack increased with an increase in the fiber volume and was especially prominent at higher values of fiber. On the other hand, in concrete without fiber addition the number of blows to produce first and damage cracks were considerably less, an average of 2.33 in both cases. The absence of fiber to produce resistance, lead to the appearance of flexural crack (Fig. 1b) which is always taken as the first as well as damage cracks. At the first blow, the specimen was still stable and intact, but immediately after the impact of the second or third blow full flexural crack appears. However, in concretes with fiber addition, the first crack is mostly taken as the appearance of hairline cracks after impact which mostly starts from weak zones where the presence of the fiber is not felt, the smooth sides of the cylinder or crevices generated during the cutting process. The maximum average number of blows observed to produce damage cracks was 50 blows at fiber addition level of 2 % which can be seen in Fig. 2. This produced a corresponding trend in the amount of impact energies for the first crack and damage cracks respectively. This could be attributed to energy absorption capacity of fibers which manifest at higher fiber addition level.

It is known that steel fiber addition has more effect on the toughness characteristics which is the energy absorbed by the specimen covered by area under the load – deformation curve, than on strength characteristics and in this case impact energy as well. The number of blows to fiber addition levels shows a considerable increase in the number of blows required to either induce crack or damage the specimen beyond addition level of 1 % and peaked at 2 %, specimens with 2 % fiber volume receiving the highest impact energies compared with other specimens with lower percentages of steel fiber.

It was also observed that failure of the specimens resulted in the separation of the fiber from the matrix despite the matrix being stronger, and in some cases the fibers could be seen vertically aligned from the matrix. This could be attributed to preferential alignment of fibers during placing of the mix in the cylindrical mould. There is no much room for maneuver so that the fibers could be placed in such a manner as to carry the load after the first crack. More so, the cylinders were cut perpendicular to the direction of the fibers.

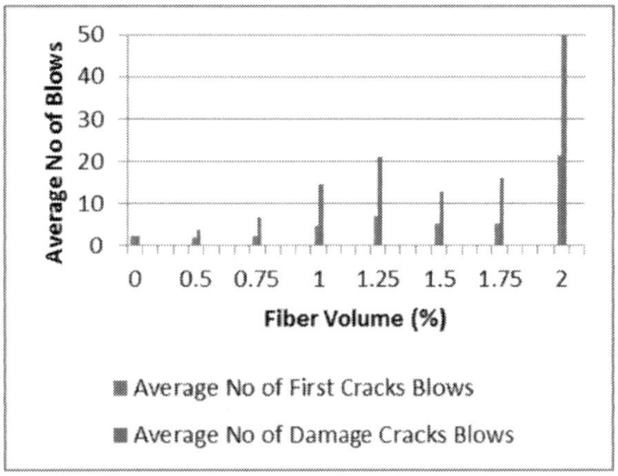

Figure 2. Relationship between numbers of blows against fiber volume.

If the cylinders had been cut in the diametral direction, probably this problem might not occur, alternatively, prisms should be used instead of cylinders. The graph of impact energy versus fiber addition levels (Fig. 3) indicated an increase the addition levels resulted in energy absorption capacity, though somewhere at 1.5% addition, there was a decline in both the first and damage cracks energy but which later rises.

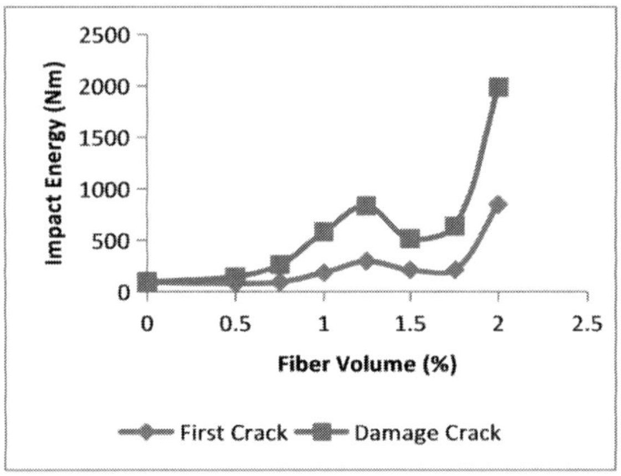

Figure 3. Relationship between impact energy against fiber volume.

4 Conclusion

An evaluation of the long – term exposure of concrete specimens under environmental weather conditions on the impact resistance of HPSFC was conducted and the following conclusion has been arrived at:

Long – term exposure to weather elements and increase in fiber percentage increases the impact resistance of the concrete slowly at the beginning but substantially at up to 2 % fiber addition. The impact energy effectiveness at first and damage cracks were 8 times and 20 times respectively at 2 %.

References

1. ACI 544.1R-96: 2001 State-of-the-Art Report on Fiber Reinforced Concrete American Concrete Institute Farmington Hills MI USA
2. Ö. Eren, K. Marar and T. Celik (1999) Effects of Silica Fume and Steel Fibers on Some Mechanical Properties of High-Strength Fiber-Reinforced Concrete *Journal of Testing and Evaluation* **27**, 380–387
3. K. .Marar, Ö. Eren and T. Celik (2001) Relationship between impact energy and compression toughness energy of high-strength fiber-reinforced concrete *Materials Letters* **47**, 297-304
4. P. S. Song, J. C. Wu, S. Hwang and B. C. Sheu (2005) Statistical analysis of impact strength and strength reliability of steel-polypropylene hybrid fiber-reinforced concrete *Construction and Building Materials* **19**, 1-9
5. K. H. Mo, S. P. Yap, U. J. Alengaram, M. Z. Jumaat and C. H. Bu (2014) Impact resistance of hybrid fibre-reinforced oil palm shell concrete *Construction and Building Materials* **50**, 499-507
6. ACI 544.2R-89: 1988 Measurement of Properties of Fiber Reinforced Concrete American Concrete Institute Farmington Hills MI USA
7. P-C. Aitcin (1998) High-Performance Concrete E & FN SPON New York NY U

Effect of Elevated Temperature on Engineering Properties of Ternary Blended No-cement Mortar

Harry Hermawan[1], Ta-Peng Chang[1], Herry Suryadi Djayaprabha[1,2] and Hoang-Anh Nguyen[3]

[1]*Department of Civil and Construction Engineering, National Taiwan University of Science and Technology, Taipei 106, Taiwan*
[2]*Department of Civil Engineering, Parahyangan Catholic University, Bandung 40141, Indonesia*
[3]*Department of Rural Technology, College of Rural Development, Can Tho University, Can Tho City 90000, Vietnam*

Abstract. This paper aims to examine the engineering properties of ternary blended no-cement mortar which subjected to the various elevated temperatures exposure. The mortars were produced by mixing ground granulated blast furnace slag (S), Type-F fly ash (F) and circulating fluidized bed combustion (CFBC) fly ash (C). The water-to-binder ratio was fixed at 0.40 and the CFBC fly ash content was fixed at 15 wt.% of the mixture that acts as the main activator. The specimens were exposed to the elevated temperatures ranging from 200 °C to 800 °C. The mass loss, compressive strength, and ultrasonic pulse velocity were determined before and after exposure to the elevated temperatures. The obtained results showed after exposed to high temperature, the mortar weight reduction was discovered in the range of 6.0–8.7% when temperature rose from 200 °C to 600 °C, and decreased significantly up to 12.4% as temperature reached 800 °C. The major strength loss occurred after 600 °C with the residual compressive strength approximately at 44.2%. At 200 °C, increased strength was found on SFC mixture and when temperature rose to 400 °C, the specimens still can resist the load reliably with the strength loss less than 8.0%. Consequently, SFC mortar generates good durability and heat resistance below 400 °C.

1 Introduction

Nowadays, the production of Portland cement is the main issue that needs to be solved due to the CO_2 emission from the cement plant was produced in the huge amount. Thus, researchers geared to develop the new alternative pozzolanic material in order to replace the Portland cement that is environmental friendly beside lower cost, the required potential energy to produce it can be reduced and did not create the emission with the substantial extent. The use of industrial wastes is one of the alternative approaches to reduce the use of Portland cement. Example of these wastes includes ground granulated blast furnace slag (GGBFS), fly ash (FA), and circulating fluidized bed combustion (CFBC) fly ash. The utilization of fly ash and slag as the basis material to replace the ordinary Portland cement (OPC) has been investigated with the major formation of ettringite (AFt) and calcium silicate hydrate (C–S–H) gel.

CFBC fly ash is the current new material that has been proposed and developed by some researcher. This particular fly ash was produced from the thermal power plant with the clean coal combustion technology that releases fly ash and bottom ash that contains physicochemical properties and acts as the main activator to trigger the binding mechanism. The aforementioned technology was considered as the environmental-friendly technology that has many advantages such as low SO_2 and NO_x emissions, high combustion efficiency, wide fuel flexibility and low combustion temperatures (typically in the range of 800–900 °C) [1]-[5]. The CFBC fly ash theoretically includes calcium oxide (CaO) and sulfur trioxide (SO_3) and mostly formed in the irregular shape. High content of free lime and $CaSO_4$ were found in the CFBC fly ash particle that induces the setting and hardening process if it contacts with the aqueous solution which characterized by self-cementing behavior. The main hydration products of CFBC fly ashes are hydrated calcium silicate (C–S–H), ettringite (AFt), gypsum and some portlandite [1]-[8].

Several studies [1]-[8] have focused on adding the CFBC fly ash in an effort to enhance the mechanical properties such as compressive strength, durability, Young's modulus, shear modulus, etc. Chen *et al.* [1] reported that the optimum value for no-cement SFC binder to obtain the best compressive strength is about 15–20 wt.% CFBC fly ash and 10–30 wt.% FA as replacement of GGBFS. For SFC pastes with the water-to-binder ratio of 0.35, the compressive strengths at 3 and 28 days can reach 47.7 MPa and 68.4 MPa, respectively. In other work, the SFC binder has been proposed by Nguyen *et al.* [6] in producing the high strength self-compacting concrete (SCC). In contrast, the utilization of CFBC fly ash in the mortar can significantly increase the sulfate resistance [2], [8]. However, several studies [4], [5] confirmed that the CFBC fly ash replacement leads to increase length change (expansion) and the amount of CFBC fly ash replacement cement was recommended to be limited below 20%.

Generally, concrete provides the good fire resistance properties due to its constituent materials (i.e., binder and aggregates). The strength of concrete has significant influence on its properties at both room and high temperatures. When the concrete was subjected to fire, the spalling phenomenon occurred where the surface cracks and breaks into smaller pieces as the concrete deteriorates at the elevated temperature [9]. Horszczaruk *et al.* [10] stated that at high temperature, there was transformation in mechanical properties of paste and aggregates within the concrete since calcium hydroxide (CH) decomposes into lime and water in the temperature range of 350–550 °C and the chemical compound calcium silicate hydrate (C–S–H) gel completely dehydrates and decomposes in the range of 700–900 °C. Nadeem *et al.* [11] reported that the major strength and durability loss of fly ash mortar occurred after 400 °C where the strength loss was up to 16% and the strength loss was even higher for about 46% and 80% at elevated temperatures of 600 °C and 800 °C, respectively. However, for the performance of using CFBC fly ash in the mortar, no more details have been reported and this suggests that there is not yet a full understanding of the impact of the elevated temperature exposure on the properties of SFC mortar. This present paper aims to investigate the durability of SFC mortar subjected to the elevated temperatures through various experiments which assess the mass loss, compressive strength, ultrasonic pulse velocity and thermal conductivity.

2 Experimental program

2.1 Materials and mix proportions

Three industrial waste by-products of ground granulated blast furnace slag (S), class F fly ash (F) and CFBC fly ash (C) were used as binder in this study. Its chemical and physical properties are given in Table 1. The commercial Type I Portland cement in accordance to ASTM C150 was used as the reference binder. Natural river sand was used as fine aggregate with the specific gravity at saturated surface dry condition (SSD) of 2.716 and the water absorption of sand is 0.78%.

The mix proportion of SFC mortar is provided in Table 2. The independent variable of CFBC fly ash content was determined. Chen *et al.* [1] pointed out that the additional of CFBC fly ash from the total binder weight of mixture (S+F) in the range of 15–20 wt.% can improve the compressive strength and trigger the hydration process. In fact, the higher amount of CFBC fly ash results in the discontinuous structure of the SFC binder due to the secondary gypsum and delayed AFt formation. Consequently, the CFBC fly ash content was fixed at 15% by total weight of slag and class F fly ash. The binders were prepared by dry mixing of pozzolanic materials with 4 different class F fly ash to binder ratios, i.e., F/(S+F) = 0, 0.1, 0.3 and 0.5 with the CFBC fly ash addition, i.e., C/(S+F) = 0.15. The water-to-binder ratio was fixed at 0.40 for all mixtures. In order to produce the SFC mortar, the sand-to-binder ratio was fixed at 2.75 in accordance with ASTMC109/C109M. Superplasticizer was added into the mixture at certain amount during the mixing process in order to obtain the desired workability.

Table 1. Properties of the binder materials.

	Portland Cement	GGBFS	Class F fly ash	CFBC fly ash
Chemical composition (%)				
SiO_2	20.42	33.54	59.98	5.22
Al_2O_3	4.95	13.00	25.38	2.21
Fe_2O_3	3.09	0.50	3.65	0.58
CaO	61.96	39.85	5.72	56.80
MgO	3.29	6.90	1.26	2.06
SO_3	2.40	1.67	0.26	32.40
K_2O	-	-	0.48	-
TiO_2	-	-	1.46	-
L.O.I	1.75	4.54	3.00	-
Physical properties				
Specific gravity	3.15	2.9	2.17	2.7
Blaine fineness (cm^2/g)	3450	6000	-	3000

Table 2. Mix proportions of SFC mortar (unit: kg/m^3).

Mixture	Cement	GGBFS	Class F fly ash	CFBC fly ash	Sand	Water	SP
OPC	578.04	-	-	-	1589.61	231.22	0.83
S100F0C15	-	493.88	-	74.08	1561.90	227.19	0.54
S90F10C15	-	441.80	49.09	73.63	1552.44	225.81	0.53
S70F30C15	-	339.51	145.51	72.75	1533.87	223.11	0.66
S50F50C15	-	239.64	239.64	71.89	1515.73	220.47	1.15

2.2 Specimen preparation and testing methods

The mixtures were casted into a 50-mm cubic mold and sealed with the plastic wrap to prevent the water evaporation then after 24 h, all specimens were removed from the molds and cured at a temperature of 25±2 °C and relative humidity level of 65% until the day of testing. The engineering properties of SFC mortar were investigated by mass loss, compressive strength, ultrasonic pulse velocity (UPV) and thermal conductivity tests. The compressive strength test was conducted in accordance with ASTM C109. Evaluation of non-destructive testing method was examined thoroughly by applying the ultrasonic waves and heat transfer through the specimen. The ultrasonic pulse velocity was conducted by UPV test instrument (PROCEQ Pundit Lab) with the nominal transducer frequency of 54 kHz in accordance with ASTM C597, and the thermal conductivity was conducted by heat transfer analyzer (ISOMET model 2104) with the specific surface probe for thermal conductivity measurement range between 0.30 and 2.0 W/m·K and temperature between −15 and 50 °C. Two series of experimental works were conducted, i.e. the specimens were tested before and after subjected to elevated temperatures. Before exposed to elevated temperatures, the compressive strengths of mortar from each mixture were measured at 3, 7, 14 and 28 days using 200 tons uniaxial compression testing machine. Moreover, the measurement of the UPV and thermal conductivity were also carried out before exposed to the elevated temperatures. The other mortar specimens were prepared to be subjected to the elevated temperature exposure of 200 °C, 400 °C, 600 °C, and 800 °C only at age of 28 days. The specimens were heated up in the high temperature electric furnace with a thermostat to control the rate of heating. The heating rate was set at 5 °C/min and the temperature increased progressively until it reached the desired temperature and maintained for 60 min. Furthermore, in order to prevent the specimens from the thermal shock, the specimens were taken after cooled down until reached the uniform room temperature. After the specimens were subjected to elevated temperatures, the measurement of mass loss, UPV and compressive strength were carried out.

3 Results and discussions

3.1 Development of mortar properties before elevated temperature exposure

The investigation of compressive strength, ultrasonic pulse velocity (UPV) and thermal conductivity were established in order to measure the development of engineering mortar properties. Fig. 1 presents the experimental results for the compressive strength test result of OPC and SFC mortars. The compressive strength of OPC mortar was obviously higher than the SFC mortar however the strength was also considerable. The compressive strengths of SFC mortar with the 0–30% FA were higher than those SFC mortar with 50% FA. The increased FA of 50% exhibited the decreased compressive strength up to 35%. Moreover, the 10% slag replacement by fly ash with the fixed CFBC fly ash content of 15% in the S90F10C15 mixture achieves the highest compressive strength of 22 MPa and 30 MPa at 3 and 28 days, respectively compared with the other SFC mixtures. This results show the good agreement with the research conducted by Chen *et al.* [1].

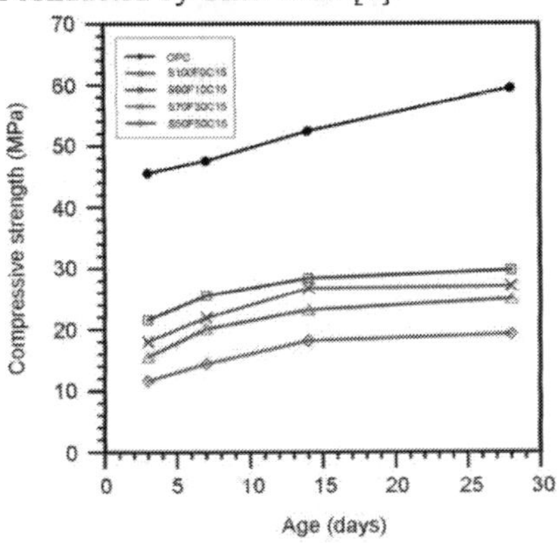

Figure 1. Compressive strength development of mortars.

The UPV results of mortars are given in Fig. 2. Generally, the UPV evaluates the compactness of specimens by measuring the passing speed and attenuation of an ultrasonic wave. Based on the test results, the higher compressive strengths resulted in higher UPV values for all mortars. The development of UPV was equivalent starting from 3 up to 28 days for OPC mortar, however the SFC mortars induced the substantial increment. The huge increment was shown by S90F10C15 and S50F50C15 mixtures at approximately 4%. As 15% CFBC fly ash fixed amount was used, SFC mortar composed of 10% FA had the UPV value comparable with the OPC mortars at 28 days. The result suggested the replacement of slag should be limited in order to obtain the dense microstructure of mortars.

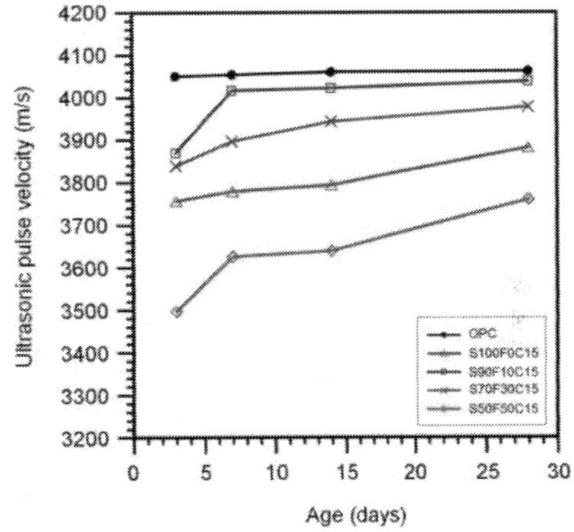

Figure 2. Ultrasonic pulse velocity of SFC mix proportions.

The thermal conductivity of the mortars at 28 days is summarized in Table 3. The mortar with longer hydration age exhibits higher value of thermal conductivity as more amount of solid phase formed. As the replacement ratio of the slag for the fly ash increased from 0 to 10%, the thermal conductivity increased from 1.21 W/(m·K) to 1.52 W/(m·K). However, the increase of the slag replacement ratio by the fly ash in the range of 10–50%, the thermal conductivity of mortars decreased simultaneously. Accordingly, specimen with the low thermal conductivity represents the weak heat transfer and high probability of porosity occurred with the high air content.

Table 3. Thermal conductivity examination at 28 days

Mixture	OPC	S100F0C15	S90F10C15	S70F30C15	S50F50C15
Thermal conductivity (W/(m·K))	1.65	1.21	1.52	1.36	1.24

3.2 Mass loss

The weight of mortar decreased with increasing temperature due to the release of water from the cement paste caused the air voids formed in the inner part of mortar. The mortar integrity disintegrates as demonstrated by the increase of weight reduction with increased temperature. In this present study, the mass losses of SFC mortar were slightly higher than the mass losses in OPC mortar except S50F50C15 mixture above 200 °C. The average mass losses of SFC mortar exposed to 200 °C, 400 °C, 600 °C, and 800 °C were 6.6%, 7.6%, 8.0% and 12.0%, respectively. When temperature rises from 200 to 600 °C, the mass loss for both mortar types is sufficiently small in the range of 6.0–8.7%, however it increased dramatically around 12.4% after exposed to 800 °C. Fig. 3 shows that the mass loss tends to increase in SFC mortar when the slag content was replaced up to 30%, moreover the increased FA of 50% contributed lower weight reduction compared with all of mortar types.

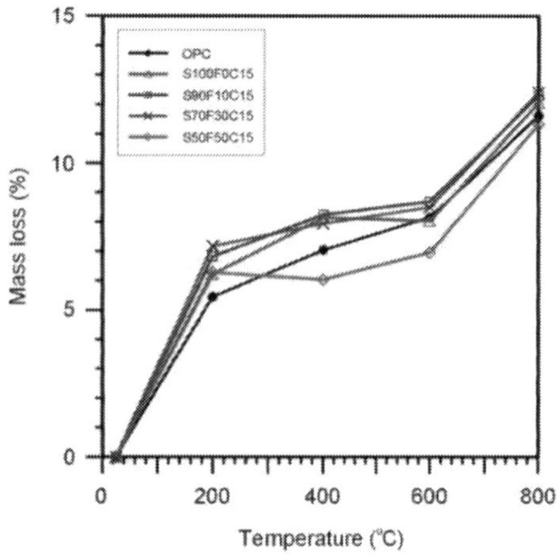

Figure 3. Mass losses of the mortar specimens.

3.3 Ultrasonic pulse velocity under elevated temperatures

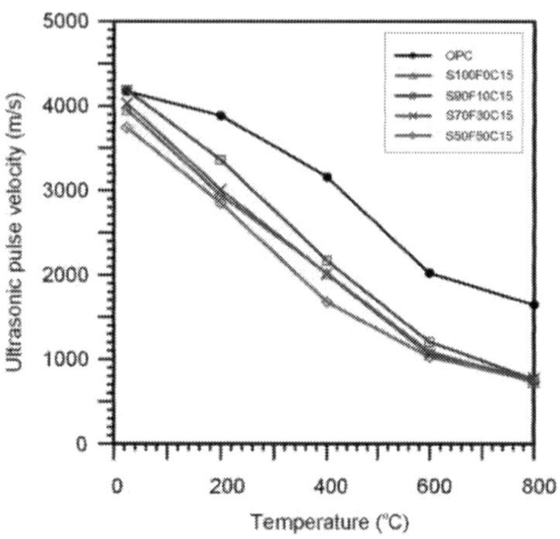

Figure 4. UPV changes in the mortar exposed to elevated temperatures.

Fig. 4 plots results from ultrasonic pulse velocity tests on the mortars after exposed to the elevated temperatures ranging from 200 °C to 800 °C at the mortar age of 28 days. At ambient temperature, the UPV result of OPC and S90F10C15 was equivalent approximately 4180 m/s. This result proved that the compressive strength of mortar with additional fly ash by 10% gained the better strength than the other SFC mortars. When the mortars were subjected to the elevated temperatures, the UPV decreased significantly. The ultrasonic pulse velocity of SFC mortars at 800 °C fell down at one point with the reduction by 80%. As temperature rises, induction of microcracks as a result of degradation of calcium silicate hydrate and portlandite and also formation of CaO crystals during the cooling period may have loosened the microstructure [12]. In general, the microstructures of the specimens that were used in the study deteriorated due to increasing temperature, resulting in decreasing UPV values, especially at 800 °C.

3.4 Residual compressive strength

The percentages of residual compressive strengths of OPC and SFC mortars at different elevated temperatures were described in Fig. 5. It should be noted that x-axis named S100, S90, S70 and S50 stands for mixes S100F0C15, S90F10C15 and so on. From Fig. 5, the enhancement in compressive strength occurred in OPC and mix S50F50C15 that exposed at 200 °C by 1.9% and 12.7%, respectively. In the scope of SFC content, S50F50C15 mixture showed the lowest compressive strength and the highest was achieved by S90F10C15 mixture by slight difference of 20%. Above 200 °C, the compressive strength reduced gradually. A declining trend in residual compressive strength was observed as temperature was elevated to 800 °C. This may be attributed to the induction of cracks as a result of evaporation of physically combined water as well as bound water from calcium silicate hydrate [12].

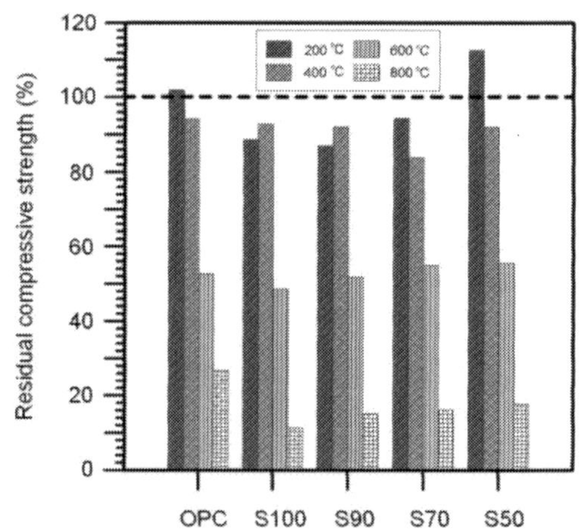

Figure 5. Residual compressive strength of OPC and SFC mortars at different elevated temperatures.

When the mortars heated up to 400 °C, mixes S100F0C15 and S90F10C15 showed a good durability as the corresponding compressive strengths exceeded the compressive strengths when it was subjected to 200 °C. The residual compressive strengths were approximately 92.4% for the aforementioned mixes. The comparable result was also found interestingly on S50F50C15 mixture. The major strength loss occurred after 600 °C. In this case, the presence of surface microcracks was initially observed in the specimens. As the high temperature of 600 °C was applied to the mortars, the dramatic reduction was initially occurred with the substantial difference up to 44.2% compared with the residual compressive strength at 400 °C. Both OPC and SFC mortars lose approximately 48.6 to 55.5% of their compressive strength at temperature of 600 °C. In the other hand, when the mortars were exposed to 800 °C, the test results showed that the strength reduction was significant and the compressive strengths were incredibly low. The residual strength of SFC mortar was measured in the range of 11.4–17.7% at 800 °C. As a result, SFC mortars still can resist the load with the acceptable strength loss and show good durability below 400 °C, however above this limit, the strength is considerably low due to the formation of cracks.

4 Conclusions

The addition of fly ash by 10–30% with the fixed CFBC fly ash amount of 15% was found to increase the compressive strength up to 30 MPa. Ultrasonic pulse velocity and thermal conductivity tests proved that SFC mortar with 90% slag content showed better properties with denser microstructure and good heat transfer. After exposed to elevated temperatures, the mass losses of mortar were found in the range of 6.0–8.7% as temperature rises from 200 to 600 °C and 12.4% when exposed to 800 °C. For all mixes, the major strength loss occurred after 600 °C with the residual compressive strength approximately by 44.2%. Furthermore, the 12.7% increased compressive strength was found at 200 °C with the utilization of slag-to-binder ratio equal to 0.5. The SFC mortars still showed good performance when subjected to 400 °C with the strength loss of 7.6% similarly with OPC mortar except mix S70F30C15. Above 400 °C, the strength decreased dramatically up to 88.6%. These results indicate the SFC mortar generates good durability and heat resistance acceptably below 400 °C. For future work, it is recommended that further investigation is carried out to examine the microstructure of the hardened paste after subjected to elevated temperatures and ensure the effects of adding various amount of CFBC fly ash on long-term strength development.

Acknowledgment

The authors would like to gratefully acknowledge the financial supports from both the Ministry of Science and Technology through the grants of MOST 103–2221–E–011–078–MY3 and National Taiwan University of Science and Technology (NTUST) to conduct this experimental study.

References

1. C.-T. Chen, H.-A. Nguyen, T.-P. Chang, T.-R. Yang, T.-D. Nguyen, Constr. Build. Mater. **76**, 264-272 (2015)
2. T.-D. Nguyen, T.-P. Chang, C.-T. Chen, Constr. Build. Mater. **63**, 40-48 (2014)
3. X.-G. Li, Q.-B. Chen, K.-Z. Huang, B.-G. Ma, B. Wu, Constr. Build. Mater. **36**, 182-187 (2012)
4. M. Chi, Constr. Build. Mater. **123**, 565-573 (2016)
5. T. Wu, M. Chi, R. Huang, Constr. Build. Mater. **66**, 172-180 (2014)
6. H.-A. Nguyen, T.-P. Chang, J.-Y. Shih, C.-T. Chen, T.-D. Nguyen, Constr. Build. Mater. **106**, 670-677 (2016)
7. H.-A. Nguyen, T.-P. Chang, J.-Y. Shih, C.-T. Chen, T.-D. Nguyen, Constr. Build. Mater. **91**, 208-215 (2015)
8. H.-A. Nguyen, T.-P. Chang, J.-Y. Shih, C.-T. Chen, T.-D. Nguyen, Constr. Build. Mater. **102**, 239-243 (2015)
9. S. A. Omer, R. Demirboga, W. H. Khushefati, Constr. Build. Mater. **94**, 189-195 (2015)
10. E. Horszczaruk, P. Sikora, K. Cendrowski, E. Mijowska, Constr. Build. Mater. **137**, 420-431 (2017)
11. A. Nadeem, S. A. Memon, T. Y. Lo, Constr. Build. Mater. **38**, 338-347 (2012)
12. N. Farzadnia, A. A. A. Ali, R. Demirboga, M. P. Anwar, Constr. Build. Mater. **43**, 469-479 (2013)

Three-Dimension Random Aggregate Generation of Numerical Concrete Model

Jianxin Ding[1] and Qingzhou Yang[2]

[1]*Changjiang Institute of Survey, Planning, Design and Research, Wuhan, 430010, China*
[2]*PowerChina Hubei Electric Engineering Corporation Limited, Wuhan, 430040, China*

Abstract. The aggregate generation of concrete is one of the important problems in concrete mesoscopic mechanics. Firstly, the mesoscopic numerical model with spherical aggregates is obtained by the method of excluding the occupied space, and fully-graded concrete model of high aggregate content can be quickly generated. Then, based on the spherical aggregate model, the generation method of random convex polyhedral aggregates is proposed. Finally, a full-graded concrete model with spherical aggregates is shown in Case 1 and a cylindrical concrete model with random convex polyhedral aggregates is shown in Case 2. The result shows that the aggregates are equally distributed in the concrete models which can be used in the study of mesoscopic numerical calculation.

1 Introduction

Concrete is a kind of typical heterogeneous composite material. Mesoscopic mechanics is one of the important methods of property research of concrete material, which has distinct advantages. The concrete mesoscopic mechanics method mainly includes: the generation and packing of aggregates, mesh discretization, and the numerical analysis of the mesoscopic mechanics model. These three aspects are the main contents of the mesoscopic mechanics, and also the major difficulties. Among them, the former two aspects are the basis of mesoscopic mechanics analysis, the generation and packing of aggregates with high content is difficult, while the aggregate content is very high in the actual hydraulic structures of mass concrete. At present the studies mainly focus on two-dimensional plane problem with circular or polygonal aggregates, while little three-dimensional research with spherical or polyhedral aggregates.

Scholars proposed various methods to generate and pack concrete aggregates, and achieved a series of research results. Wittmann earlier proposed a generation method based on natural aggregate shape, which promotes the development of the concrete mesoscopic modeling [1]. Yue generated two-dimensional asphalt concrete mesoscopic structure model using digital image processing technology [2]. Ma proposed the method of excluding the occupied space to generate three-dimensional sphere and polyhedron aggregate model [3]. Cusatis developed the three-dimensional lattice discrete particle mesoscopic model, the spherical aggregate generated according to the aggregate volume fraction, and the geometrical topology algorithm is adopted to pack the aggregates [4]. Wang put forward one two-dimensional modeling approach of concrete aggregates with arbitrary shapes based on a pre-fabricated mesh [5].

In this paper, the generation method of the random aggregates of concrete is studied. Firstly, the method of excluding the occupied space is used to generate and pack spherical aggregates, and the concrete mesoscopic mechanics model with high content of aggregates can be quickly obtained. Then, a novel method is proposed for the generation of the random convex polyhedral aggregates. The generation program is compiled. Finally, A 450 mm×450 mm×450 mm full-graded concrete cube model with random spherical aggregates is shown in Case 1 and a cylindrical concrete model with random convex polyhedral aggregates is shown in Case 2. The results show that the aggregates are equally distributed in the concrete models which can be used in the study of mesoscopic numerical calculation.

2 The generation of random spherical aggregates

In concrete mesoscopic mechanics research, the aggregate is mostly assumed as spherical or polyhedral aggregate (circular or polygonal aggregate in two-dimensional model). Due to the intrusive contact judgment between aggregates is simple, the achievements mainly focus on the circular and spherical shape for aggregates. The generation and packing of three-dimensional spherical aggregates is relatively simple, but the actual aggregate content is low in a lot of papers, which did not consider the actual aggregate gradation, and most of the actual concrete aggregate content is about 40% ~ 60%, while in mass concrete of hydraulic structures, the content of aggregates is even higher, the generation and packing of concrete aggregates with full-graded and high content of the concrete mesoscopic mechanics is one of the problems to be solved. Ma proposed "the method of excluding the occupied space", which can more effectively finish high content aggregate generation and packing in certain space [3]. Qing studied the three-dimensional mesoscopic mechanics model of concrete based on CT slice [6]. This paper adopts the method of excluding the occupied space to generate concrete aggregates, and all aggregates are regarded as coarse aggregates, while the fine aggregates are supposed to be mortar.

As the method of excluding the occupied space, the specimen area is divided into cube cell with a step length less than the minimum aggregate radius, when packing aggregates, the element whose maximum distance from nodes to the aggregate center is less than the radius plus the minimum aggregate radius, the element will be excluded. When determining the position of the rest of the aggregates, the element excluded will not be considered. The method can improve the efficiency of the three-dimensional random spherical aggregate generation [3]. The algorithm is shown in Figure 1.

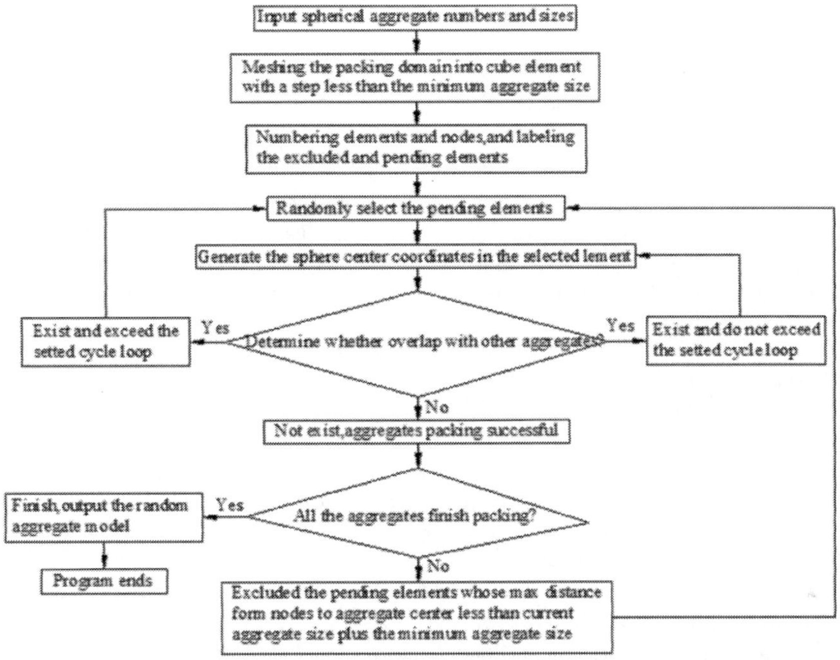

Figure 1. The program flow chart of 3 D random spherical aggregate generation and packing algorithm.

3 The generation of random convex polyhedral aggregates

The shape of the artificial crushed aggregates is mainly convex polyhedron. In order to reflect the irregular shape of the aggregates, the random convex polyhedral particle is generated in the ellipsoid (Figure 2). Reference to the stone particles of rockfill dam generation method proposed by Ma [7], the random vertices are generated on the ellipsoid surfaces, the number of vertices is stochastic, and then the convex polyhedron is generated by the vertices according to the certain algorithm, the specific method is as follows:

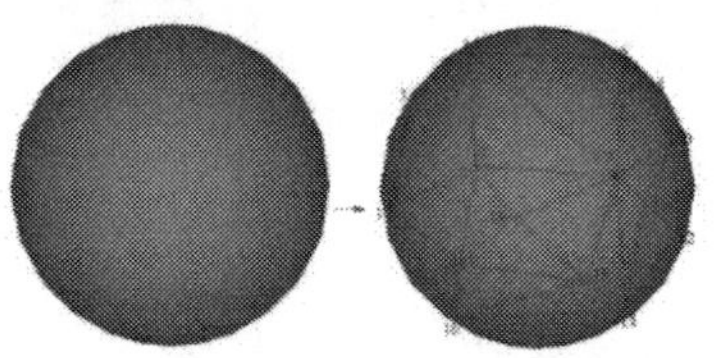

Figure 2. The ellipsoidal aggregate model.

Firstly, the radius of the ellipsoid is obtained on the basis of the generated spherical aggregate:

$$\begin{cases} a = r_1 + (r_2 - r_1) * RAND(1) \\ b = r_1 + (r_2 - r_1) * RAND(2) \\ c = r_1 + (r_2 - r_1) * RAND(3) \end{cases} \quad (1)$$

where: a 、 b and c are the length of the three semi-axis of the ellipsoid; r_1 is the lower limit of a certain size group; r_2 is the upper limit of the size group, *RAND* (1)、*RAND* (2) and *RAND*(3) are independent random numbers distributed equally in the interval [0, 1].

In order to ensure that the shape of the aggregates is fully random, the number of vertices is equally distributed:

$$n = n_{min} + (n_{max} - n_{min}) * RAND \quad (2)$$

where: *RAND* is the random number which is equally distributed in the interval [0, 1]. In this paper, the $n_{\min}$ is set to 10 and $n_{\max}$ is 18.

The vertex coordinates of the convex polyhedron is calculated by spherical coordinates:

$$\begin{cases} x_i = x_0 + a\sin\theta_i\cos\varphi_i \\ y_i = y_0 + b\sin\theta_i\sin\varphi_i \\ z_i = z_0 + c\cos\theta_i \end{cases} \tag{3}$$

where: (x_i、y_i、z_i) is the coordinate of the vertice i in the ellipsoid surface; (x_0、y_0、z_0) is the center coordinate of the ellipsoid; θ_i is the dip angle in the Spherical Coordinate System, which is of the uniform distribution in the interval [0, π]. φ_i is the direction angle in the Spherical Coordinate System and equally distributed in the interval [0, 2π].

In this paper, a novel method is proposed for the generation of the convex polyhedral aggregates with randomly selected vertices. After the number of the vertices and the coordinates of the vertices are determined, three vertices in the vertices set are randomly selected, then a plane is determined by the three vertices, and the equation of the plane is calculated. If the other vertices in addition to the three vertices are on the same side of the plane, the plane is an effective surface for the convex polyhedron, a convex polyhedron aggregate is generated until all the effective surfaces are find out.

There are sharp angles and thin deformed aggregates when using the above algorithm to generate random convex polyhedrons. It needs to improve the generation of convex polyhedral vertices. Specifically, the main control is the generation intervals for θ_i and φ_i. When the number of vertices of the convex polyhedron aggregate is determined, the generation of the interval [0, π], [0, 2π] are divided into n sub-intervals, the θ_i and φ_i are equally distributed in the interval $[\frac{\pi}{n}(i-1), \frac{\pi}{n}i]$ and $[\frac{2\pi}{n}(i-1), \frac{2\pi}{n}i]$, the shape of the generated convex polyhedral aggregate is more regular.

4 Examples

Case 1: The 450 mm×450 mm×450 mm full-graded concrete cube model, the proportion of small stone, medium stone, big stone and large stone is 2:2:3:3. In the mesoscopic numerical model, the average size of 10 ~ 20 mm fine aggregate is 15 mm, the average size of 20 ~ 40 mm medium aggregate is 30 mm, the average size of 40 ~ 80 mm big aggregate is 60 mm, and the average size 80 ~ 150 mm large aggregate is 115 mm. Numbers of aggregate are shown in Table 1.

Table 1. The numbers of spherical aggregates in concrete model.

Aggregate size	10~20 mm / fine stone	20~40 mm / medium stone	40~80 mm / big stone	80~150 mm /large stone
Numbers of aggregates	5503	688	129	18

Figure 3 shows the full-graded spherical aggregates of concrete mesoscopic mechanics model. It can be seen that the spherical aggregates are equally filled in the cube model, which proved that the generation algorithm is effective.

Figure 3. The full-graded spherical aggregate concrete model.

Case 2: A cylindrical concrete model with the diameter $\varphi = 100$ mm and the high $H = 200$ mm, there are 158 small stone particles of 10 ~ 20 mm and 20 medium stone particles of 20 ~ 40 mm. The numbers of aggregate are shown in Table 2. A three-dimensional numerical concrete model of convex polyhedral aggregates is randomly generated.

Table 2. The numbers of polyhedral aggregates in concrete model.

Aggregate size	10~20 mm / fine stone	20~40 mm / medium stone
Numbers of aggregates	158	20

Figure 4 shows the randomly generated convex polyhedral aggregate model in the cylinder. It can be seen that the random convex polyhedral aggregates are equally filled in the cylindrical specimen, and the aggregate shape is close to the natural shape.

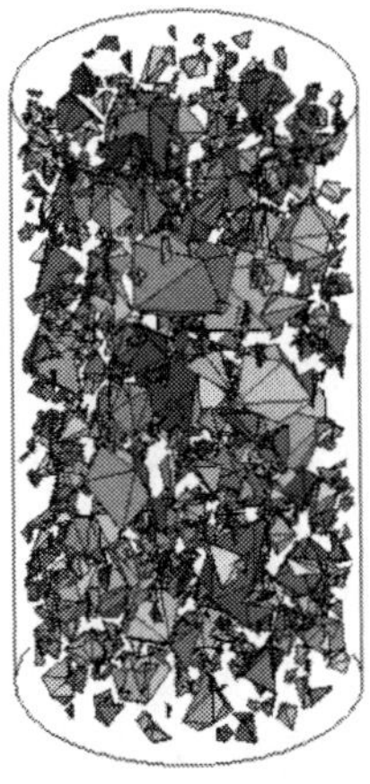

Figure 4. The random convex polyhedral aggregate concrete model.

5 Conclusions

This paper studies on three-dimension random aggregate generation of numerical concrete model, the research results are as follows:

(1) The generation of spherical aggregates of concrete mesoscopic model is realized.

(2) Based on the spherical aggregates, the convex polyhedral aggregates of concrete mesoscopic model is generated.

(3) The aggregates are equally distributed in the concrete model and close to natural aggregate shape.

At the same time, the aggregate is simplified as a sphere or a polyhedron, and the concrete model with sphere and convex polyhedral aggregates is obtained. In the future, study on the mesh discretization and the mesoscopic numerical simulation of the concrete model will be carried out.

References

1. F. H. Wittmann, P. E. Roelfstra and H. Sadouki, Mater. Sci. Eng. **68**, 239(1985).
2. Z.Q. Yue, S. Chen and L.G. Tham, Comput. Geotech. **30**, 375(2003).
3. H.F. Ma, H.Q. Chen and B.K. Li, J. Hydraul. Eng. **10**, 27(2004).
4. G. Cusatis, D. Pelessone and A. Mencarelli, Cement. Concrete. Comp. **33**, 881(2011).
5. B. Wang, H. Wang, Z.Q. Zhang and M.J. Zhou, Chin. J. Comput. Mech. **34**, 591(2017).
6. W. Qin and C.B. Du, Eng Mech. **29**, 186(2012).
7. G. Ma, W. Zhou, X.L. Chang and C.B. Zhou, Chin. J. Geotech. Eng. **33**, 746(2011).

Building Retrofit to Improve Energy Performance from Office to Accommodation. Case study: Tower Building, Nottingham, UK

Nissa Aulia Ardiani[1], Suhendri[1], Mochamad Donny Koerniawan[1], and Rachmawan Budiarto[2]

[1]*Department of Architecture, Institut Teknologi Bandung, Bandung, Indonesia*
[2]*Department of Nuclear Engineering and Engineering Physics, Universitas Gadjah Mada, Yogyakarta, Indonesia*

Abstract. Designed in Brutalism style by architect Andrew Renton, Tower Building has 17 floors for academic and lecturers' office function. As the highest point in University of Nottingham, this tower has been built for almost five decades. The aim of this project is to propose the tower retrofit from current function to accommodation purpose. Improvement in terms of function and building energy performance by applying sustainable building technologies are the objectives of the retrofit. The advanced plan of building retrofit was proposed after assessing the current building performance and determined the problems. The proposed building design was based on building performance simulation result, literature, and precedent studies. Thereafter, several technologies and design ideas were applied for further investigation, to examine how is the strategies works in the building retrofit. Integrated double skin façade with BIPV, bio composite material for internal wall, double low-e glass for external wall, as well as green lung to improve natural ventilation and create public space were proposed for the building.

1 Introduction

Sustainable building design can be approached in two ways, reducing energy consumption and supplying energy by renewables. In a comprehensive implementation of a sustainable building, the approaches are applied from the early phase of a project. Among the project phases, construction phase is the most energy intensive [1]. Thus, energy reduction strategies and renewable energy supply in this process are significant to minimize environmental impact.

Nevertheless, due to energy intensive characteristic of a new construction process, retrofit seems to be a prospective method. Menassa and Baer [2] have summarized some benefits from sustainable retrofit. A sustainable retrofit could span the lifetime of a building, give a better economic value for the owner, and thus decrease the lifecycle cost of the building. Retrofit also makes building more energy efficient [3]. However, Burton [4] mentions that retrofitting or refurbishment is still avoided by developers and architects due to its difficulty level. Also, a retrofit project is high risk because there are lots of uncertainties in the project [5]. Hence, the owner or developer should systematically analyze the possible results of a retrofit project, to make sure that the project is really feasible [6]. It is worth noting that the decision has a large number of possibilities and that it is a multi-objective problems [7].

Realizing the possible benefits of building retrofit, the Department of Architecture, University of Nottingham creates a case study for retrofitting the Tower Building. It is being used by the department as classrooms, offices, and laboratories and, in the case study, planned to be a postgraduate student accommodation. There is a big difference in terms of building typology and function if it is proposed to be a student accommodation. This revolution in building typology also should be sustainable. The design should resolve some of the difficulties in sustainable energy building design and in the choice of building materials and parameters. It also needs to solve realistic energy technology problems and provide clear interpolations of the significance of climatic parameters and their significance for the planning and design of building. Therefore, the designers should consider the influence of building parameters on the energy and carbon performance of buildings, understanding the need for effective zoning and internal form in the overall performance of buildings, and utilize thermal modelling and analysis. The opportunities and limitations of passive design solutions proposed should be assessed in terms of its annual energy use and CO2 emissions.

Considering that building retrofit in Indonesian buildings has not been intensively studied, the case study would be beneficial. There are numbers of old colonial buildings that could be retrofitted so it can be functioned for a different typology and reduce the energy consumptions.

2 Methods

This study consists of several stages which are: 1) Develop a building model to identify the building performance in the current condition as an office (before retrofit). After that, the main problems from current building were determined and anticipated by changing the function and schedule from office to accommodation. 2) Propose building design, layout, and strategies to reduce the energy consumption based on simulation result, precedent studies, and theme analysis. 3) Decide the main strategies that would be applied in the building for the retrofit to investigate the result from the simulation.

In this study, building layout, buildings materials, function, and activity were used as parameters and was simulated by using Autodesk Ecotect. Autodesk Revit and Autodesk Green Building Studio were used to make and simulate the detailed design proposal. To calculate the possibility of BIPV application in the southern façade, PVSyst software was also applied.

2.1 Site location

Tower building is located in East side of University Park campus in the engineering complex (see Figure 1). The building is one of the landmarks and the proposed function should improve the 'image' of the building as it can be seen from the main road of Nottingham city. Surrounded by important building features in campus, this location is suitable for the accommodation purpose. QMC hospital across the road, George Green Library, Engineering Science Learning Center, computer center, and bus stop are nearby and accessible within walk distance.

Figure 1. The location of Tower Building in the University of Nottingham from google maps.

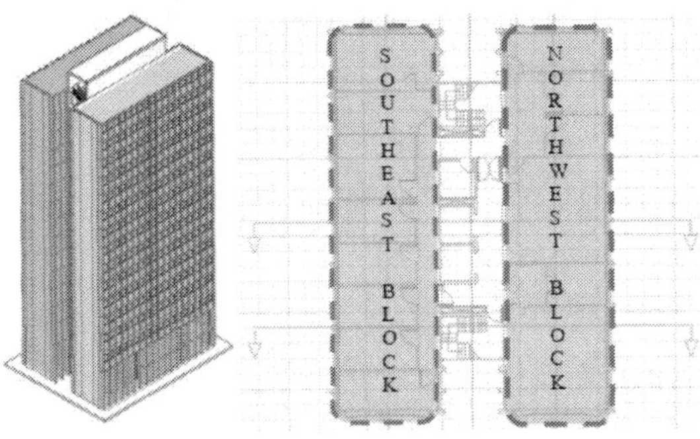

Figure 2. Axonometry (left) and the typical floor plan (right).

The Tower building consist of two masses that separated by the core for service area. The typical office and classes floor plan of the current building can be seen in the Figure 2 above. This layout was then simulated as benchmark for further study of retrofit.

2.2 Materials

Table 1. Material properties before retrofit.

Type	Materials	Thermal Properties	
		Thickness (m)	*U-Value (W/m^2K)*
Wall	Plaster + Concrete + Plaster	0.248	1.10
Partition	Framed Plasterboard	0.114	2.47
Ceiling	Fiberfax Paper + Air Gap + Plaster Board	0.3	2.49
Floor	Concrete + Carpet Underlay + Carpet	0.41	0.93
Windows	Single Glass	0.06	6.00

Table 2. Material properties after retrofit.

Type	Materials	Thermal Properties	
		Thickness (m)	*U-Value (W/m^2K)*
Wall	Concrete + Plaster	0.248	1.10
Partition	Bio Composite Panel + Insulation + Bio Composite Panel	0.10	0.46
Ceiling	Air Gap + Insulation Glass Fiber Quilt + Plaster Board	0.21	0.50
Floor	Concrete + Carpet Underlay + Carpet	0.41	0.93
Windows	Double Low E Glazed + Air Gap	0.042	2.71

Table 1 and 2 describe the material properties applied in both the current building and the building design retrofit proposal. Materials were chosen by considering its thermal properties as well as the sustainability.

2.3 Simulation setting

Daily occupancies for the office in the weekday are 100% from 8am - 12pm, 50% from 12pm - 1pm, 100% from 1pm - 6pm, 30% from 6pm - 8pm, and the building will be emptied afterwards until the next day. On the weekend, the building is always empty. In the simulation setting in Autodesk Ecotect, internal design condition for clothing was 1.00, humidity 60%, air speed 0.5 m/s, and lighting level 300 lux. Number of occupants in one floor in accordance with room layout was 13 people with sedentary activity. Sensible gain was set to 23 W/m2, air change rate was 0.5, and wind sensitivity was 0.25 air changes per hour. For the HVAC system, full air conditioning was used with efficiency 95%. Thermostat range was from 18-26 0C with operation hours from 8am to 8pm.

On the other hand, daily occupancies for the accommodation was different. Internal design condition and infiltration rate were same, however the schedule and the occupancy in the weekday are 50% from 8am - 5pm, 70% from 5pm - 7pm, 90% from 7pm - 8pm, and the building will be 100% from 8pm - 8am on the next day. The internal design condition for clothing, humidity, air speed, lighting level, air change rate, and wind sensitivity were identical with the office setting. Number of occupants in one floor in accordance with room layout was 8 people with resting activity, as well as sensible gain was set to 8 W/m2. the HVAC system, full air conditioning was used with efficiency 95%. Thermostat range was from 18 - 26 ^{0}C with operation 24 hours for every single day.

2.4 Proposed strategies and design

From the literature reviews and precedent studies, the technologies and ideas that would be applied in the Tower building are :

1) Integrated double skin façade with BIPV
2) Bio composite material for internal wall
3) Double low-e glass for external wall
4) Green lung to improve natural ventilation and create public space.

As can be seen in the Figure 3, the building is divided into six zones. Ground floor up to third floor are for public area (see Figure 4), while the upper floors are for private area. There are several types of accommodation depend on the occupant. For hotel or short stay, the proposed single hotel room with bathroom and sleep box can be seen in Figure 5. For master, undergraduate, and PhD student, there are three different choices of layout which are standard room with toilet, standard room with toilet sharing, and *ensuite* (see Figure 6).

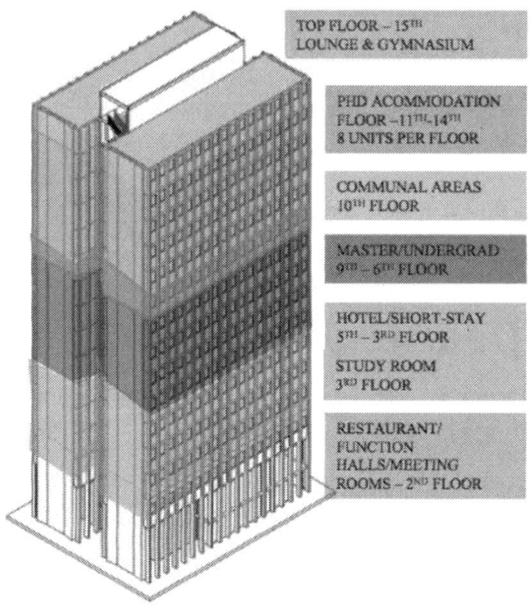

Figure 3. Zones and functions division in the design proposal of the retrofit.

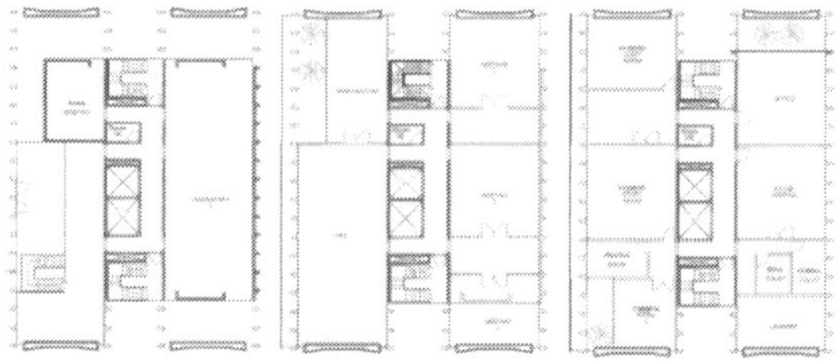

Figure 4. Floor plan proposal for public area.

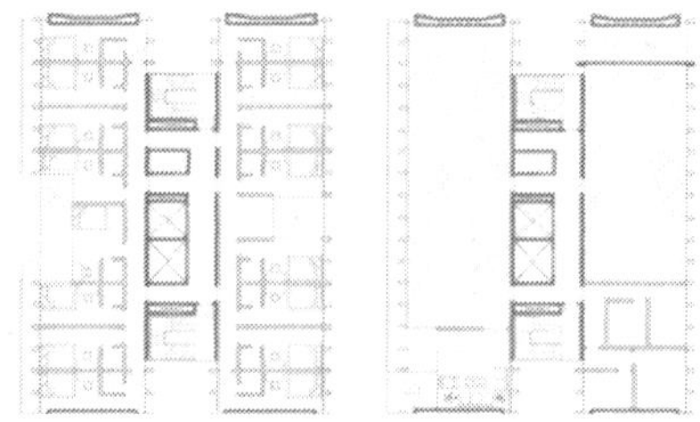

Figure 5. Floor plan proposal for short term accommodation.

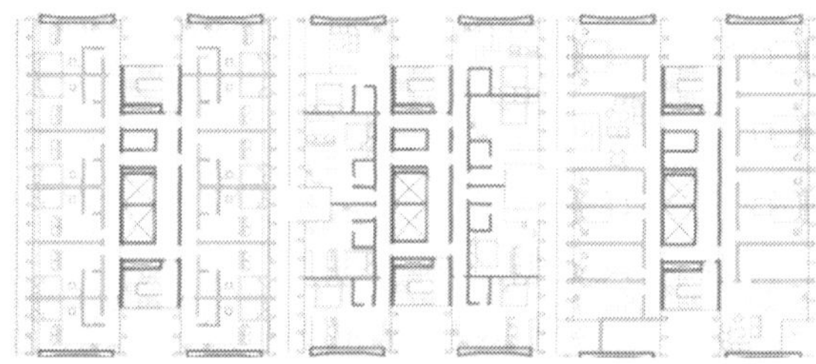

Figure 6. Floor plan proposal for long term accommodation.

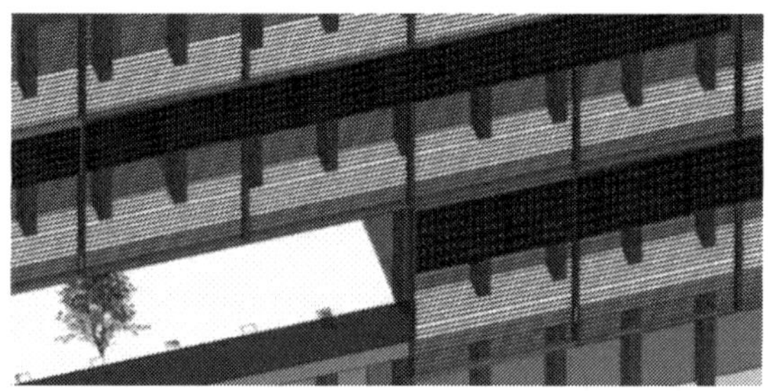

Figure 7. The application of ventilated double skin façade and BIPV system in the southern facade.

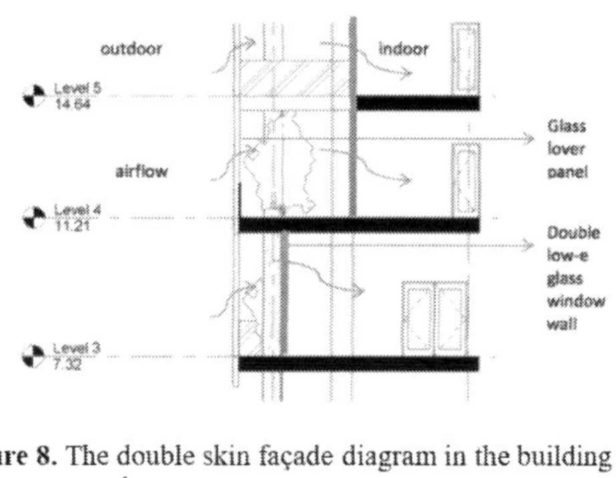

Figure 8. The double skin façade diagram in the building design proposal.

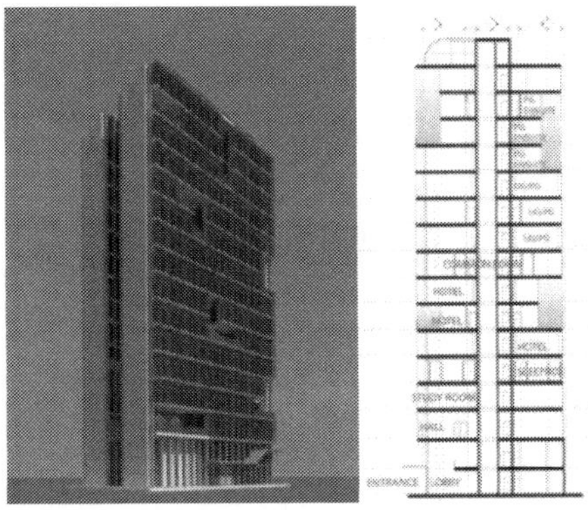

Figure 9. Perspective (left) and building section (right) of Tower Building design proposal.

Figure 7 shows the strategy to configure vertical solar photovoltaic panel in the façade. The idea is that PV panels will be integrated to the building to generate electricity. This is effective to be placed in high-rise building which its side receives high amount of solar radiation. Implementation of façade systems for Tower building is a combination between ventilated double skin façade and BIPV system (see Figure 8). The system is placed in the southern façade, because the maximum solar radiation is coming from this direction.

Figures 9 show that the designed green lung could provide community space and brings natural environment ambient in the tower. The location of the green lung which is near the outer side of building makes it possible to bring natural air from outside flows into interior. The figures also show the idea of integrating double skin façade and green lung in the building at the southern side. In some area, the glass louver panel is open for the green lung and the wall is pushed back 1.5 meters from previous wall position. The holes in double skin façade panel provide natural environment and private balcony for several rooms in the building.

3 Result and discussion

From the simulation in the Ecotect, it was revealed that the current Tower Building with office function requires a total of 35.4 MWh energy for heating from October to July (see Figure 10). This number was quite low because the schedule setting for most of the occupants were stayed inside the building only in office hours (8am - 5pm). However, with the materials applied in the current building, the energy needed for cooling in the summer time (total 48.2 MWh) was immensely high compared to the heating. The results also show that the highest heat losses were from building fabrics.

On the other hand, when the simulation setting and schedule was changed into accommodation without changing the current building materials, the energy needed for heating was increased significantly with total of 378.5 MWh (see Figure 11). This was already predicted because of the schedule of the operation of accommodation that was almost 24-hours. The highest heat losses were from ventilation because the wall materials might have not enough insulation to prevent the cold weather outside. Meanwhile, as can be seen in the Figure 11, the cooling energy in the summer was lower than the cooling energy for the office function.

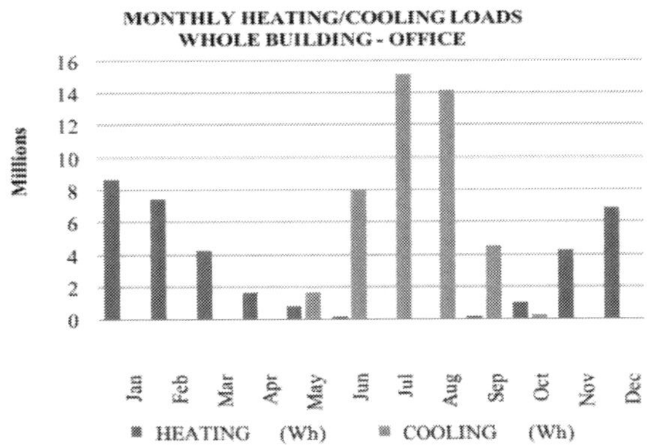

Figure 10. The energy performance of current building as an office.

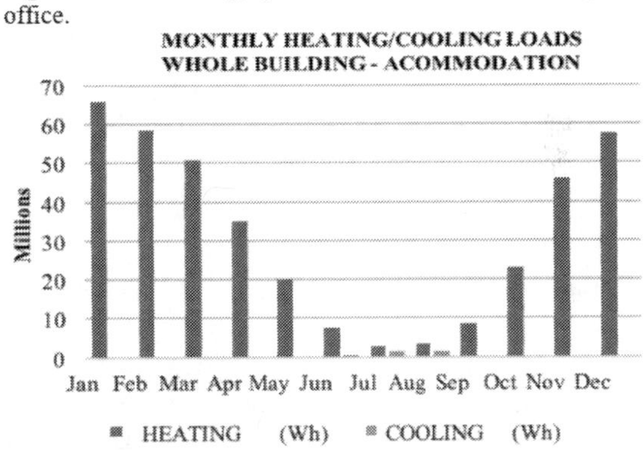

Figure 11. The energy performance of current building as an accommodation.

From the simulation of Tower building before retrofit, the strategies implemented in the project is to improve energy performance to reduce heating and cooling loads especially in the middle floor. For office function, heat gains mostly co me from internal appliances due to occupation. Nevertheless, physical factors of the building also affect the heat gain, such as building fabric and ventilation. This is also the case for accommodation function, except for the percentage of contribution from building fabric and ventilation. For office function, the contribution of heat gain from building fabric and ventilation are 63.80% and 35.80% respectively. Whereas, the contribution in the accommodation function are 45.90% and 53.90% respectively. Thus, heat gain and losses are improved by changing building fabric and ventilation. A passive solar heating and mixed mode air conditioning also used in the project.

3.1 Building performance after retrofit

As described in chapter methodology, after simulating the current building with two different function, several strategies and ideas were installed in the building. The new layouts, materials, and scheduling were then simulated again to investigate whether it could reduce the energy demands by enacting current building with accommodation simulation result as a benchmark. It was then affirmed that after retrofit, total heating loads per year is reduced to 207.7 MWh where before retrofit was 378.5 MWh. However, total cooling loads was increased from 4 MWh to 25.5 MWh (see Figure 12).

Figure 13 below shows the integration of BIPV performance in the glass louver panel. The simulation in PVsyst software was done to calculate maximum electricity that could be harvested. With 247.5 m^2 total area of vertical PV panels in the Southern façade, the result shows that it could generate 27.7 MWh per year with average 75.87 kWh per day.

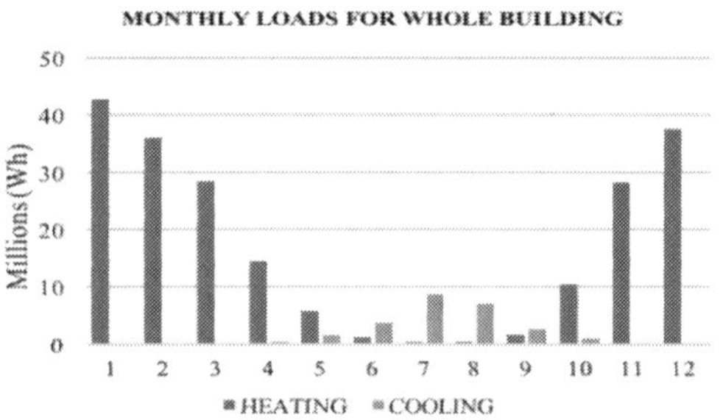

Figure 12. The energy performance after retrofit.

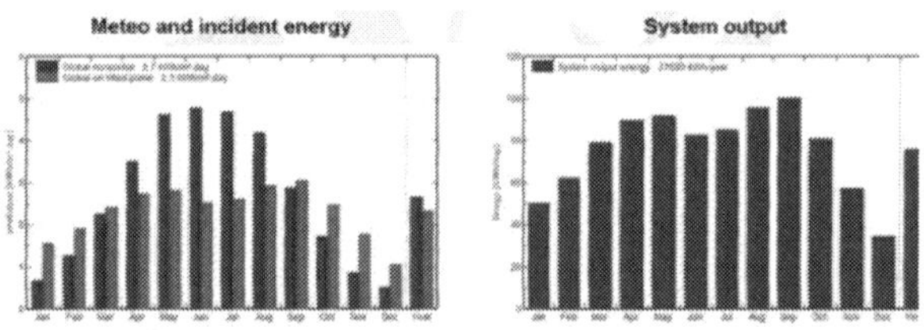

	Gl. horiz. kWh/m².day	Coll. Plane kWh/m².day	System output kWh/day	System output kWh
Jan.	0.67	1.54	50.27	1558
Feb.	1.26	1.91	62.30	1744
Mar.	2.25	2.43	79.30	2458
Apr.	3.50	2.75	89.56	2687
May	4.62	2.82	91.82	2846
June	4.79	2.53	82.54	2476
July	4.69	2.62	85.33	2645
Aug.	4.20	2.93	95.62	2964
Sep.	2.89	3.07	100.0	3001
Oct.	1.73	2.49	81.07	2513
Nov.	0.87	1.77	57.61	1728
Dec.	0.52	1.06	34.56	1071
Year	2.67	2.33	75.87	27693

Figure 13. The result of PVsyst simulation to calculate BIPV application.

4 Conclusions

Application of sustainable technology in material, passive and active design could enhance performance of the building. Simulations conducted in this project revealed that after retrofit, overall building performance is improved by the strategies applied. However, because of the function change from office to accommodation, energy loads are still high. To replace the energy needed for running the building, sustainable energy harvesting is highly possible. In this detail project, BIPV in façade could generate 3% or 27.7 MWh annual electricity. If the specification and efficiency of the PV panel was increased, it could generate more than 80 MWh per year. Figure 14 shows how the Tower building will look like from the Clifton boulevard after the retrofit.

Figure 14. Tower Building after retrofit.

For old buildings in Indonesia with the same case from office to accommodation or vice versa, retrofit could be one alternative solution to reuse the building without demolish it. However, further studies would be needed to show whether the effect of retrofit will be significant.

Acknowledgement

This building retrofit study was part of the Sustainable Research Project course in the MSc Sustainable Building Technology Program at the University of Nottingham. We gratefully acknowledge the funding from USAID in Sustainable Higher Education Research Alliances (SHERA) program through Centre for Development of Sustainable Region (CDSR).

References

1. S. Sattary and D. Thorpe, *Optimizing embodied energy of building construction through bioclimatic principles*, in proceedings of 28th Annual ARCOM Conference, Edinburgh, Association of Researchers in Construction management, pp. 1401–1411 (2012)
2. C. C. Menassa and B. Baer, *A framework to assess the role of stakeholders in sustainable building retrofit decisions*, Sustain. Cities Soc., **10**, pp. 207–221 (2014)
3. R. Yang and L. Wang, *Multi-objective optimization for decision-making of energy and comfort management in building automation and control*, Sustain. Cities Soc., **2**, no. 1, pp. 1–7 (2012)
4. S. Burton, *Sustainable retrofitting of commercial buildings: cool climates*. Routledge (2014)
5. C. C. Menassa, *Evaluating sustainable retrofits in existing buildings under uncertainty*, Energy Build., **43**, no. 12, pp. 3576–3583 (2011)
6. L. Klotz and M. Horman, *Counterfactual analysis of sustainable project delivery processes,* J. Constr. Eng. Manag., **136**, no. 5, pp. 595–605 (2009)
7. E. Asadi, M. G. Da Silva, C. H. Antunes, and L. Dias, *Multi-objective optimization for building retrofit strategies: A model and an application*, Energy Build., **44**, no. 1, pp. 81–87 (2012)

Long-term Application of Carbon Fiber Composite Cable Tendon in the Prestressed Concrete Bridge - Shinmiya Bridge in Japan

Hue Thi Nguyen[1], Hiroshi Masuya[2], Tuan Minh Ha[2], Saiji Fukada[2], Daishin Hanaoka[3], Kazuhiro Kobayashi[4] and Eiji Koida[5]

[1]*Graduate School of Natural Science and Technology, Kanazawa University, Ishikawa, Japan*
[*]*ThuyLoi University, 175 Tay Son, Dong Da, Hanoi, Vietnam*
[2]*Institute of Natural Science and Technology, Kanazawa University, Ishikawa, Japan*
[3]*Kanazawa Institute of Technology, Ishikawa, Japan*
[4]*P.S Mitsubishi Construction Co., Ltd., Ishikawa, Japan*
[5]*Tokyo Rope Manufacturing Co., Ltd., Tokyo, Japan*

Abstract. Carbon fiber reinforced plastic (Carbon Fiber Composite Cable, CFCC) has the outstanding features in comparison with regular steel. In October 1988, CFCC was applied as the tensioning material in main girders of new Shinmiya Bridge in Ishikawa, Japan. This was the first bridge in Japan and in the world, which CFCC tendons were used in the prestressed concrete bridge to counter salt damage. To investigate the serviceability and durability of the main girders and CFCC, three full-scale test girders were fabricated in 1988. At the same time, a bending experiment was conducted on one girder to investigate the ultimate behavior, load carrying capacity of the PC girder, as well the strain behavior of the CFCC. Besides, two PC girders were placed next to the main girders of the bridge in the same conditions. One of them was used for a destructive test after six years of the construction time (1994). In this study, another test specimen that was exposed to the actual corrosive environment after nearly 30 years was subjected to a destructive test by bending load. The load carrying capacity of the girder was clarified, and the durability of the PC girders using CFCC tendon was confirmed.

1 Introduction

The bridges in the coastal area of the Noto Peninsula, Hokuriku region (including Toyama, Ishikawa and Fukui Prefecture) are significantly affected by the flying salinity on the seasonal wind from the Northwest in Japan. As consequences, the deterioration in these bridges due to salt damage is a remarkable problem (see Figure 1). The old Shinmiya Bridge was a reinforced concrete bridge, which had been damaged by salt corrosion. Therefore, the construction of a new bridge became necessary after twelve years of the construction time.

Figure 1. Salt damage due to flying salinity in Hokuriku region.

However, research of Carbon Fiber Reinforced Plastic (CFRP) had started in architectural engineering and civil engineering field from the 1980s in Japan, CFRP was applied as a new material in an alternative to the conventional steel [1]. Carbon Fiber Composite Cable (CFCC) was manufactured and developed by Tokyo Rope Manufacturing Company, Ltd. It is a type of CFRP so CFCC had the similar characteristics of CFRP. CFCC has the outstanding features in comparing with normal steel, it has not only excellent characteristics such as corrosion resistance, lightweight, high tensile strength, high elastic modulus but it was also low relaxation loss, high fatigue strength, low linear expansion and easily coiled flexibility [2].

Therefore, CFCC was selected as the first material for tendons in pre-tensioned prestressed concrete slab system of new Shinmiya Bridge in October 1988. This was the first bridge in Japan and in the world using CFCC tendons in main girders against the salt damage [3]. After that, many of bridge structures were built with this technique in Japan [4]. Furthermore, CFRP was studied and applied not only in Japan but also in other countries in the world [5]. In Canada, the Bendding Trail Bridge was built as the first bridge prestressed with CFCC tendons and Leadline strands in November 1993 in Calgary, Alberta [6]. Abdelrahman and Rizkalla (1997) also conducted the other study about the flexural behavior of concrete beams prestressed by CFRP reinforcement in Canada. The T beams with 6.2 m in length and 330 mm in depth were used, and the CFRP bars had a diameter of 8 mm, the tensile strength of 1970 MPa, and elastic modulus of 147 GPa was applied [7]. In the United States, the Bridge Street Bridge was built in Southfield, Michigan in 2001. It was the first bridge in the USA for using CFCC tendon in T beam and the monitoring systems will be used until 2020 to confirm the long-term value [8]. Besides, the first CFRP application for the a prestressed girder in road-bridge in Belgium was in 2004. Carbon fiber polymer wires were used with 5 mm in the diameter, 2450 MPa in tensile strength and 160 GPa in Young's modulus for a T beam [9]. And more than 140 structures applied CFCC as reinforced material as shown by Enomoto et al. (2012), which confirmed that CFCC was suitable for the bridge construction [10].

CFCC has the excellent properties for prestressed concrete bridges but CFCC has not been widely applied in the world due to the cost and uncertain durability. However, in 2012, Grace et al. showed that despite the higher initial cost of CFRP purchasing, the life cycle cost of the bridge was not high when compared to traditional steel reinforced bridges. The cost of CFCC will become the least expensive option after 20–40 years of service time[11]. Regarding the durability, there was only a limited data on the long-term value of prestressed concrete bridges using CFCC tendon, especially the structure has full-scale and was exposed to the actual corrosive salt environment.

In the first bridge in the world using CFCC tendons-Shinmiya Bridge, the aim of investigation the durability and serviceability of the main girders using CFCC tendon was performed from the construction time (1988). Three test girders with the same size of the main girders were fabricated in 1988. At the same time, in order to investigate the ultimate behavior and load carrying capacity of the PC girder using CFCC as tendon, a bending experiment was conducted on one girder [12], and the deflection of the main girder, and the strain behavior of the CFCC were shown [13]. Two other PC girders (mountainside, seaside) were placed next to the main girders of the bridge in the same conditions for confirming the long-term quality. Six years after the construction time (1994), the destructive test was carried out using the seaside girder, the comparison of the load carrying capacity with the value in the time of construction was done, and the durability of the CFCC tendon have been confirmed [14]. In this study, a destructive bending test was conducted on the mountainside girder, which was exposed to an actual corrosive salt environment for nearly 30 years (2017), and the results of the load carrying capacity and durability have been evaluated and shown.

2 Description of Shinmiya Bridge

The Shimiya Bridge was built at the Shika-machi, Hakui-gun, Ishikawa Prefecture, Japan in October 1988. Figure 2 and Figure 3, respectively show the side view and cross-section of Shinmiya Bridge. The bridge had 6.1 m in length and 7.0 m in the effective width.

The main bridge has a structure, in which 24 main girders as shown in Figure 3 with the I-shaped cross-section (JIS A 5313, S106-325) shown in Figure 4. For the aim of confirming the long-term quality, two full-scale test girders on both sides of the bridge (mountainside, seaside) were fabricated, and transversely tightened together with the main girders in the same environment. When six years passed (1994) after construction, the test girder on the seaside was taken out. The destructive experiment of this girder by bending load was conducted. And transmission length test, salinity measurement, tensile test and chemical composition test were done after removing of tension by CFCC [14]. In this research, other test girder on the mountainside was removed from the main bridge and the destructive experiment of the girder by bending load was conducted.

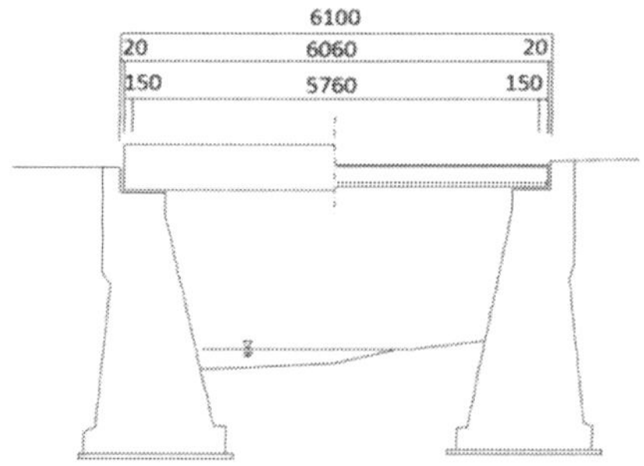

Figure 2. Side view of Shinmiya Bridge (unit:mm).

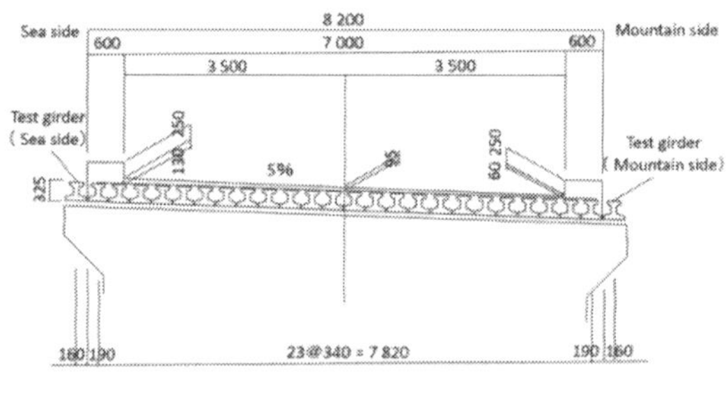

Figure 3. Cross-section of Shinmiya Bridge (unit:mm).

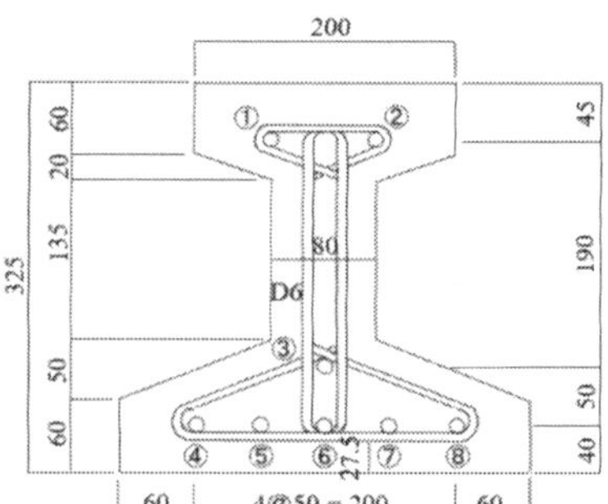

Figure 4. Cross-section of main girder (unit: mm) (o: CFCC).

With respect to the test specimen, the detailed dimension of the mountainside girder was shown in Figure 2, Figure 3 and Figure 4. The length of the test girder was 6060 mm, the height of it was 325 mm while the width of the top flange was 200 mm, and the width of the bottom flange was 320 mm. The compressive strength of concrete in the past was recorded of 59.8 N/mm2. The stirrup used epoxy coated rebar of D6 as a method against salt damage. Eight straight CFCCs (Carbon Fiber Composite Cable) with a diameter of 12.5 mm, and seven strands were employed for tendons. The surface of CFCC tendons were twisted and roughened to enhance the adhesion effect with concrete

3 Bending test

3.1. Outline of experiment

The destruction test by bending load was conducted according to the same method of the test, which was done at six years passed after construction. The outline and the test setup of bending experiment are shown respectively in Figure 5 and Figure 6. The load was applied at the two points in the middle of the span with a distance of 1000 mm and two steel rollers supported for the girder. Measuring items including load, displacement, and strain at the center and at the support positions of the main girder were arranged as the Figure 7.

The loading procedure was the same as t that was done in the loading test at six years passed after construction. First, the load was applied up to 35.4 kN (design load 35.3 kN) and returned to 0 kN. Secondly, the load was loaded to crack initiation load and returned to 0 kN again. Finally, load was loaded to the ultimate load as the final stage.

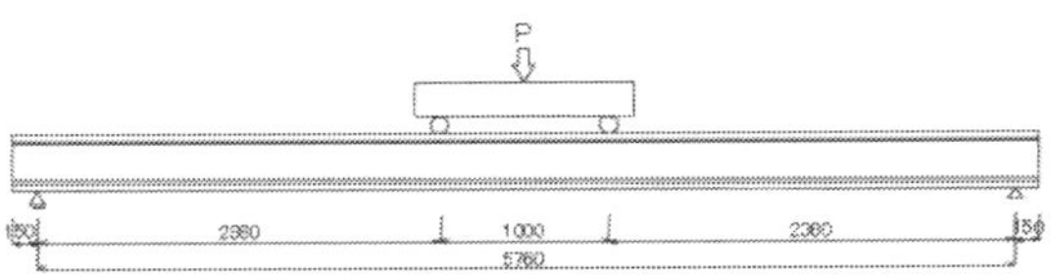

Figure 5. The outline of experiment.

Figure 6. The setup of bending experiment.

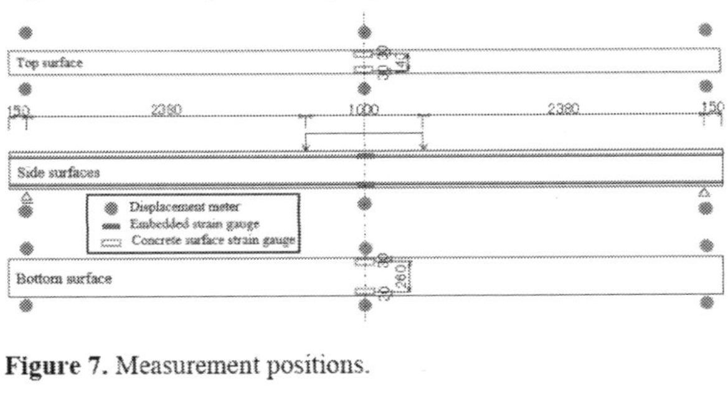

Figure 7. Measurement positions.

3.2 Flexural behavior

The situation of the PC girder after failure is shown in Figure 8. The relationship between load and displacement at the middle of the span, the relationship between load and strain, and the crack pattern obtained by the destructive test are shown in Figure 9, Figure 10 and Figure 11, respectively. In addition to the results of this loading test, crack initiation load and bending fracture load at the time of construction, six years after construction, and design values are summarized in Table 1.

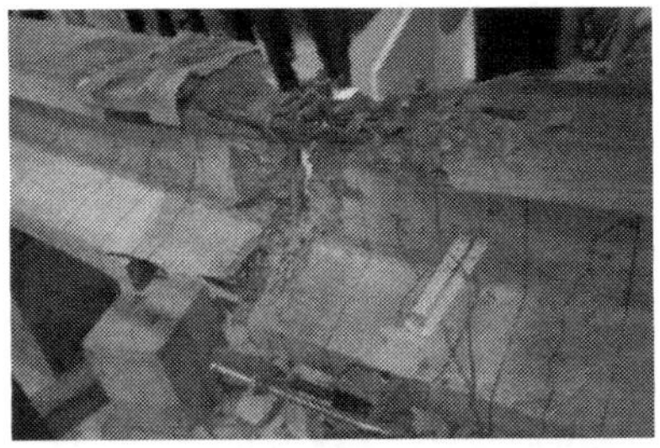

Figure 8. The situation of the girder specimen after failure.

Table 1. Comparison of crack initiation load and ultimate load.

	Design value	1988	1994	2017
		At the time of construction	After 6 years	After 29 years
Cracking initiation load (kN)	68.3	70.6	98.3	82.8
Ultimate load (kN)	131.2	132.3	167.1	157

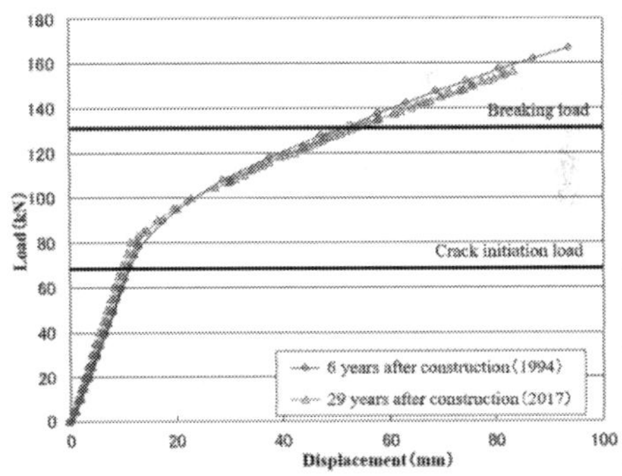

Figure 9. Relationship between load and displacement at mid-span.

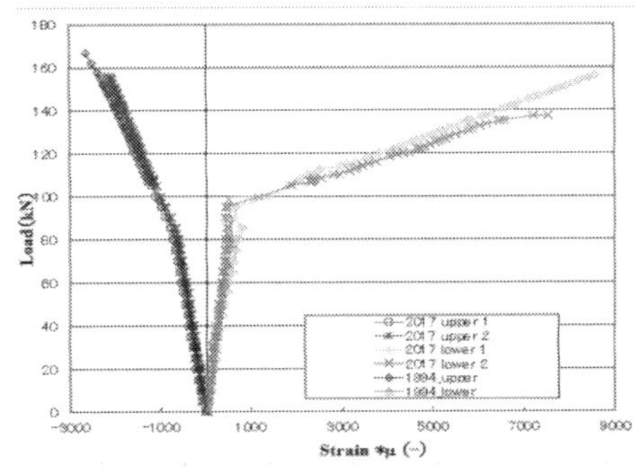

Figure 10. Relationship between load and strain of concrete.

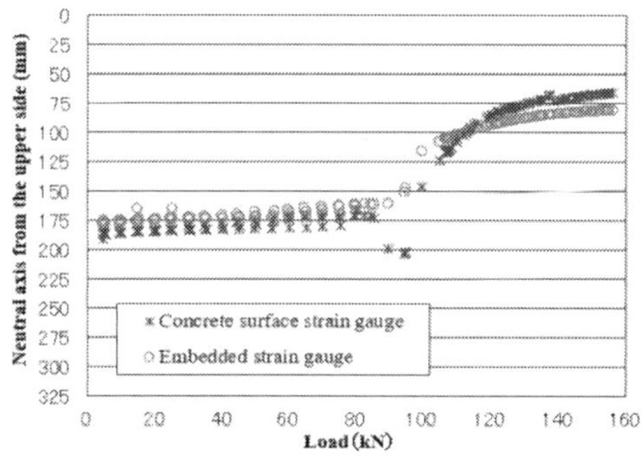

Figure 11. The neutral axis depth.

The failure mode imposed in this loading test was the crushing of concrete in the compressive area after the bending crack propagation (see Figure 8). This occurred because the CFCC had no yield phenomenon, the girder was designed to be a destructive form, which the top flange of the concrete collapsed before CFCC ruptured. In this experiment, the destruction was as the same with design. This destruction was similar to that obtained in the tests at the time of construction time and six years after construction.

The situation of the PC girder surface after the concrete collapse and CFCC after the destruction test were observed. It was found that the CFCC in the upper side was twisted and ruptured after crushing of the concrete. However, at the CFCC of the lower side, failure due to tension could not be confirmed.

3.3 Load-displacement relationship

From the relationship of load-displacement shown in Figure 9, when the load was unloaded after loading up to design load and crack initial load, almost no residual defection was observed. The relationship of load-displacement obtained by this examination was overlapped with the results at the time of six years passed after construction.

This relationship was the linear until the initial crack occurred at the load of 82.8 kN (approximately 50 percent of the final load). In comparing with the applied load of the six years after construction time, the initial crack load decreased by 15.7% but this number was also higher than the design value (68.3 kN). In addition, the ultimate load and the mid-span displacement were 157 kN and 84 mm, respectively. The value of final load reduced by about 6% as compared with the result obtained in the bending test in 1994 (six years after construction time). However, as shown in Table 1 this result was superior to the value at the construction time (132.3 kN) and the design value (131.2 kN). These results revealed that there was no problem in term of the load carrying capacity of the main girder.

3.4 Strain distribution

As can be seen from the Figure 10, the relationship between applied load and strain obtained by this experiment was close to the results in 1994.

Figure 12 shows the results of neutral axis depth calculated from the concrete surface strain gauge and embedded strain gauge. It assumed the linear distribution of strain in the cross-section for the neutral axis calculation. In the elastic stage, the neutral axis depth was almost similar between two results. After 82.8 kN, the values of neutral axis were a slight difference but both of them had the trend toward the compression zone with an increase of the applied load. This difference may be due to one of the concrete surface strain gauges on the lower side that was broken before ultimate load and the strain distribution on the cross-section was not linear. In term of the result calculated from embedded strain gauge, the strain of concrete at the middle span was 0.0011 in compression and 0.0060 in tensile at failure. The neutral axis depth of test girder was approximately 81 mm at 156.2 kN.

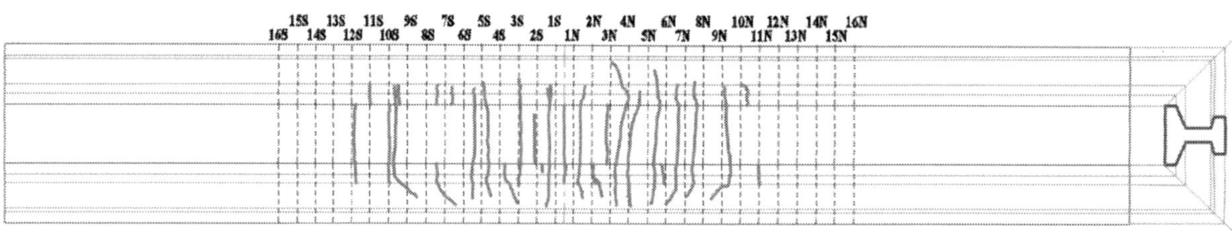

Figure 12. The neutral axis depth.

4 Conclusions

The first destructive experiment was conducted on the full-scale PC girder using CFCC tendons, which was exposed for approximately 30 years in the actual corrosive salt environment. The durability and load carrying capacity of PC girder were clarified. The remark conclusions were given follow.

- The mode of failure in this study was the crushing of concrete followed by the rupture of CFCC tendon in the compression zone. The result was similar to the failure modes in the construction time (1988) and six years after construction time (1994).
- The cracks occurred mainly in the middle of girder zone. It began in the bottom of flange toward the loading points.
- From the results of the comparison regarding the load-displacement relationship, load-strain relationship obtained by the destructive bending test and the experiment at six years after construction, it has become clear that there was no significant

https://doi.org/10.1051/matecconf/201820602011

difference in the shape of relationship curves. Although, the ultimate load in this experiment was about 6% lower than the data at six years after construction. However, this value was also higher than the design value. It was judged that there was no problem in load carrying capacity after nearly 30 years.

- From the results of previous studies and this outcome, it can said that CFCC is suitable to replace the regular steel in bridges, especially the prestressed concrete bridges constructed in corrosive environments to combat salt damage.
- Furthermore, with the reasonable cost and the durability confirmed, CFCC should be studied and applied more in the field of architectural engineering and civil engineering worldwide

Acknowledgment

This research was conducted as part of the Cross-ministerial Strategic Innovation Program (SIP). (R & D subject: Resolution of Early-aged Deterioration Mechanisms in Concrete Bridges and Development of Total Management System Based on Evaluation for Material and Structure Quality Performance, Chairman: Kazuyuki Torii). We are deeply grateful to everyone concerned and supported

References

1. JSCE, Recommendation for design and construction of concrte stuctures using continuous fiber reinforcing materials,(1998).
2. H. Kimura, N. Shiratori, and H. Takagi, Concr. Eng. Annu. Rep. 12-1PP (1990) (in Japanese)
3. N. Santoh, H. Kimura, T. Enomoto, T. Kiuchi, and Y. Kuzuba, Fiber-Reinforced-Plastic (FRP) Reinforcement Concr. Struct. Proc. Int. Symp. ACI, SP-138, pp. 895 - 911 (1993)
4. V. M. Karbhari, Int. Technology Research Institute, (1998).
5. P. B. Potyrała, Thesis, (2011)
6. S. H. Rizkalla and G. Tadros, Concr. Int., vol. 6, no. 4, pp. 37–42, (1994)
7. A. A. Abdelrahman and S. H. Rizkalla, ACI Struct. J., no. 94, pp. 447–454, (1998)
8. N. F. Grace, F. C. Navarre, R. B. Nacey, W. Bonus, and L. Collavino, PCI Journal, 47 5, pp. 20-35., (2002)
9. W. D. Corte and P. Van Bogaert, Compos. Part B Eng., vol. 36, no. 2, pp. 91–98 (2005)
10. T. Enomoto, N. F. Grace, and T. Harada, Proc., Conf. FRP Compos. Civ. Eng., vol. Rome Italy (2012)
11. N. F. Grace, E. Jensen, C. Eamon, T. Enomoto, and X. Shi, Proc. CICE 2012 6th Int. Conf. FRP Compos. Civ. Eng., pp. 1–8 (2012)
12. T. Yamashita, T. Kiuchi, H. Inukai, and T. Iwasaki, Prestress. Concr. Vol. 31, No.2, pp. 71-78 (1989) (in Japanese)
13. T. Futakuchi, S. Komada, T. Kiuchi, and K. Matsumoto, Proc. 2nd Symp. Prestress. Concr. Tech. Assoc. pp. 307-310 (1991) (in Japanese)
14. H. Kanda, T. Kiuchi, and K. Matsumoto, Proc. 5th Symp. Prestress. Concr. Tech. Assoc., pp. 529–532, (1995) (in Japanese)

Analysis on the Characteristics of Ordinary and More Serious Accidents in Building Projects in China from 2005 to 2016

Hui Zhang , Xuewei Ji and Aizhi Wu

Beijing Academy of Safety Science and technology, Beijing 100070, China

Abstract: In China, the frequent safety accidents in building projects seriously affect the sustainable development of construction projects. The paper makes a statistical analysis on the characteristics of ordinary and more serious accidents in building projects from the types of accidents, the trends of development and the characteristics of geographical distribution from 2005 to 2016 in China by using the histogram, trend chart, pie chart and table. It can be concluded that in the recent 12 years, the security situation of construction projects generally fluctuates in a big way and is still not optimistic. Among the five major types of accidents, the electric shock and the lifting accidents have been maintained at a relatively stable and low incidence, and the sloughing accidents have accounted for more than half of the proportion. Accidents concentrated in East China, Central China, North China and Southwest China. The economic development in all provinces and cities in China is positively related to the accident. The greater the total regional economy, the rate of economic growth and the demand for infrastructure construction, the more deaths and accidents in construction projects in the corresponding regions.

1 Introduction

The large-scale construction projects in China have employed a large number of workers which are the largest working groups in the world [1]. Since 2001, macro-economy has entered a new round of ascent cycles, and the number of construction projects has also skyrocketed in China. In the Construction Engineering, which is a branch of China's construction industry, the data published by the National Bureau of Statistics shows that, the added value of China's construction industry amounted to 17.8591 trillion yuan from 2003 to 2012 [2], the number of employees increased from 38.5247 million in 2011 to 43.5679 million in 2015, and the total output value increased from 11.663332 trillion yuan in 2011 to 18. 757 trillion yuan in 2015 [3]. However, in fact, China's construction projects also include civil engineering, pipeline and equipment installation and renovation works. With the expanding scope and scale of engineering construction in our country, the natural environment, construction types and structures confronted by construction are more diversified, increasing the engineering construction technology challenge [4], as a result that the number of accidents and casualties have not been effectively contained, causing great economic losses to the construction and negatively affecting the social harmony and stability. Therefore, combining the data of construction safety accidents occurred in China in recent years, this paper analyzes the type characteristics, geographical features and development trends of the accidents, which have reference and guidance significance to the control and prevention of construction accidents in China.

Relevant experts and scholars in the statistical analysis of construction safety accidents in China have made a series of achievements: Li Yanjie, Yang Liqiong and others [5] count the accidents occurred in China's construction industry from 2007 to the first half of 2017, analyze the regulation of the accident from the accident type, the occurrence time and the occurrence season, basing on the overall status of the accidents which have occurred, and reached a conclusion. Wu Yongfeng [1] analyzes the security incidents occurred in China from 2004 to 2012, through charts, histograms, pie charts and other forms, finds the law of accident and puts forward specific suggestions on the safety management of the industry. Zhang Tong and others [6] analyze building accidents in China from 2003 to 2012, from the correlation between the characteristics of the accident and economic development trends, and put forward corresponding improvement measures for the prevention and control of the construction accidents. Hao Zhong and Wei Yanxiao [7] analyzed the rules and characteristics of the 448 accidents from the degree of injury, type of injury, working type of accident and so on, and put forward measures and suggestions to prevent and control occupational injuries, basing on the rules and characteristics of the accident. Yuan Shuang [8] collectes 8293 safety accidents occurred in China's municipal housing project from 2003 to 2012., with "The People's Republic of China Ministry of Housing and Urban-Rural

Construction Injury Express Letters" as the core data source, extracts the time, region, grade, etc. from the accident to establish an absolute index system, at the same time, obtains regional construction output and practitioners'number from the China Statistical Yearbook and local statistical yearbook in 2003- 2012, and calculats the output index of 10 billion yuans and the death rate of 100,000 employees, establishing a relative index system to conduct a comprehensive evaluation on the fluctuation characteristics and development trend of the management level of safety accident. Wang Dan, Cui Li Miao and others analyze the construction accidents information in China from 2011 to 2015 by using SPSS in terms of the month, time period, provinces and types of accidents, and propose measures to reduce the construction accidents in China to provide a reference for safety management of building enterprise. Based on the theory of safety system engineering, safety management engineering and safety evaluation, Sun Jian [9] analyzes the characteristics of safety accidents that may occur during the construction operations and discusses the safety management and comprehensive evaluation of construction projects from the perspective of man-machine-environment, using analytic hierarchy process and gray comprehensive evaluation method. Shui Zhengxiang [10] automatically collectes and manages 318 typical cases of major accidents in China from 2005 to 2014 and extractes accident keywords by using software and analyzes the distribution characteristics from time, accident type and region by using statistical analysis method to find the law of the accident occurrence, with "The notification of production safety accident in housing municipal engineering " issued by Ministry of Housing and Urban-Rural Development of the People's Republic of China as the main source of data.

Based on the data of construction accidents obtained from the Accident Inquiry System of the State Administration of Work Safety, this article counts 520 ordinary and more serious construction accidents in China from 2005 to 2016, including the construction of bridges, railways and ports, housing construction projects and underground tunnel construction, and explores the space-time characteristics and development trend of the accidents from time, type and geographical area to guide the construction and prevention of ordinary and more serious construction accidents in China.

2 The basic characteristics of ordinary and more serious construction accidents

According to the types of accidents including falling, collapse, object strike, lifting injury, electric shock and others, this section calculated the number of accidents and deaths from ordinary and more serious construction accidents in China during 2005-2016 to get the trend of the total number of accidents deaths. The statistical results are shown in Table 1.

It can be seen from Table 1 and Figure 1 and Figure 2 that the number of accidents and deaths of ordinary and more serious construction accidents in China from 2005 to 2016 have not shown the increasing or decreasing trend regularly and have fluctuated obviously. In recent 12 years, the trend of construction engineering in China has the following characteristics:

Table 1. The statistics table of ordinary and more serious construction accidents in China from 2005 to 2016.

Year	The number of accidents	Accident reduction rate (%)	The number of deaths	The decreasing rate of deaths (%)
2005	40	——	152	——
2006	46	-15.0	168	-10.5
2007	36	21.7	148	11.9
2008	52	-44.4	214	-44.6
2009	40	23.1	169	21.0
2010	55	-37.5	245	-45.0
2011	47	14.5	187	23.7
2012	42	10.6	184	1.6
2013	44	-4.8	170	7.6
2014	51	-15.9	194	-14.1
2015	31	39.2	118	39.2
2016	36	-16.1	222	-88.1
Total	520	——	2171	——

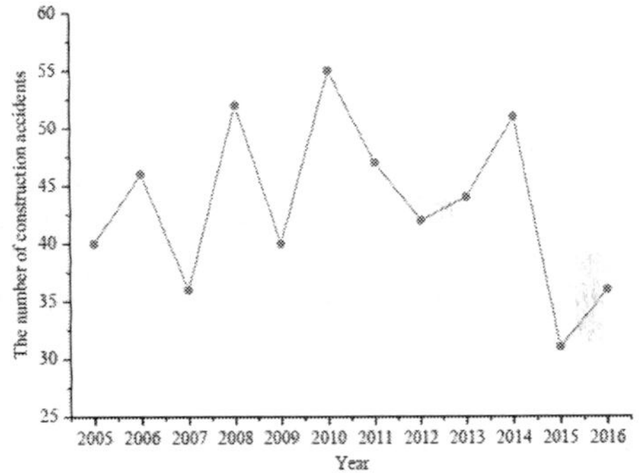

Figure 1. The trend of the number of accident.

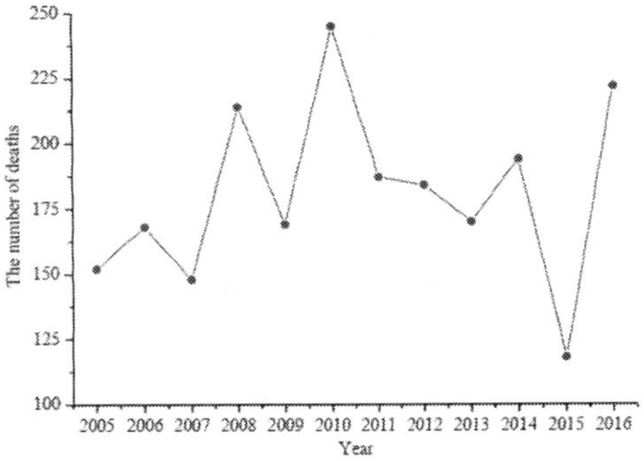

Figure 2. The trend of the number of deaths.

(1) In the ordinary and more serious construction accidents in China from 2005 to 2016, the trend of the total number of accidents and deaths is almost the same. In other words, the total number of deaths in the year also increased when the total number of ordinary and more serious construction accidents increased in a year was larger than that in the previous year; When the total number of accidents in a year decreased, the total number of deaths corresponding to the year also decreased compared with that of the previous year. However, with the exception of the statistics in 2013, the number of accidents increased during the year but that of deaths decreased compared to 2012.

(2) The total number of accidents and deaths in 2010 was the highest in 12 years between 2005 and 2016, with a 37.5% increase in accidents and a 45% increase in deaths compared to 2009, showing that in 2010, the construction safety inChina is extremely poor. Compared with the data in 2010, accidents from 2005 to 2010 can be assumed to show a increasing trend year by year.

(3) The incidence of accidents between 2010 and 2016 can be approximated as showing a downward trend. In particular, a good construction safety was shown in 2015, with a drop of 39.2% in the number of accidents and deaths in the year, which is the lowest annual accident count and death tolls in the past 12 years. Although showing a slight rebound in 2016 compared with 2015, these two indicators generally remained at a relatively low level.

The number of construction accidents has fluctuated rather than showing a steady decline since 2005. But the number of accidents and deaths in 2015 and 2016 have dropped significantly compared with the previous 10 years, which showed the results of strengthening construction safety management in China in recent years. However, the occurrence of accidents has not been fundamentally deterred, which is the key that restricts the increasing of the safety management and construction in China.

3 Type characteristics of the accident

"Lifting injury, falling, objects strike, electric shock, collapse" is known as the five major construction safety injuries. In the ordinary and more serious construction accidents in China from 2005 to 2016, in addition to object strike, the four major types of injuries occupied the major part of the accident. In order to analyze the annual variation trend of different accident types and the changes of death toll better, this section classified and counted the ordinary and more serious construction accidents in China in the past 12 years. The chart of accident types the annual trend of the different types of accidents are shown in Figure 3 to 6.

According to the statistical results, only considering collapse, object strike, lifting injury, electric shock and falling, 307 ordinary and more serious construction accidents have occurred, with 1,322 deaths in the past 12 years. Among these, there were 197 collapse accidents and 858 deaths in them, accounting for 64% and 65% of the total respectively. There were 84 falling accidents and 372 people died in them, accounting for 27% and 28% respectively. There were 15 lifting injuries and 56 people died, accounting for 5% and 4% respectively. There were 11 electric shock accidents and 36 deaths, accounting for 4% and 3% respectively.

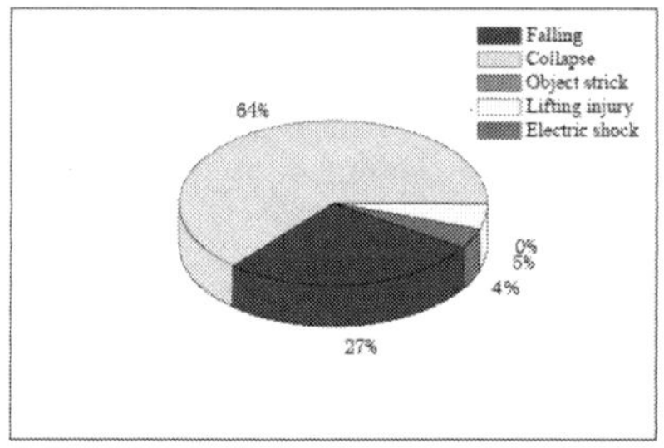

Figure 3. Scale of accident type.

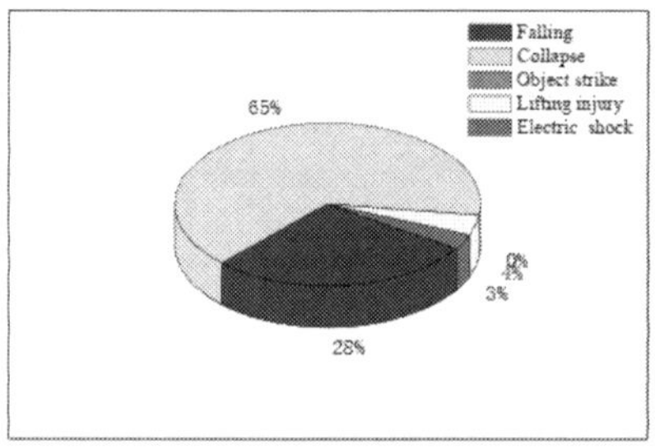

Figure 4. Annual trend of the different accidental types.

It can be seen from Figure 3 and Figure 4 that the collapse accidents all took the first place, far exceeding those of the other three Kinds of accidents, no matter the number of accidents or the number of deaths, among the five major types of construction accidents during 2005-2016. Fallings, lifting injuries and electric shock accidents came in second, third and fourth respectively. Thus, collapse accident is the most common form of injury and the easiest event to happen in construction, but also the focus of accident prevention.

As can be seen from Figure 5 and Figure 6, the number of deaths, involving falling accidents, lifting injuries and electric shock accidents, showed the same trend with the number of accidents in each year. Collapse accidents were the most volatile, especially in 2015 and 2016. There were 22 collapse accidents in 2015 with 88 deaths and 16 collapse accidents with 136 deaths in 2016. It can be seen that although the number of accidents increased slightly in 2015 as compared with 2014, the total number of deaths decreased, indicating that the seriousness of accidents in 2015 was not high. And although the number of accidents in 2016 markedly decreased, the total number of deaths increased significantly, indicating that most of the accidents occurred in 2016 were destroyous accident and tremendous devastating accident, indicating thatthe construction safety situation in China remains very serious. From the trend chart, it can be seen that, for electric shocks and lifting accidents from 2005 to 2016, the fluctuations in the number of accidents and the deaths are both very small and at a relatively low and stable level. The annual average number of accidents and deaths in lifting accidents are 2 and 5, and those in electric shocks are 1 and 3. Due to the fact that collapse accidents occupy a larger proportion in all types of accidents than before, the focus of construction safety management should be on the prevention of collapse accidents and falling accidents in the future.

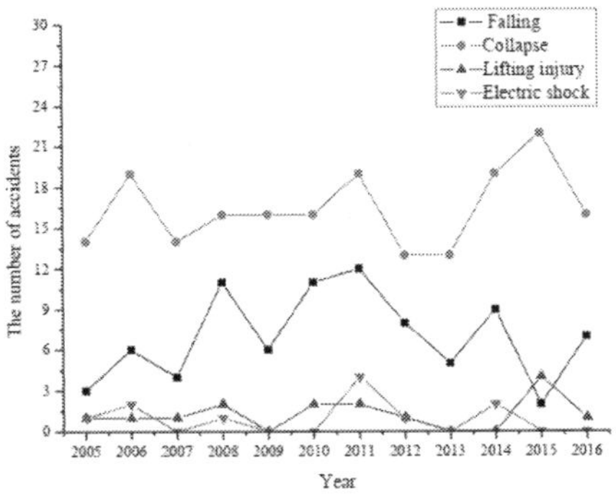

Figure 5. Annual trend of various accidental types.

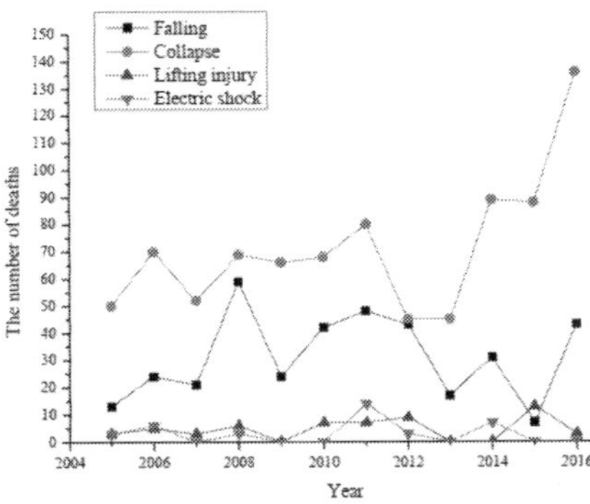

Figure 6. Trend of death toll of each accident type.

4 Regional characteristics of the accident distribution

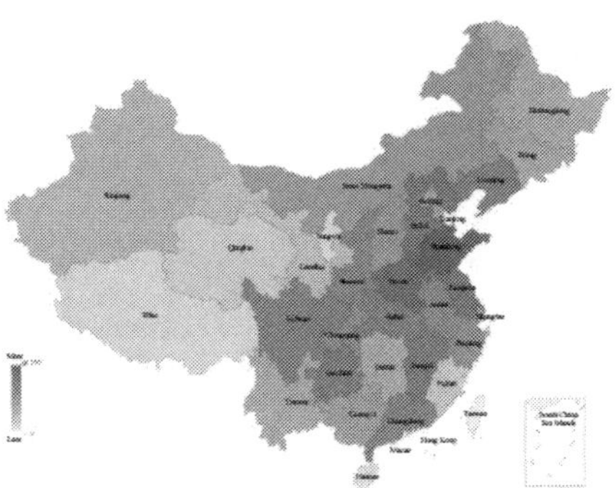

Figure 7. The distribution of the number of deaths in construction in different provinces from 2005 to 2016.

The statistics of the deaths of ordinary and more serious construction accidents in various provinces (cities) in China are shown in Figure 7.

It can be seen that from 2005 to 2016, the total number of ordinary and more serious construction accidents in some provinces, including Shandong Province, Hebei Province, Henan Province, Jiangsu Province, Guizhou Province, Sichuan Province, Guangdong Province, Zhejiang Province, Liaoning Province, Hubei Province, Chongqing Municipality and Shaanxi Province, are decreasing from year to year and are also above 20. A total of 319 ordinary and more serious construction accidents occurred in these 12 provinces (cities), with a total of 1,260 deaths, accounting for 61.3% and 87.2% of the national total respectively.

The statistical results of the death toll of ordinary and more serious construction accidents in various provinces in China from 2005 to 2016 are shown in Figure 7. As can be seen from the figure, the distribution of construction accidents in China shows obvious and regional characteristics. Among them, the security situation in some areas is severe, including Shandong, Jiangsu, Zhejiang in East China, Henan, Hubei, Hebei in North China, Sichuan, Guizhou, Chongqing in Southwest China, Guangdong in South China, Shaanxi in Northwest China and Liaoning in Northeast China. The number of accidents and the number of fatalities of ordinary and more serious construction accidents in these 12 provinces from 2005 to 2016 were plotted as a histogram. The results are shown in Figure 8.

Figure 8 show that the total number of deaths from 2005 to 2016 in the 12 provinces was significantly higher than those in other regions, with the total number of deaths in some provinces of the 12 provinces, including Shandong, Hebei, Henan, Jiangsu, Guizhou, Guangdong, Liaoning, Hubei, reaching over 100, were 149, 113, 127, 116, 105, 100, 103, 100. The average number of deaths per accident was about 4, while the largest average number of deaths per accident which are were 9.82 in Jiangxi Province, 6 in Shanghai, 5.75 in Qinghai and 5 in Tibet. This shows that the security accidents in China are not only concentrated in areas with rapid economic development but also in areas with relatively backward economy, where the security situation is also harsh due to geographical culture, local topographical conditions and inadequate supervision by local governments.

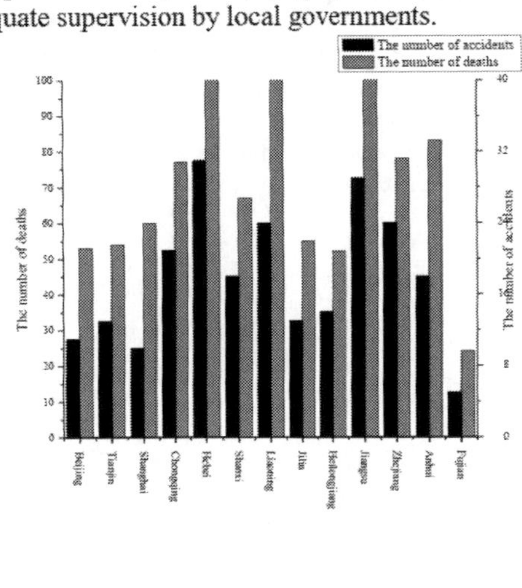

Figure 8. The total number of construction accidents and deaths in the ordinary and more serious construction accidents in different areas.

In addition, from 2005 to 2016, the top four provinces are Shandong, Henan, Jiangsu and Hebei no matter from the number of accidents or the total number of deaths in 2005-2016. According to the data released by the National Bureau of Statistics of China, the GDP of the whole country reached 68,905.21 billion yuan in 2015, of which the GDP of Jiangsu Province reached 7,011.638 billion yuan, the GDP of Shandong Province reached 6,300.233 billion yuan, the GDP of Henan Province Reached 3,700.216 billion yuan, and the GDP of Hebei Province reached 2,980.611 billion yuan. Thus it can be seen that the large number of construction accidents in China are concentrated in the provinces with large population where the economy is more developed and the rate of development is increasing. The larger the total regional economy, the growth speed of economic and the demand for infrastructure construction, the greater the number of deaths and incidents of construction accidents. Therefore, the future work should focus on strengthening the safety supervision of construction in these 12 provinces, especially in the four provinces of Shandong, Henan, Jiangsu and Hebei.

5 Conclusion

(1) The security situation of construction in China did not show a steady trend of improvement year by year, with the number of accidents and deaths fluctuating greatly. Among them, from 2005 to 2016, both the number of accidents and the death toll in 2010 were the largest in nearly 12 years. Both indicators dropped significantly in 2015 reaching the lowest point in 12 years. The number of construction accidents in 2016 slightly increased, but the death toll rebounded significantly.

(2) From the five major types of accidents such as collapsing, falling, electric shock, object striking and lifting injury, it can be found that among the ordinary and more serious accidents in China from 2005 to 2016, the collapse accidents accounted for more than half of them; electric shock and lifting injuries have been maintained at a relatively stable, low incidence. In the future, it is possible to improve the intrinsic safety of the system from the technical point of view during the construction operation, and to control the accidents that occured frequently to improve the safety of the construction work.

(3) As can be seen from the geographical features of the construction accidents, the 61.3% of the ordinary and more serious accidents in China are concentrated in Shandong Province, Hebei Province, Henan Province, Jiangsu Province, Guizhou Province and Sichuan Province , Guangdong, Zhejiang, Liaoning, Hubei, Chongqing and Shaanxi provinces. From the number of accidents and the total number of deaths, the top four are Shandong Province, Henan Province, Jiangsu Province and Hebei Province. According to the data released by the National Bureau of Statistics of China, the GDP of these four provinces in 2015 was 6.3002 trillion yuan, 3.700216 trillion yuan, 7.011638 trillion yuan and 2.980611 trillion yuan respectively. It can be seen that there is a positive correlation between the economic development and the accident of each province (city) inChina. The large number of construction accidents in China are concentrated in the provinces with large population where the economy is more developed and the rate of development is increasing. The larger the total regional economy, the growth speed of economic and the demand for infrastructure construction, the greater the number of deaths and incidents of construction accidents.

Reference

1. YF. Wu, MH. Yuan, Value Engineering **33(21)**, 96-98 (2014).
2. XM Li, China High-tech Enterprise **20**, 67 (2013).
3. D Wang, LM Cui, YL Zhang, Engineering Economics **3 (27)**, 38-41(2017).
4. YR Xing. *Research on the causal relationship between construction accidents [D]* .Beijing: School of Construction and Management, Tsinghua University (2010).
5. YJ Li, LQ Yang, X Huang, Engineering Science and Technology **22**, 132-133 (2013).
6. T Zhang, DL Sun, Y Jiang, Construction Safety **8**, 18-21 (2013).
7. Z Hao, YX Wei, China Standardization **2**, 245-248(2017).
8. S Yuan. *Analysis on the characteristics and countermeasures of safety accidents of municipal buildings in China [D]*. Chongqing: College of Construction Management and Real Estate, Chongqing University (2014).
9. J Sun. *Construction enterprise safety management and evaluation [D]*. Jiangxi: Jiangxi University of Science and Engineering (2009).
10. ZX Gong. *Research on the ordinary and more serious accidents in China [D]*. Jiangxi: School of Architecture and Civil Engineering, Nanchang University (2015).

Design Experiment for Leveraging Material Value Coal Ash as Material Mixture of Paving Block

Harris Setyo Fernanto , *Djoko Sihono* Gabriel

Department Industrial Engineering, University of Indonesia, Depok, Indonesia

Abstract. The coal-fired thermal power plant is the main source of coal ash production. Coal ash collected at the bottom of the furnace (boiler) is called coal bottom ash (CBA) and fly ash remaining combustion called coal fly ash (CFA). In Indonesia, the utilization of bottom ash and fly ash generally is still limited to landfill material due to the value of the material too low. The objective of this research is to leverage the value of the utilization of bottom ash and fly ash materials as a mixer of construction industry products. Design of Experiment (DoE) was conducted on paving block product using various experimental factors such as composition of sand, cement, bottom ash, fly ash, and curing period of product. This Design of Experiment (DoE) applied orthogonal array to finding the best mixed compositions that affect the quality of the end result. After it, the proof was performed with a compressive strength test and water absorption test to meet the standard requirements. Testing compressive strength of paving block was done at 7 days, 14 days and 28 days to get the best result.

1 Introduction

Coal is one of the energy sources used in Indonesia, based on the 2015 estimate of power plants is an industry that ranks first as a coal user as a fuel. Once used in a burning medium (boiler), coal turns into fly ash (fly ash) and ash that falls into the bottom of the combustion chamber (bottom ash) [1]. The old paradigm regarding the utilization of fly ash and bottom ash as landfill hoarders due to too low the value of the material so it is not interesting to use again. Therefore, a new paradigm is considered to consider the leverage of fly ash and bottom ash materials by mixing with other materials to produce better construction products. Both types of materials are classified as waste electricity production, but research utilization is still relatively small. As for existing research, it only shows that fly ash and bottom ash can be used as a substitute material for cement and sand [2], but no one has shown the right experimental method to replace the material in producing the best quality, so research needs to be done to increase the value of performance on the value of economic efficiency.

This research aims to obtain the fly ash & bottom ash utilization as a paving block industrial product mixer which gives better material value and also seeks evidence of compliance with the Standard National Indonesia (SNI), and then calculates the performance that can be gained from the utilization of fly ash and bottom ash.

This research is limited to the use of fly ash and bottom ash for the construction industry which is used as raw material for paving block product substitution, with explanation of performance test result using laboratory test.

2 Material and Methods

1.1 Materials

The concrete prepared and tested in this study was composed of Ordinary Portland cement (OPC) ASTM C 150-04a, crushed sand (mixing stone ash & bangka sand), fly ash and bottom ash. Class F fly ash and Bottom ash were obtained from PT. Indonesia Power, Steam Power Plant Station, Lontar, Indonesia. The properties of CBA & CFA showed in Table.1 CBA was dried in an oven at a temperature of 110 C for three hours before use to ensure it is in dry state.

Table 1. Properties of fly ash and bottom ASH

FA	SiO_2	Al_2O_3	Fe_2O_3	TiO_2	CaO	MgO	K_2O	Na_2O	SO_3	MnO_2	P_2O_5	LOI
%	34,29	16,62	15,38	0,73	18,18	7,52	1,35	2,97	1,63	0,17	0,25	0,36

BA	SiO_2	Al_2O_3	Fe_2O_3	TiO_2	CaO	MgO	K_2O	Na_2O	SO_3	MnO_2	LOI
%	34,39	10,02	18,41	0,65	21,16	9,70	0,90	0,24	0,66	0,22	3,54

1.2 Methods - Design of experiments

Various samples were prepared using different methodologies based on experimental design to predict the interaction component of the concrete mix. Experimental design is an important tool that can be used for product or process development and improvement [3]. In this study the effect of five control factors on the response, compressive strength of concrete, was investigated. All continuous variables were studied at 3 levels. Control factors levels used in this designed experiments were: % of cement mixing (5%, 20%, 30%), CFA (0%, 15%, 30%), sand (15%,40%,85%), CBA (0%,70%,85%) and length of curing period/age (7, 14, 28 days). Table 2 below summarizes factors considered along with number of levels and level settings.

Minitab 18.0 software is used create and design the DOE with orthogonal array method that resulted running in Table 3, summarizes the list of 27sets of mixtures with 1 set replication that were prepared. Experimental results along with design matrix are tabulated. The order in which the experiments were performed was selected at random so that this design is completely randomized. The compressive strength of each of the samples was measured and evaluated for each case as the response variable.

Table 2. DOE factors

DOE Factors

Factor	Level 1	Level 2	Level 3
% Cement	5%	20%	30%
% CFA	0%	15%	30%
% Sand	15%	40%	85%
% CBA	0%	70%	85%
Curing Period (Days)	7	14	28

Water content was adjusted until the required flow capability was achieved. The fresh mixture was then poured into paving mould of size (21cm x 10.5 cm x 6 cm), its top surface was leveled and kept covered with a wet cloth for two days in the laboratory condition after which the specimens were removed from the moulds and transferred to curing condition of 95% relative humidity at 22 C. The curing condition was achieved by keeping the paving block in partially water filled closed storage boxes and keeping the boxes in an air conditioned room as mentioned in [4].

Some nuisance (noise) factors were also identified, these are variables that probably have some effect on the response, but which are of little or no interest to the experimenter. They differ from held-constant factors in that they either cannot be held entirely constant, or they cannot be controlled at all. They include all composite batch-to-batch variation, and ambient temperature of the room and vibration of table while compaction and mixing. Other factors were predicted to exert some effect on the responses, but for the current study these factors are not of interest, so they have been held constant at a specified range. Examples of these factors include type of all composite, operator, water content in mix and cement to sand ratio. Table 4 outlines the noise factor and noise management strategy of each one. A standard Hobart mixer was used for the mixing process.

Table 3. Orthogonal array result

Minitab - Paving Block Experiment.mpj - [P...

	C1 Cement	C2 CFA	C3 Sand	C4 CBA	C5 Curing Periode
1	1	1	1	1	1
2	1	1	1	1	2
3	1	1	1	1	3
4	1	2	2	2	1
5	1	2	2	2	2
6	1	2	2	2	3
7	1	3	3	3	1
8	1	3	3	3	2
9	1	3	3	3	3
10	2	1	2	3	1
11	2	1	2	3	2
12	2	1	2	3	3
13	2	2	3	1	1
14	2	2	3	1	2
15	2	2	3	1	3
16	2	3	1	2	1
17	2	3	1	2	2
18	2	3	1	2	3
19	3	1	3	2	1
20	3	1	3	2	2
21	3	1	3	2	3
22	3	2	1	3	1
23	3	2	1	3	2
24	3	2	1	3	3
25	3	3	2	1	1
26	3	3	2	1	2
27	3	3	2	1	3

Current Worksheet: Paving Block Experiment.mtw

Table 4. Noise factor and management strategies

Factors	Anticipated Effect	Noise Management Strategy
Operator experience and skill	High	Use one experienced operator
Water content	High	Water in the mix was limited to 0.5
Curing condition	High	Use same procedure for all experiment
Cement to sand ratio	High	kept at 1:3
Batch to batch variation	Low	Block & Randomization
Temperature of room effect	Low	Randomization

3. Result and discussion

3.1 Compressive strength

Results from the sets of experiments were investigated in more details for the compressive strength in Table 5. Fig. 1 shows the Pareto Chart of the factors affecting the compressive strength. Using Cement% mixing and CBA% have the highest effect which will be illustrated further when discussing effect of each factor. The interaction of the age/curing time with CFA% and CBA% also has a major effect. Other factors can be seen from the Pareto chart in descending order of influence. As stated earlier there is no interaction between wet mixing and sonication time since not all the sonication times can exist for sand mixing, similarly between wet mixing and whisking and other 3rd and 4th order interaction terms."

Fig. 2 illustrates the main effect plot of each of the factors studied. The effect of a factor is defined to be the change in response produced by a change in the level of the factor. Main effect is related to the primary factors of interest in the experiment. It can be seen that there is a non-linear relation between increasing the CFA & CBA and the gain in sample compressive strength compared to the control, indicating that adding more CFA & CBA does not necessarily result in a higher strength. As the age of the concrete increase, the compressive strength ratio also increases.

Table 5. Compprresive strength

Exp No.	Cement (%)	CFA (%)	Sand (%)	CBA (%)	Curing Periods (Days)	Compressive Strength (MPa) R1	Compressive Strength (MPa) R2	Mean	SNI Standard
1	25	0	75	0	7	4.13	5.52	4.83	N/A
2	25	0	75	0	14	5.53	5.1	5.32	N/A
3	25	0	75	0	28	8.51	8.75	8.63	Quality D
4	8.33	25	66.67	0	7	4.89	6.12	5.51	N/A
5	8.33	25	66.67	0	14	6.72	8.61	7.67	N/A
6	8.33	25	66.67	0	28	8.76	8.27	8.52	Quality D
7	2.45	14.63	41.46	41.46	7	17.12	19.04	18.08	Quality B
8	2.45	14.63	41.46	41.46	14	22.45	23.87	23.16	Quality B
9	2.45	14.63	41.46	41.46	28	27.13	26.52	26.83	Quality B
10	13.8	0	27.58	58.62	7	21.35	21.53	21.44	Quality B
11	13.8	0	27.58	58.62	14	27.04	29.11	28.08	Quality B
12	13.8	0	27.58	58.62	28	35.56	37.08	36.32	Quality A
13	16.67	12.5	70.83	0	7	20.56	20.44	20.50	Quality B
14	16.67	12.5	70.83	0	14	25.65	26.13	25.89	Quality B
15	16.67	12.5	70.83	0	28	33.34	35.07	34.21	Quality B
16	14.82	22.22	11.11	51.85	7	10.49	9.94	10.22	Quality D
17	14.82	22.22	11.11	51.85	14	14.78	15.42	15.10	Quality C
18	14.82	22.22	11.11	51.85	28	18.97	21.24	20.11	Quality B
19	16.23	0	45.94	37.83	7	15.82	17.1	16.46	Quality C
20	16.23	0	45.94	37.83	14	19.34	20.67	20.01	Quality B
21	16.23	0	45.94	37.83	28	24.45	25.66	25.06	Quality B
22	20.68	10.34	10.34	58.64	7	19.22	19.51	19.37	Quality B
23	20.68	10.34	10.34	58.64	14	27.67	26.41	27.04	Quality B
24	20.68	10.34	10.34	58.64	28	32.54	34.42	33.48	Quality B
25	30	30	40	0	7	5.66	5.95	5.81	N/A
26	30	30	40	0	14	4.33	5.74	5.04	N/A
27	30	30	40	0	28	7.56	8.63	8.10	N/A
28	15	0	85	0	7	5.94	6.13	6.04	N/A
29	15	0	85	0	14	7.47	8.01	7.74	N/A
30	15	0	85	0	28	10.96	12.04	11.50	Quality D

The best Compressive strength resulted from unproportional mixture with dominantly CBA is much higher than sand, and Cement higher than CFA (0%) as per experiment no 10 - 12. This is also influenced by other factors like curing period as illustrated in the interaction plot. This interaction was not captured by previous studies.

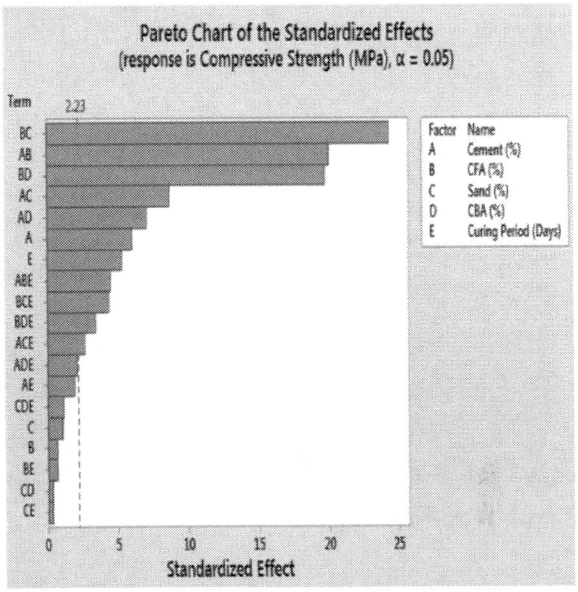

Fig. 1. Pareto chart of compressive strength.

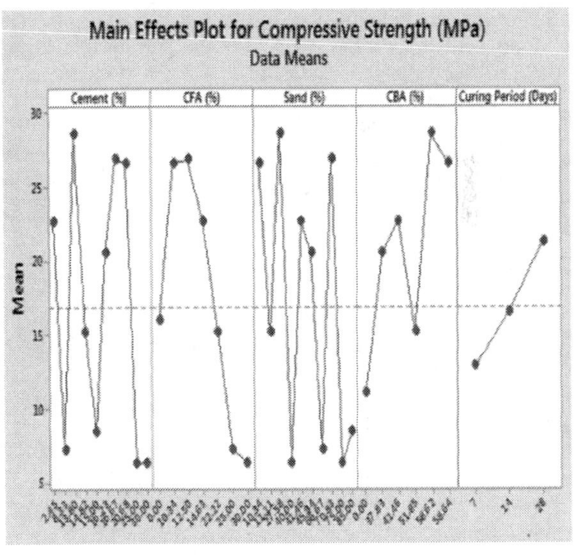

Fig. 2. Main effect plot for compessive strength.

The findings of this study were investigated further by considering the interaction plot for compressive strength as illustrated in Fig. 3. In some experiments the difference in response between the levels of one factor is not the same at all levels of the other factor. When this occurs there is an interaction between the factors. Any crossing indicates an interaction between variables.

https://doi.org/10.1051/matecconf/201820602013

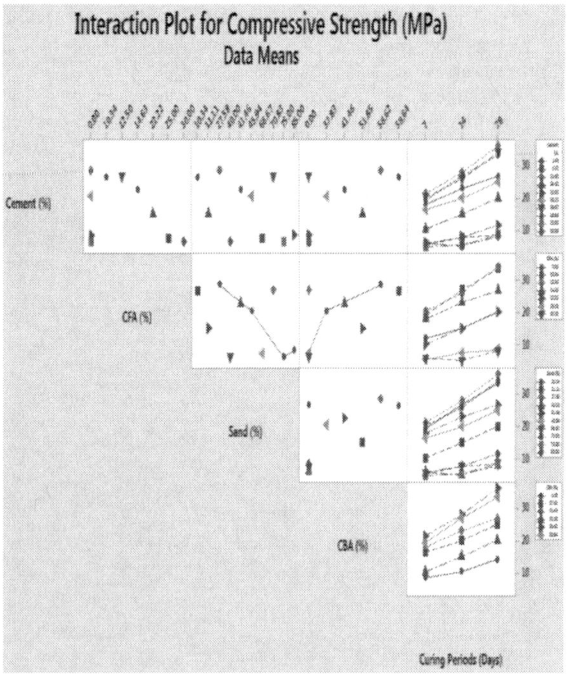

Fig. 3. Interaction plot for compessive strength.

To capture the above behavior, responses were modeled using response surface methodology as discussed earlier. Eq. (1) shows the model result for the compressive strength in terms of the process variables along with the goodness of fit. Quadratic terms, linear effects and interaction discussed earlier was captured through the model.

Compressive Strength (MPa)
28.158 + 4.833 Cement (%) + 0.335 CFA (%) -
0.528 Sand (%)
+ 6.00 Curing Periods (Days)
+ 35.46 Cement (%)*CFA (%)
+ 29.24 Cement (%)*Sand (%)
+ 14.14 Cement (%)*CBA (%)
+ 1.846 Cement (%)*Curing Periods (Days)
+ 51.37 CFA (%)
*Sand (%) + 31.82 CFA (%)*CBA (%) + 0.394 CFA (%)
*Curing Periods (Days) + 0.39 Sand (%)*CBA (%) -
0.181 Sand (%)
*Curing Periods (Days) + 9.49 Cement (%)*CFA (%)
*Curing Periods (Days) + 10.68 Cement (%)*Sand (%)
*Curing Periods (Days) + 5.12 Cement (%)*CBA (%)
*Curing Periods (Days) + 10.93 CFA (%)*Sand (%)
*Curing Periods (Days) + 6.58 CFA (%)*CBA (%)
*Curing Periods (Days) - 1.49 Sand (%)*CBA (%)
*Curing Periods (Days)
...... (1)
$R2 = 99.44\%$
With anova (Analysis of Variance) result is:

Analysis of Variance

Source	DF	Adj SS	Adj MS	F-Value	P-Value
Model	19	2764.01	145.474	93.19	0.000
Linear	4	105.71	26.427	16.93	0.000
Cement (%)	1	56.07	56.073	35.92	0.000
CFA (%)	1	0.67	0.668	0.43	0.528
Sand (%)	1	1.86	1.864	1.19	0.300
Curing Periods (Days)	1	43.33	43.329	27.76	0.000
2-Way Interactions	9	1140.15	126.683	81.15	0.000
Cement (%)*CFA (%)	1	623.31	623.312	399.30	0.000
Cement (%)*Sand (%)	1	117.07	117.073	75.00	0.000
Cement (%)*CBA (%)	1	76.64	76.638	49.09	0.000
Cement (%)*Curing Periods (Days)	1	5.75	5.753	3.69	0.084
CFA (%)*Sand (%)	1	922.34	922.336	590.86	0.000
CFA (%)*CBA (%)	1	605.17	605.166	387.67	0.000
CFA (%)*Curing Periods (Days)	1	0.65	0.647	0.41	0.534
Sand (%)*CBA (%)	1	0.19	0.188	0.12	0.736
Sand (%)*Curing Periods (Days)	1	0.15	0.155	0.10	0.759
3-Way Interactions	6	46.40	7.733	4.95	0.013
Cement (%)*CFA (%)*Curing Periods (Days)	1	31.41	31.405	20.12	0.001
Cement (%)*Sand (%)*Curing Periods (Days)	1	10.99	10.991	7.04	0.024
Cement (%)*CBA (%)*Curing Periods (Days)	1	7.08	7.084	4.54	0.059
CFA (%)*Sand (%)*Curing Periods (Days)	1	29.39	29.388	18.83	0.001
CFA (%)*CBA (%)*Curing Periods (Days)	1	18.23	18.225	11.68	0.007
Sand (%)*CBA (%)*Curing Periods (Days)	1	1.89	1.889	1.21	0.297
Error	10	15.61	1.561		
Total	29	2779.62			

In the end, result of compressive strength will be compared with Standard National Indonesia (SNI). Mostly the result show that the compressive strength get the quality B of SNI with average compressive strength 20 – 35 MPa. For the best result showed by experiment No. 12 with 36.32 MPa that the including quality A (average 40 MPa).

3.2 Water absorption

The first procedure is Dry the specimen in a ventilated oven at a temperature of 105 °C to 115°C till it attains substantially constant mass. And then, cool the specimen to room temperature and obtain its weight (M1) specimen too warm to touch shall not be used for this purpose. Immerse completely dried specimen in clean water at a temperature of 27+2°C for 24 hours. In the end, remove the specimen and wipe out any traces of water with damp cloth and weigh the specimen after it has been removed from water (M2).

Water absorption, % by mass, after 24 hours immersion in cold water in given by the formula Eq. (2) Shows,

$$W = \frac{M_2 - M_1}{M_1} X\ 100 \qquad \ldots\ \ldots (2)$$

The final result shows by Table.6 and Fig 4. Water absorption for the mixtures ranged between 1.98% and 10.72%. The maximum water absorption for paving block under severe weather exposure is 17% whereas the same for paving block under moderate weather exposure is 22% [5]. Most of the paving block are found to have more water absorption values in this investigation. However these can be used in inner walls and situations of minimal weather exposure. Fig. 4 with implicit indicates the relationship between water absorption and BA/Cement ratio. It can be observed that the water absorption values for the mixtures with BA:FA ratio of 1:5 are the lowest. This is because bottom ash is not a water absorber as the percentage of bottom ash water absorption is only 0.3–6.1% [5]. Yet fly ash is water absorbent material that increases the water absorption capacity of hardened matrix when it is used in the mix [6].

Table 6. Water absorpstion

Exp. No	Cement (%)	CFA (%)	Sand (%)	CBA (%)	Wet Mass (Kg)		Dry Mass (Kg)	% Water Absorpstion		Mean	SNI Standard
					R1	R2	R1	R1	R2		
1	25	0	75	0	3.62	3.74	3.12	16.03	4.76	10.39	N/A
2	8.33	25	66.67	0	3.42	3.48	3.14	8.92	9.43	9.18	Quality D
3	2.45	14.63	41.46	41.46	3.56	3.65	3.44	3.49	3.99	3.74	Quality B
4	13.8	0	27.58	58.62	3.62	3.58	3.54	2.26	1.70	1.98	Quality A
5	16.67	12.5	70.83	0	3.68	3.7	3.48	5.75	5.41	5.58	Quality B
6	14.82	22.22	11.11	51.85	3.68	3.71	3.41	7.92	8.48	8.20	Quality D
7	16.23	0	45.94	37.83	3.62	3.66	3.38	7.10	7.65	7.37	Quality C
8	20.68	10.34	10.34	58.64	3.41	3.38	3.32	2.71	3.05	2.88	Quality A
9	30	30	40	0	3.63	3.64	3.36	8.04	13.40	10.72	N/A
10	30	0	70	0	3.68	3.66	3.39	8.55	6.71	7.63	Quality C

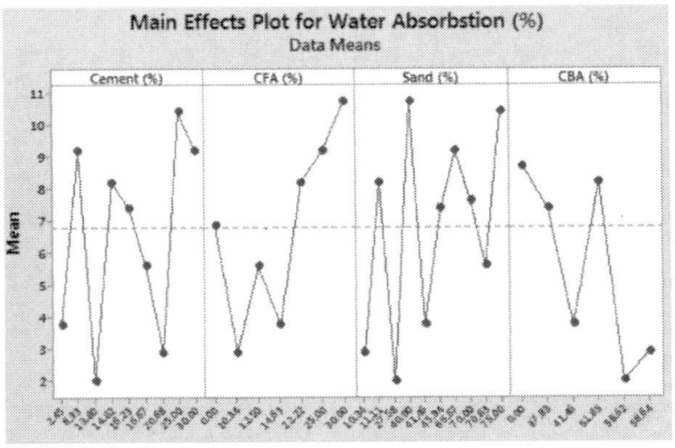

Fig. 4. Main effect plot for water absorption.

4 Conclusion

Mixed results in literature about the influence of CBA & CFA on the mixture properties of paving block are very well justified since varying the procedure significantly affects the result. Depending on the procedure adopted, CBA & CFA can be effectively dispersed in the mix resulting in higher strength. This research compared the degree of variability that exists in concrete blocks from CBA & CFA dispersions when prepared using different protocols for the same cement to water ratio. Results clearly illustrates that for the purpose of data comparability, there is a need to provide the exact details of all steps involved in a dispersion protocol and an optimum protocol can be standardized.

From the data above, results of Compressive Strength & Water Absorptions show that experiment No.12 with the mixtures component Cement 13.8%, CFA 0%, Sand 27.58%, and CBA 58.62% in curing period 28 days give the best result and recommendation for production of paving block, of course also with concern the noise factor, but for the both influence composition CFA & CBA that showed exp. 24 with Cement 20.68%, CFA 10.34%, Sand 10.34%, and CBA 58.64% in 28 days. In the case experiment for industrial manufacturer, Orthogonal Array is very helpful because the company can reduce time and also reduce cost for to get the best result.

For the future work of the experiment, next researcher can mixture with the other sand types, add addictive fluid to composite and resetting DoE factor to get the best result. Researcher too can verify result of experiment with SEM (Scanning Electron Microscope)

Acknlowledgement

Authors wish to acknowledge the Department of Industrial Engineering of University Of Indonesia.

References

1. R. Siddique,. "Utilization of Coal Combustion by-Products in Sustainable Construction Materials. Resources. Conservation and Recycling". (2010). 54.pp 1060– 1066.
2. M. Rafieizonooz, J. Mirza, M.R. Salim, M.W. Hussin, E.Khankhaje. "Investigation of Coal Bottom Ash and Fly Ash in Concrete as Replacement for Sand and Cement. Construction and Building Materials". (2016) 116. 15–24.
3. Douglas Montgomery, Design and Analysis of Experiments, John Wiley & Sons,NJ, 2009.
4. Hashim Abdul Razak, Sivakumar Naganathan, Siti Nadzriah Abdul Hamid, Performance appraisal of industrial waste incineration bottom ash as controlled low-strength material, J. Hazard. Mater. 172 (2009) 862–867
5. Craig H. Benson, Sabrina Bradshaw, User Guideline for Coal Bottom Ash and Boiler Slag in Green Infrastructure Construction. Recycled Materials Resource Centre, University of Wisconsin Madison, USA, 2011.
6. S. Naganathan, N. Subramaniam, K.N. Mustapha, Development of bricks from thermal power plant bottom ash and fly ash, Asian J. Civil Eng. 1 (2012) 275– 287. M. Young, The Technical Writer's Handbook. Mill Valley, CA: University Science, 1989.
7. Badan Standar Nasional. (1996). Bata beton (*Paving block*). SNI. 03-0691-1996.
8. Concrete Manufacturers Association (2009). Concrete Block Paving, Book 1- Introduction. Midrand: Author. Februari 5, 2017. www.cma.org.za
9. Concrete Masonry Association of Australia. "Permeable Interlocking Concrete Pavements Design and Construction Guide". Artarmon NSW, Australia: Author, 2010
10. Gabriel, D.S. "How to Increase Plastic Waste Acceptance for Mechanical Recycling: An Introduction to Material Value Conservation and Its Phenomenon". Key Engineering Materials. (2016). 705. 362-367.
11. Interlocking Concrete Pavement Institute. (2003). "Structural Design of Interlocking Concrete Pavement for Roads and Parking Lots". Tech spec No. 4. Washington, DC: Author. 14 Mei, 2017. http://www.panablock.com/content_images/Tech%20Spec%204.pdf
12. R. Siddique. "Utilization of Coal Combustion by-Products in Sustainable Construction Materials.

https://doi.org/10.1051/matecconf/201820602013

Resources. Conservation and Recycling". (2010). 54. 1060– 1066
13. R. Siddique. "Compressive Strength, Water Absorption, Sorptivity, Abrasion Resistance and Permeability of Self-Compacting Concrete Containing Coal Bottom Ash". Construction and Building Materials. (2013). 47. 1444–1450.
14. M. Singh & R. Siddique,). Effect of Coal Bottom Ash as Partial Replacement of Sand on Properties of Concrete. Resources. Conservation and Recycling. (2013). 72. 20–32.

Development of Implementation Guidlines for Maintenance, Repairation and Determine Material Spesifications for Mechanical Component Design in Government Own Green Building Based On Work Breakdown Structure

Dwi Endah Aryaningrum, *Yusuf* Latief, *Leni Sagita* Riantini, *Budi* Susilo

Department of Civil Engineering, University of Indonesia, Jakarta Indonesia

Abstract. The development of green buildings has become a global issue as global climate change is gradually becoming a serious concern for the future. Government buildings can be categorized as buildings that must comply with green building requirements, because they consume energy, water and other resources with large quantities. The absence of guidelines in maintenance and repair of government buildings will have an impact on the security and comfort of building users. In addition, the absence of guidance regarding the material specifications used will also result in a material stock system for maintenance and repair being ineffective. This study discusses the development of implementation guidelines for maintenance, reparation and determine material specifications for mechanical components design in government own green buildings based on work breakdown structure. This research will detail starting from the sub work section, work package, alternative design, implementation requirements, activities, resources and material specifications. With the implementation guidelines for maintenance, reparation and determine material specifications for mechanical components design in government own green buildings based on work breakdown structure (WBS) expected maintenance and reparation of government buildings can be implemented maximally, according to specifications and functions.

1 Introduction

The main purpose of green building developement is to create a healthy environment based on resource use and efficient ecological design [1]. In Singapore, a building is considered green if it meets the requirements under the Green Mark Scheme, which adopts assessment criteria that include energy efficiency, water efficiency, environmental protection, indoor environmental quality, features and other green innovations [2]. In Indonesia, matters concerning green building are regulated in Peraturan Menteri Pekerjaan Umum dan Perumahan Rakyat Republik Indonesia No. 02 / PRT / M / 2015. Based on the regulation, government buildings can be categorized as buildings that must comply with green building requirements.

One component in a building that can support the concept of green building is a mechanical component. Mechanical work is a work that includes mechanical things, which essentially require a mechanical principle in its application. Mechanical systems applied in the building for the purpose of supporting activities carried out in the building, including in terms of comfort and security for each activity and the culprit in the building.

Mechanical problems that often arise in government buildings such as elevator damage, pipeline leakage and ineffective material stock systems for maintenance and reparation. This problem is caused by the absence of work implementation guidelines that explain the rules, standards used, material specifications and implementation requirements for maintenance and reparation of mechanical components.

Preparation of guidelines for maintenance and reparation of mechanical components refering to Peraturan Menteri Pekerjaan Umum No. 24 / PRT / M / 2008 will be developed structurally in the form of Work Breakdown Structure (WBS) as well as the need for material specifications for the design of mechanical components referring to green buildings.

With the implementation guidelines for maintenance, reparation and determine material specifications for mechanical components design in government own green buildings based on work breakdown structure (WBS) expected maintenance and repair of government buildings can be implemented maximally, according to specifications and functions.

2 Research objectives

The objectives of this research are :

1. Identify the sub work section and work package of maintenance and reparation mechanical components.
2. Identify alternative designs/methods, activities, implementation requirements and resources based on the sub work sectio and work package of maintenance and reparation mechanical components.
3. Identify material specifications for each resource on maintenance and reparation mechanical components.
4. Establish implementation guidelines for maintenance, reparation and specification of materials for mechanical component design in government own green buildings based on work breakdown structure.

3 Literature review

3.1 Green building at government building

According to Green Building Council Indonesia (GBCI), the concept of green building is planning, development, operation and maintenance that pay attention aspects in protecting, saving, reducing the use of natural resources, maintaining the good quality of the building and the air quality in in the room and pay attention to the health of its inhabitants who are all based on the principle of sustainable development. This study refers to the GBCI benchmark for existing building [3].

State buildings are buildings for official purposes that are state / regional property and held with financing sources derived from APBN, and / or APBD funds or other acquisitions (Article 1 point 1) [4].

3.2 Work breakdown structure

The WBS is a hierarchical decomposition of the total scope of work to be carried out by the project team to accomplish the project objectives and create the required deliverables. The WBS organizes and defines the total scope of the project and represents the work specified in the current approved project scope statement [5].

Maintenance and Repair Work

The scope of maintenance work includes the type of cleaning, weaning, inspection, testing, repair and / or replacement of building materials and equipment and other similar activities under the guidance of operation and maintenance of buildings [6].

The scope of building maintenance work includes the repair and / or replacement of building parts, components, building materials, and / or infrastructure and facilities based on document of technical plan of building construction, taking into consideration the document of construction implementation [6].

4 Methodology

To answer the research objectives, this research is using descriptive research with qualitative approach. Archive analysis and literature study are the methods used in data collection. Expert interview are conducted to validate the data. The sequence of the research can be examined in Figure 1. The Research Flow.

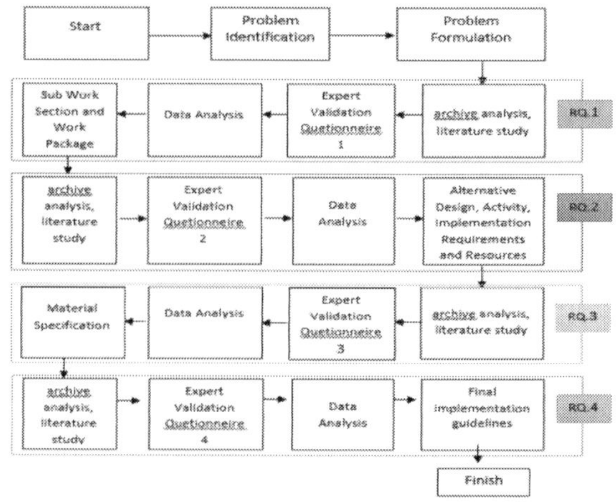

Fig. 1. The research flow.

5 Result and discussion

The results of this study are as follows :

1. To answer RQ.1 identify sub work section and work package of maintenance and reparation mechanical components for government own green building in Figure 2. Work Breakdown Structure (WBS) Diagram.

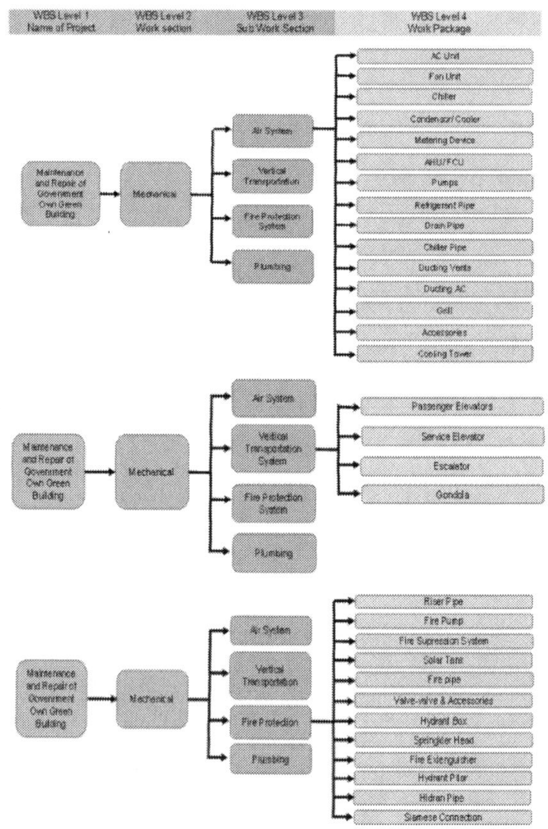

https://doi.org/10.1051/matecconf/201820602014

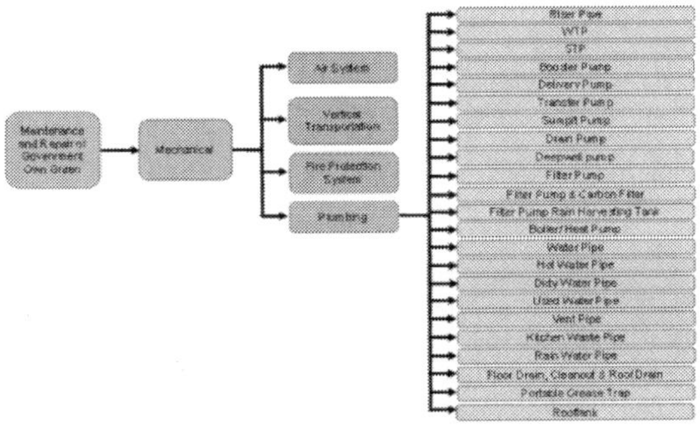

Fig. 2. Work breakdown structure (WBS) diagram.

Level 1 is for the name of project, maintenance and reparation of government own green building. Level 2 is work section, Mekanikal. Level 3 is for sub work section, air system, vertical transportation, fire protection system, and plumbing. Level 4 is work package.

2. To answer RQ.2 there is alternative designs/methods, activities, implementation requirements and resources based on sub work section and work package of maintenance and reparation mechanical components for government own green building. In table 1. shown alternative design/ method and activities of chiller as a sample in this study.

Table 1. Alternative design/ methode and activities of chiller

WBS Level 3 Sub Work Section	WBS Level 4 Work Package	Alternative Design/ Method	**WBS Level 5** Activities
Air Sistem	Chiller	Water Cooled Chiller	Inspection
			Maintenance
			Repair
			Testing
		Air Cooled Chiller	Inspection
			Maintenance
			Repair
			Testing

From the table above shows that alternative design/ method of work package chiller are water cooled chiller and air cooled chiller. Level 5 is activity, which consists of inspection, maintenance, reparation and testing.

In mechanical work section, data collection from archive analysis and literature study obtained 96 alternative design/ method based on work package at Level 4. Expert validation results, obtained 83 alternative design / method corresponding to the work package of maintenance and reparation mechanical components for government own green building.

Each activity based on alternative design/ method will be details with implementation requirement and resources.

Table 2. Implementation requirement and resources of air cooled chiller

WBS Level 2 (Work Section)	: Mechanical
WBS Level 3 (Sub Work Section)	: Air System
WBS Level 4 (Work Package)	: Chiller
Alternative Design/ Method	: Air Cooled Chiller

WBS Level 5 (Activity)	Requirement	WBS Level 6 (Resources)
Inspection	-Chiller check periodically. -Compression check of the compressor that measured at the high pressure side (discharge) and the low pressure side (suction). -By using pressure gauge, oil pressure as the lubricat parts moving in the compressor are measured periodically. While lhe level of oil that can be seen in sigh glass visually should be considered and should not bel lower than factory required. -Check the refrigerant pressure on the condenser and cooler on a regular basis to determine whether heat transfer is good or not. -Check the fan motor that serves to move the cooling air (especially for condenser). - routine checking of metering device. - inspection of water inlet and outlet against leakage. -Colling coil check against leakage and corrosion. -Check operation of the actuator control zone. -Inspection of the water damper operation. -Inspection of fresh air damper operation. -filter check. -inspection of refrigerant, to use non-CFC refrigerant. -analyze whether chiller maintenance/ reparation is necessary.	Man : -Mechanical Technician Equipment : -Checklist -pressure gauge
Maintenance	- Prepare the equipment and mateials needed for chiller maintenance. - Reset the metering device when the refrigerant flow is interrupted. - Clean the dirt on the compressor valve and reset. - add compressor oil periodically - The condenser machine needs to be cleaned regularly on the cooling fluid side. The air-cooled condenser requires cleaning of the fins on the air side. - For air-cooled condensers, external airflow needs to be maintained sufficiently and unobstructed, and there is no short circuit between the airflow out of the condenser and the air flow to enter the condenser. - Clean the surface of cooling coil. - clean the actuator control zone. - lubricate the air damper bearing. - lubricate the fresh air damper bearing. - adding / replacing a non-CFC refrigerant that has a small Ozone Depleting Potential (ODP) value (ODP <1).	Man : -Mechanical Technician Equipment : -plastic/brass brush -water pump -wipes Material : -chemical liquid -refrigerant -oil
Reparation	- Prepare equipment and materials needed for chiller repair. - Damaged filters must be replaced in order to function properly. - Worn and rusty compressors that cause leaks should be replaced. - An unworkable condenser should be replaced. - Metering device that is not working must be replaced. - Evaporators that are not working should be replaced.	Man : -Mechanical Technician Equipment : -chiller set repair equipment Material : -filters -compressors -condensers -metering

		devices -evaporators
Testing	- Prepare equipment and materials needed for chiller testing. - Periodically, recommissioning or retro commissioning with performance improvement targets (KW / TR) on chiller equipment at maximum 3 years with minimum efficiency for water cooled chiller (recip / screw chiller) is 1.27 KW/TR.	Man : -Mechanical Technician Equipment : -chiller set testing equipment

3. To answer RQ.3 Identify material specifications for each resource on maintenance and reparation mechanical components for government own green building. In table 3. shown material specification for air cooled chiller.

Table 3. Material specification for air cooled chiller

Alternative Design/ Method	Material Specification
Air Cooled Chiller	1. Refer to Standar Nasional Indonesia (SNI) 6390:2011 [7]. 2. Certified by ISO 14001 about environmental management system. 3. Certified by ISO 9001 about quality management system 4. Use non-CFC/ non-HCFC refrigerant (ODP=0). 5. Regional Production.

Material specifications are determined to regulate material use, so that material stock systems become more effective. Material specification that use In this case, material specification that use in air cooler chiller are refer to Standar Nasional Indonesia (SNI) 6390-2011. It also must certified by ISO 14001 and ISO 9001, use non-CFC/ non-HCFC refrigerant (ODP=0) and local fabrication.

Beside certified SNI / ISO / ecolab, for other material specifications refers to requirements that prioritize the expenditure of environmentally friendly materials such as regional production, recyclable materials, reused materials, renewable materials and other eco-categorized environmental friendly materials.

4. To answer RQ.4 Establish implementation guidelines for maintenance, reparation and specification of materials for mechanical component design in government own green buildings based on work breakdown structure. Table 4 shows the implementation guidelines that provide detailed information about WBS levels, alternative designs/ methods, activities, implementation requirements, resources and material specifications required to complete mechanical work in government own green buildings.

Table 4. Implementation guidelines for air cooled chiller

CHILLER

Description :
Chiller work package includes the provision of manpower, materials, equipment and tools needed for the work perform

AIR COOLED CHILLER

WBS Code	**Description**	
WBS Level 2	**Work Section**	**Mechanical**
WBS Level 3	**Sub Work Section**	**Air System**
WBS Level 4	**Work Package**	**Chiller**
	Alternative Design/ Method	**Air Cooled Chiller**

WBS Level 5 (Activity)	**WBS Level 6 (Resources)**	**Specification Material**
Inspection	Manpower : Mechanical Technician Equipment : Checklist, pressure gauge	Refer to Standar Nasional Indonesia (SNI) 6390:2011 [8]. Certified by ISO 14001 about environmental management system. Certified by ISO 9001 about quality management system Use non-CFC/ non-HCFC refrigerant (ODP=0). Regional Production.
Maintenance	Manpower : Mechanical Technician Equipment : Plastic/brass brush, water pump, wipes Material : Chemical liquid, refrigerant, oil	
Reparation	Manpower : Mechanical Technician Equipment : Chiller set repair equipment Material : Filters, compressors, condensers, metering devices, evaporators	
Testing	Manpower : Mechanical Technician Equipment : Chiller set testing equipment	

Implementation Requirement :

Inspection :
- Chiller check periodically.
- Compression check of the compressor that measured at the high pressure side (discharge) and the low pressure side (suction).
- By using pressure gauge, oil pressure as the lubricat parts moving in the compressor are measured periodically. While lhe level of oil that can be seen in sigh glass visually should be considered and should not bel lower than factory required.
- Check the refrigerant pressure on the condenser and cooler on a regular basis to determine whether heat transfer is good or not.
- Check the fan motor that serves to move the cooling air (especially for condenser).
- routine checking of metering device.
- inspection of water inlet and outlet against leakage.
- Colling coil check against leakage and corrosion.
- Check operation of the actuator control zone.
- Inspection of the water damper operation.
- Inspection of fresh air damper operation.
- filter check.
- inspection of refrigerant, to use non-CFC refrigerant.
analyze whether chiller maintenance/ reparation is necessary.

Maintenance :
- Prepare the equipment and mateials needed for chiller maintenance.
- Reset the metering device when the refrigerant flow is interrupted.
- Clean the dirt on the compressor valve and reset.
- add compressor oil periodically
- The condenser machine needs to be cleaned regularly on the cooling fluid side. The air-cooled condenser requires cleaning of the fins on the air side.
- For air-cooled condensers, external airflow needs to be maintained

sufficiently and unobstructed, and there is no short circuit between the airflow out of the condenser and the air flow to enter the condenser. - Clean the surface of cooling coil. - clean the actuator control zone. - lubricate the air damper bearing. - lubricate the fresh air damper bearing. adding / replacing a non-CFC refrigerant that has a small Ozone Depleting Potential (ODP) value (ODP <1).
Reparation : - Prepare equipment and materials needed for chiller repair. - Damaged filters must be replaced in order to function properly. - Worn and rusty compressors that cause leaks should be replaced. - An unworkable condenser should be replaced. - Metering device that is not working must be replaced. Evaporators that are not working should be replaced.
Testing : - Prepare equipment and materials needed for chiller testing. - Periodically, recommissioning or retro commissioning with performance improvement targets (KW / TR) on chiller equipment at maximum 3 years with minimum efficiency for water cooled chiller (recip / screw chiller) is 1.27 KW/TR.

6 Conclusion

The implementation guideline is the required reference in every work. Development of Implementatiion Guidelines for maintenance and reparation work for mechanical components in government own green building based on work breakdown structure facilitate the users in implementing the work and the determination of material specifications for the design make the material stock more effective.

References

1. Bon-Gang Hwang, Xianbo Zhao, Lene Lay Ghim Tan, (2015) "Green building projects: schedule performance, influential factors and solutions", Engineering, Construction and Architectural Management, Vol. 22 Issue: 3, pp.327-346.
2. Hwang Bon-Gang, Jack See Tan, "Sustainable Project Management for Green Construction : Challenges, Impact and Solutions", World Construction Conference, Sri Lanka, June 2012.
3. A Guide to The Project Management Body of Knowledge PMBOK GUIDE Sixth Edition.
4. Peraturan Presiden Republik Indonesia Nomor 73 Tahun 2011, "Pembangunan Bangunan Gedung Negara".
5. A Guide to The Project Management Body of Knowledge PMBOK GUIDE Sixth Edition.
6. Peraturan menteri Pekerjaan Umum Nomor 24/PRT/M/2008, "Pedoman Pemeliharaan dan Perawatan Bangunan Gedung".
7. Standar Nasional Indonesia (SNI) 6390:2011, "Konservasi Energi Sistem Tata Udara Bangunan Gedung".
8. Rajasa Miralia. Pengembangan Standar WBS (Work Breakdown Strukture) Untuk Estimasi Biaya Pekerjaan Mekanikal dan Elektrikal Bangunaan Gedung Berbasis Risiko. Fakultas Teknik Universitas Indonesia. 2017.
9. Peraturan Gubernur Provinsi Daerah Khusu Ibukota Jakarta Nomor 38 Tahun 2012, "Bangunan Gedung Hijau"
10. Sana Azeem, Malik Asghar Naeem, Abdul Waheed, Muhammad Jamaluddin Thaheem, (2017) "Examining barriers and measures to promote the adoption of green building practices in Pakistan", Smart and Sustainable Built Environment, Vol. 6 Issue: 3, pp.86-100.
11. Suzaini M. Zaid, Amir Kiani Rad, Nurshuhada Zainon, (2017) "Are green offices better than conventional?: Measuring operational energy consumption and carbon impact of green office in Malaysia", Facilities, Vol. 35 Issue: 11/12, pp.622-637.

Effect of Temperature and Asphalt Binder Type on the Properties of Foamed Asphalt Mix

Martin Bitolog[1], Mouhamed Bayane Bouraima[1], Xiao-hua Zhang[2], Christian Magloire Ndjegwes[1] and Yanjun Qiu[1]

[1]*School of Civil Engineering, Southwest Jiaotong Univ., Highway Engineering Key Laboratory of Sichuan Province, Chengdu China.*
[2]*Sichuan Provincial Transport Department Highway Planning, Survey, Design and Research Institute, Chengdu China.*

Abstract: In this study, an effect of temperature on both indirect tensile strength (ITS) test and Marshall Stability Test of foam asphalt mixture using Shell #70 A asphalt is firstly investigated. A detailed investigation of the mechanical properties of foam asphalt mixture was then carried out using two different asphalt binders. The experimental work compares the mechanical properties of foam asphalt mixes including water stability test, indirect tensile strength (ITS) test, and freeze-thaw splitting test. The test results indicate that the foam asphalt mixture is temperature dependent and also temperature sensitive material. The foaming properties of asphalt have an important effect on the properties of the foam asphalt mixture for the strength and water stability of the foam asphalt mixture.

1 Introduction

With the rapid development of transport infrastructure in China [1], the use of foamed bitumen as a stabilizing agent in pavement since 1956 [2] is improving steadfastly and internationally. Foamed technology has been used by numerous road and highway stabilization and recycling projects [3-6] after its production process is modified by injecting steam to cold water into hot bitumen in 1976 and also introducing new machineries in the market in mid-1990 [7].

The use of foam bitumen stabilization is imperative because of its high field performance properties and its environmentally friendly benefits [8-11]. Many factors such as bitumen type, grade, and additives; in particular, antifoaming agents often are added to bitumen produced by solvent precipitation processes affected its foaming characteristics.

Asphalt can be foamed at high temperature by adding approximately 2% amount of water to hot bitumen resulting in the formation of liquid asphalt binders, which is called foam asphalt [12]. Foamed bitumen advantages have grown in recent years and consist of (a) improving mix strength and durability, (b) increasing spectrum of aggregates usable, (c) reducing binder content, (d) improving workability, (e) propensity of using labor-intensive methods, (f) energy savings during the mix processing and others [13-16]. Foam asphalt technology has thus been widely used in asphalt pavement structure during the last decade all over the world [17-20]. China has mainly used it to use to recycle asphalt pavement [21]. The effects of asphalt binder properties on foamed asphalt mix properties were studied and reported by in references [22] and [23] respectively. They have concluded through their investigation that asphalt binders with higher expansion ratio and longer half-life tend to produce better quality foamed asphalt mixes

Use of new aggregate in foam asphalt mixture is also a good choice for new asphalt pavement structure. However, there is a difference between foam asphalt mixtures with foam asphalt recycled mixture in terms of the strong performance, it is imperative to research the strength characteristic in the laboratory. Mechanical performance of cold foamed bitumen mixes was investigated through laboratory [24] using bitumen source, foaming additive, fiber reinforcement and cement effect. The mechanical properties included dynamic modulus and indirect direct tensile strength of dry and conditioned specimens. The results show that bitumen with low foamability provided better mechanical performance than that with the better foamability.

The objectives of this study were to first investigate the effect of temperature on both indirect tensile strength (ITS) test and Marshall Test of foam asphalt mixture using Shell #70 A asphalt and secondly to see the effect of two different asphalt binder type on the mechanical properties of foam asphalt mixture.

2 Materials and test program

Zhonghai # 90 A grade asphalt and Shell #70 A bitumen currently used in China were investigated. For each source, penetration, viscosity, and softening point were measured. They used to produce foam asphalt at 150°C. Water content is 4% of asphalt mass when producing foam asphalt. Physical properties of base asphalt for the two sources and their respective foaming characteristics are shown in Table 1. The mixture design gradation is given in Table 2. The optimal water content of foam asphalt mixture is determined by Marshall 75 times compaction test. The optimal foam asphalt content is 3.5%.

3 Test program

3.1 Water Stability

The residual Marshall stability is used as the index to evaluate the water stability according to the Chinese Technical Specification for Construction of Highway Asphalt Pavement. Standard Marshall Test and immersion Marshall Test were conducted according to Chinese specification T0709-2011, and then the residual stability can be determined from the Marshall stability and the immersion Marshall Stability according to formula (1)

$$S = \frac{S_1}{S_0} \times 100 \quad (1)$$

where S is the residual stability, %; S_1 is the immersion Marshall stability, kN; and S_0 is the Marshall stability, kN.

3.2 Indirect tensile strength test

For the indirect tensile test, Marshall specimens were subjected to compressive loads in equipment fabricated for this purpose. The tests were conducted at 25°C, 50 mm/min loading rate. The test procedure adopted was as per Chinese specifications JTJ T0716-2011. The split strength can be obtained according to formula (2).

$$R_T = \frac{0.006287 P_T}{h} \quad (2)$$

in which, R_T is a split strength (MPa), h: displacement (mm), and P_T is the load when specimen destroyed (N).

3.3 Comparison of freeze-thaw splitting

The tensile strength ratio is defined as the tensile strength of specimen after freeze-thaw conditioning compared to the tensile strength of specimen at ambient temperature. During the freeze-thaw conditioning, the specimen is immersed in water under vacuum state for 15 min, then froze at −18°C in the refrigerator for 16 h, and then immersed in 60°C water for 24 h. The splitting strength tests were carried out according to the Standard Test Methods of Bitumen and Bituminous Mixtures for Highway Engineering, and the maximum strength values were recorded. The tensile strength ratio (TSR) can be calculated based on Eq. 2.

$$TSR = \frac{R_{T2}}{R_{T1}} \quad (3)$$

Where TSR: freeze-thaw strength ratio, %, R_{T1}: strength without freeze and thaw, MPa, R_{T2}: Strength after freeze and thaw, MPa.

4 Test results and discussion

4.1 Comparison of test indexes under different temperature

The effect of temperature on both indirect tensile strength (ITS) test and Marshall Test of foam asphalt mixture using Shell #70 A asphalt is investigated. The test results of dry and wet indirect tensile strength test at 25°C are shown in Table 3. Table 4 shows the results of dry wet indirect tensile strength test under different temperatures. The Marshall test results at 40°C and 60°C are shown in Table 5.

Table 1. Physical property of the two asphalt binders

Test items	Unit	Shell #70 A	Zhonghai # 90 A
Penetration (25°C)	0.1 mm	65	84
Ductility (10°C)	cm	34	135
Softening point	°C	47.5	48
Dynamic viscosity (135°C)	Pa.s	1..71	1.73
Density (15°C)	g/cm³	1.02	1.03
Expansion ratio		6	14
Half-life	s	5	8

Table 2. Foam asphalt mixture gradation

Passing percentage (%) of each mesh sieve (mm)											
26.5	19	16	13.2	9.5	4.75	2.36	1.18	0.6	0.3	0.15	0.075
100	96.4	92.6	82.5	76.6	53.5	37.7	28.8	22	15.8	13.4	11

Table 3. 25°C dry and wet indirect tensile strength test

Test item	25°C dry ITS	24h 25°C soaking wet ITS
Average value	0.417	0.353
Dry ITS/wet ITS	84.7%	

Table 4. 15°C and 25°C dry wet t indirect tensile strength test

Test item	15°C dry ITS	15°C soaking wet ITS	25°C dry ITS	25°C soaking wet ITS

Average value	0.4853	0.4669	0.417	0.353
Dry ITS/wet ITS	96.8%		84.7%	

Table 5. 60°C and 40°C Marshall test

Test item	60°C standard Marshall stability (35min)	40°C immersion Marshall stability (35min)
Average value	5.96	9.1

Note: ITS stands for indirect tensile strength.

Table 6. Water stability test results

Asphalt binder type	40°C standard Marshall stability (kN,35min)	40°C Immersion stability (kN,48h)	Residual stability (%)
Shell #70 A	9.1	8.89	97.7
Zhonghai # 90 A	9.23	9.21	99.8

Table 7. 25°C Indirect tensile test results

Shell #70 A asphalt			Zhonghai # 90 A asphalt		
Dry ITS (MPa)	Wet ITS (MPa)	Dry/Wet ratio	Dry ITS (MPa)	Wet ITS (MPa)	Dry/Wet ratio
0.417	0.353	84.6%	0.485	0.415	85.5%

Table 8. Comparison of freeze-thaw splitting test results

Test item	Shell #70 A asphalt		Zhonghai # 90 A asphalt	
	Before freeze-thaw	After freezing and thawing	Before freeze-thaw	After freezing and thawing
	0.3924	0.303	0.3819	0.3648
Wet and dry splitting ratio	77.2%		95.5%	

Through Tables 3 to 5, it can be seen that the properties of the foam asphalt mixture are temperature dependent and are also temperature sensitive materials. The dry-wet ITS ratio of foam asphalt mixture at 15°C increased obviously than that of25°C, and the dry ITS increased 40.4% while that of wet ITS increased by 60.6%

4.2 Mechanical properties of foam asphalt mixture.

The compared performance properties of foam asphalt mixture from two kinds of asphalt binder including water stability, indirect tensile strength, and freeze-thaw splitting was carried. The results of water stability tests were shown in Table 6.

It can be seen from Table 6 that the water stability of these two foam asphalt mixtures meets the specification requirement. And the residual stability of Zhonghai # 90 A asphalt mixture is greater than that of Shell #70 A, which means that the water stability of Zhonghai # 90 A asphalt mixture is better than that of Shell #70 A asphalt mixture. There is no much difference from Shell #70 A asphalt and Zhonghai # 90 A asphalt, after 48 hours of immersion Marshall Attenuation. The above test data demonstrate that the Marshall stability is not sensitive to the performance of the foamed asphalt and that the Marshall stability can meet the specified requirements when the foaming performance is poor. Table 7 presents the results of the indirect tensile test.

From the Table 7, it can be seen that the dry splitting strength of Zhonghai # 90 A asphalt is 16.3% higher than shell asphalt while its wet splitting strength is 17.5% higher. Whether it is dry or wet splitting, the Zhonghai # 90 A asphalt has a better strength ability compare to Shell #70 A asphalt. The results of the freeze-thaw splitting test for the two sources of asphalt are shown in Table 8.

It can be seen that after a freeze-thaw cycle, the wet splitting strength of both Shell #70 A and Zhonghai # 90 A foam asphalt mixtures decreased in varying degrees. After the freeze-thaw cycle, the water stability of Zhonghai # 90 A foam asphalt mixture is better than that of Shell #70 A foam asphalt mixture, and after a freezing-thawing cycle the water stability of Shell #70 A foamed asphalt mixture was significantly decreased

5 Conclusion

The aim of this study is to first investigate the effect of temperature on both indirect tensile strength (ITS) test and Marshall Test of foam asphalt mixture using Shell #70 A asphalt and secondly to see the effect of two different asphalt binder type on the mechanical properties of foam asphalt mixture. The main conclusions drawn from the current research work are presented as follows:

- Foam asphalt mixture is temperature dependent and

are also temperature sensitive materials

- The water stability of Zhonghai # 90 A asphalt mixture is better than that of Shell #70 A asphalt mixture.
- Foam asphalt mixture with Zhonghai # 90 A asphalt has a better indirect tensile strength compared to that of the Shell #70 A asphalt.
- The water stability of Zhonghai # 90 A foam asphalt mixture is better than that of Shell #70 A foam asphalt mixture after the freeze-thaw cycle.

References

1. Bayane, B. M., Yanjun, Q. (2017). Transport infrastructure development in China. Journal of Sustainable Development of Transport and Logisti2cs, 2 (1), 29–39. doi:10.14254/JSDTL.2017.2-1.3.
2. Csanyi LH, 1957. Foamed asphalt in bituminous pavements. Design procedures for soil modification or stabilization. IN, USA: Division of Engineering and Asset Management, Office of Geotechnical Services; 2013, February.
3. Wirtgen Cold Recycling Technology, 2012. The Wirtgen Cold Recycling Manual, Wirtgen GmbH, Windhagen, Germany.
4. Geoff J, 2013. Design and performance of foamed bitumen stabilized pavements: progress report 1. Australia: Austroad Ltd.
5. Abel, F., 1978. Foamed Asphalt Base Stabilization. Proc.,6th Annual Asphalt Paving Seminar, Colorado State Univ,Dec.7-8,1978.
6. Acott, S.M., 1979. Sand Stabilisation Using Foamed Bitumen. In: 3rd Conference on Asphalt Pavements for Southern Africa, 3rd, 1979, Durban, pp.155-172.
7. Mohammad LN, Abu-Farsakh MY, Wu Z, Abadie C, 2003. Louisiana, experience with foamed recycled asphalt pavement base materials. USA: Transportation Research Board, TRR 1832; 2003: pp. 17–24.
8. Saleh, M. F. New Zealand Experience with Foam Bitumen Stabilization. In Transportation Research Record: Journal of the Transportation Research Board, No. 1868, Transportation Research Board of the National Academies, Washington, D.C., 2004, pp. 40–49.
9. Kaveh Ostad-Ali-Askari, Mohammad Shayannejad. The Study of Mixture Design for Foam Bitumen and the Polymeric and Oil Materials Function in Loose Soils Consolidation. Journal of Civil Engineering Research 2015, 5(2): 39-44 DOI: 10.5923/j.jce.20150502.04.
10. Saleh, M. F. Effect of Aggregate Gradation, Mineral Fillers, Bitumen Grade, and Source on Mechanical Properties of Foamed Bitumen-Stabilized Mixes. Transportation Research Record: Journal of the Transportation Research Board. Issue Number: 1952. 2006.
11. Saleh, M. F. Cost evaluation of foam bitumen and other stabilization alternatives. International Journal of Pavement Engineering, Vol. 00, No. 0, 2006, 1-5. DOI: 10.1080/10298430601149585
12. Csanyi, L. H, 1963. Foamed Asphalt for Economic Road Construction. Civil Engineering, Vol. 32, No. 6, pp. 62–69.
13. Namutebi, M., Birgisson, B., & Bagampadde, U, 2011. Foaming effects on binder chemistry and aggregate coatability using foamed bitumen. International Journal of Road Materials and Pavement Design, 12(4), 821–847.
14. S. W. Du et al., "The Mechanical Properties of Foam Asphalt Mixture", Applied Mechanics and Materials, Vols. 71-78, pp. 1150-1153, 2011.
15. Gaudefroy, V., Olard, F., Cazacliu, B., Roche, C.L., Beduneau, E., & Antoine, J.P, 2007. A laboratory investigation on the mechanical performances of foamed bitumen mixes using half-warm aggregates. TRB 2007 Annual Meeting CD-ROM.
16. Muthen, K.M, 1998. Foamed asphalt mixes–mix design procedure (Contract Report CR-98/077). Pretoria, South Africa: SABITA Ltd & CSIR Transportek.
17. Ruenkrairergsa, T., Phromsorn, C., Silarom, P. and Ketnoot, W., 2004. Engineering Properties of Foamed Bitumen Mixtures in Thailand. Proceedings of the 8th Conference on Asphalt Pavement for Southern Africa (CAPSA '04), ISBN No. 1-920-01718-6, 12-16 September 2004, Sun City, South Africa.
18. Chiu, C.T. and Huang, M.Y., 2002. A Study on Properties of Foamed Asphalt Treated Mixes. Chung Hua University, Department of Civil Engineering, No.30 Tung Shiang, Hsin Chu,300, Taiwan
19. Sakr, H.A. & Manke, P.G., 1985. Innovations in Oklahoma Foamix Design Procedures. In: Asphalt materials, mixes, construction and quality. Washington, DC: Transportation Research Board. (Transportation Research Record; 1034), pp 26-34.
20. He, Z. and Lu, W., 2004. Research on the Properties of Road Building Materials Treated with Foamed Bitumen in China. Proceedings of the 8th Conference on Asphalt Pavement for Southern Africa (CAPSA '04), ISBN No. 1-92001718-6, 12-16 September 2004, Sun City, South Africa..
21. LI Xiu-jun, SHI Fang-Zhi, ZHANG Yong-ping. Journal of Tongji University (Natural Science); 2009 (01):47-50.
22. Jenkins KJ. Mix design considerations for cold and half-cold bituminous mixes with emphasis on foamed bitumen. Ph.D. Thesis, University of Stellenbosch, South Africa; 2000.
23. Pengcheng Fu, David Jones, John T. Harvey, 2011. The effects of asphalt binder and granular characteristics on foamed asphalt mix strength. Construction and Building Materials, (25): 1093-1101.
24. Gilberto Martinez-Arguelles, Filippo Giustozzi, Maurizio Crispino, Gerardo W. Flintsch, 2015. Laboratory investigation on mechanical performance of cold foamed bitumen mixes: Bitumen source, foaming additive, fiber-reinforcement and cement effect. Construction and Building Materials. Volume (93): 241-248.

Numerical Simulation of Temperature Distribution in the Glass Fiber Epoxy Composites during Microwave Curing

X Zhang , X L Chang, R L Ma, L Zhang, X D Chen and Q Zhang

Department of Power Engineering, Xi'an Hi-Tech Institute, Xi'an, 710025, China

Abstract. A three-dimensional coupled model of electromagnetic field, temperature field and curing degree field was established. Based on this model, the simulation of microwave curing process of glass fiber epoxy ring was realized, and the temperature distribution at different time was obtained. Numerical results indicate that the temperature difference within the composite ring is mainly formed in the initial stage during microwave curing.

1. Introduction

Advanced polymer matrix composites have the advantages of high strength-to-weight and stiffness-to-weight ratios, as well as good corrosion resistance and fatigue resistance performance, which have been widely used in aerospace and aircraft[1, 2]. Currently, the main curing technique for fabricating polymer composites is conventional thermal curing technique, which has some disadvantages such as high operational costs and long processing cycle3. Microwave curing technique, however, has been shown to be a promising alternative to solve the existent problems with some superiorities including fast heating rate, selective and volumetric heating, energy saving, easy controllability and high curing efficiency4 5.

During microwave curing process, the electromagnetic field and temperature distribution in different positions of the composites at different time are difficult to obtain. In addition, the conventional method by experiments will consume a lot of manpower, material and financial resources. In view of these constraints, it is very important to carry out the numerical simulation of microwave curing process. The realization of numerical simulations on microwave curing processes of epoxy composites will provide a feasible method for the study of microwave curing technique.

In this paper, the modeling and numerical simulations of multiphysics coupling on microwave curing processes of epoxy composites are carried out based on electromagnetic, heat transfer and cure reaction kinetics theories, a three-dimensional coupled model of electromagnetic field, temperature field and curing degree field is established through analysis of microwave curing process of glass fiber epoxy composites, and the model includes two internal heat sources.

2. Numerical modeling

The microwave curing process of epoxy matrix composites is a nonlinear multiphysics coupling process involving two internal heat sources. In order to establish the model, the electromagnetic field, temperature field and curing degree field need to be simulated at the same time. The Maxwell equations, heat conduction equation and curing kinetics equation are combined to obtain the temperature distribution in the composites.

2.1. Maxwell equations

The theoretical basis for solving the electromagnetic field distribution in the composite microwave curing process is Maxwell equations

$$\nabla \times \mathrm{E} = -\partial \mathrm{B} / \partial t \quad (1)$$

$$\nabla \times \mathrm{H} = \mathrm{J} + \partial \mathrm{D} / \partial t \quad (2)$$

$$\nabla \times \mathrm{D} = \rho \quad (3)$$

$$\nabla \times \mathrm{B} = 0 \quad (4)$$

Where *E, H, D, B, J,* ρ respectively represent the electric field strength, magnetic field strength, electric flux density, magnetic flux density, current density and charge density. According to the formulas (1) to (4), together with appropriate boundary conditions, the distribution of the electromagnetic field inside the microwave oven and inside the composites can be solved.

2.2. Heat conduction equation

According to the law of conservation of energy, the three-dimensional transient heat conduction equation of the composites during microwave curing is given as

$$\frac{\partial}{\partial x}(K_x\frac{\partial T}{\partial x})+\frac{\partial}{\partial y}(K_y\frac{\partial T}{\partial y})+\frac{\partial}{\partial z}(K_z\frac{\partial T}{\partial z})+P+Q_V=\rho_c C_{pc}\frac{\partial T}{\partial t} \quad (5)$$

Where P and Qv are the two internal heat sources in the composites microwave curing process. Among them, P is the microwave energy absorbed by the composites per unit volume at a certain moment. The microwave energy absorbed by the composites can be calculated according to the Poynting theorem[6]:

$$P = 2\pi f\varepsilon\varepsilon'' E^2 \quad (6)$$

Where P is the microwave energy absorbed by the composites (W/m^3); f is the microwave frequency, the value is 2.45GHz; ε is the free space permittivity, the value is 8.854×10^{-12}F/m; ε″ is the dielectric loss factor of the material; E is the electric field strength(V/m). In the numerical simulation of composites microwave curing process, the electric field strength E is obtained by solving Maxwell equations.

Qv is the reaction heat released by the curing reaction of the resin matrix, which is expressed as follows

$$Q_V = \rho_r V_r H_r d\alpha/dt \quad (7)$$

Substituting (6) and (7) into (5), the three-dimensional transient heat conduction equation for the microwave curing process can be expressed as

$$\rho_c C_{pc}\frac{\partial T}{\partial t}=\frac{\partial}{\partial x}(K_x\frac{\partial T}{\partial x})+\frac{\partial}{\partial y}(K_y\frac{\partial T}{\partial y})+\frac{\partial}{\partial z}(K_z\frac{\partial T}{\partial z})+2\pi f\varepsilon\varepsilon'' E^2+\rho_r V_r H_r\frac{d\alpha}{dt} \quad (8)$$

2.3. Curing kinetics equation

The phenomenological curing reaction kinetics model as follows

$$d\alpha/dt = k(T)(1-\alpha)^n$$
$$k(T) = A\times e^{(-\Delta E/RT)} \quad (9)$$

Where α is the curing degree, n is the reaction order, k is the reaction rate constant, A is the frequency factor, and ΔE is the activation energy of the resin curing reaction.

2.4. boundary condition

The sample will conduct convective heat transfer with air, and the boundary equation can be expressed as

$$-k\left(\frac{\partial T}{\partial n}\right)_W = h(T_W - T_a) \quad (10)$$

Where k is coefficient of heat conduction, n is Surface normal direction, T_W is The temperature of the boundary, T_a is ambient temperature, h is Convective heat transfer coefficient.

3. Simulation Process

The cavity of the microwave curing oven used in this study is hemispherical with a rated of power 700W and a operating frequency of 2.45GHz7. The same glass fiber/epoxy composite ring is selected as in Ref8, whose thickness, width and inner diameter are 3mm, 6mm and 150mm respectively. The PTFE ring mould wrapped with composites is placed on the bottom of the cavity of the glass rotating center. Physical and thermochemical parameters of glass fiber/epoxy composite components and PTFE mould as shown in table 1. The reaction order is 0.916 and reaction activation energy is 5.46×104J/mol.

In the microwave curing oven, the inner wall of the cavity and the waveguide are made of metal. These inner walls can be regarded as perfect electric conductors, so the tangential component of the electric field is zero. The initial temperature of the ring is set at 20℃, and the initial curing degree is 0.001. During curing, the microwave directly heats the composite ring to cause it cured gradually, and the PTFE mould merely serves to fix and support the composites. In the microwave curing process, as the temperature of the composite ring gradually increases, part of the heat of the composites is transferred to the PTFE mould in the form of heat conduction.

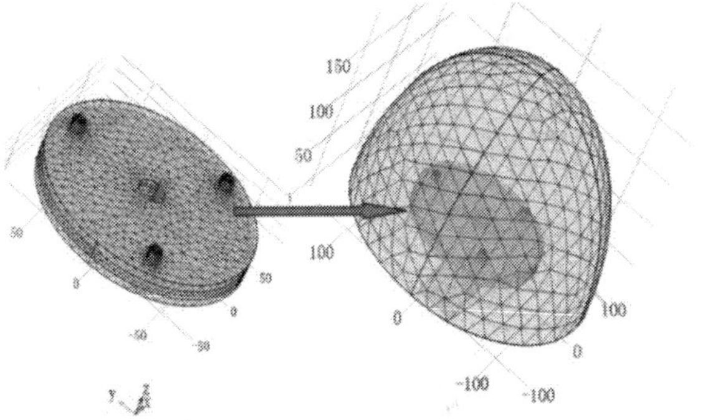

Figure 1. The overall mesh of model.

The finite element method is used to simulate the microwave curing process, and the model is meshed by tetrahedral meshing. Due to the more attention to the distribution of the electric field at the ring of the composites, the mesh at the ring and its mould is refined. The overall meshing situation is shown in Figure 1. A total of 83,286 grid units and about 429,815 degrees of freedom are divided.

In order to simplify the calculation, the basic assumptions made in the numerical simulation are as follows: (1) suppose the resin does not flow during the curing of the composite; (2) assuming that the physical parameters such as density, specific heat and thermal conductivity of the composite are constant; (3) do not take into account the deformation of the composite during curing; (4) in view of the dielectric loss glass turntable and PTFE mould are low, without considering the effect of microwave heating on the two kinds of medium in numerical simulation.

Table 1. Physical and thermochemical parameters

value	parameter		
	density (kg/m^3)	specific heat capacity J/(kg·K)	thermal conductivity W/(m·K)
fiber	2.54×10^3	7.37×10^2	1.250
resin	1.14×10^3	1.24×10^3	0.165
PTFE	2.19×10^3	1.05×10^3	0.244

4. Results and discussion

4.1. Electric field distribution

The ring mould wrapped with composites is horizontally placed at the center of the microwave curing oven turntable. The height of the lower end side of the ring from the bottom of the cavity is 0.012m. For this reason, the electric field distribution at the height of z = 0.012m is selected for analysis. Figure 2 and Figure 3 are respectively the electric field distribution cloud at this altitude under no-load and load states.

In the no-load state, the electric field distribution in the microwave curing oven is not uniform enough, but it is worth noting that due to the spherical symmetry of the cavity shape, the electric field distribution also shows obvious symmetry characteristics. Observations also found that the electric field intensity in the vicinity of the waveguide was significantly higher than that in other regions, and due to the spherical reflection effect of the cavity, another approximately symmetrical strong electric field region appeared at the center of the cavity bottom.

When the composite ring is placed at the bottom of the cavity, the electric field distribution in the cavity changes obviously compared with no-load, which indicates that the addition of lossy medium will have a great influence on the electric field distribution. It is considered that that the addition of ring enhances the absorption and reflection of the microwaves, so that the number of electromagnetic field modes inside the cavity of the oven increases, and the electric field is redistributed. From Fig. 3 shows that the electric field intensity inside the cavity under load is significantly higher than that under no-load conditions, and the electric field intensity on the ring is higher than the electric field strength in the surrounding area, mainly due to the electric field distribution has a characteristic to concentrate on areas with lossy media.

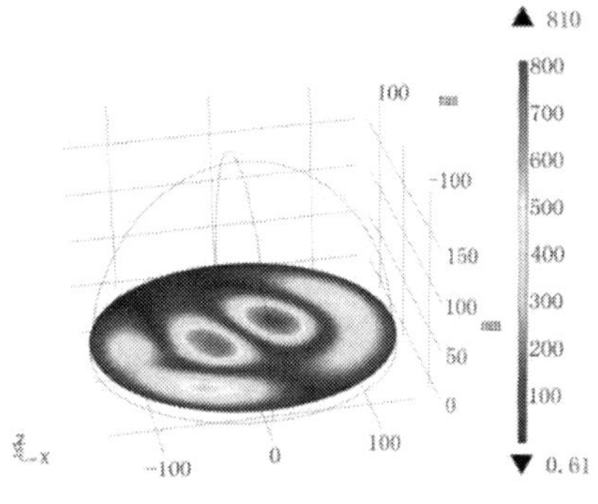

Figure 2. Electric field distribution cloud chart under no-load state(z=0.012m)

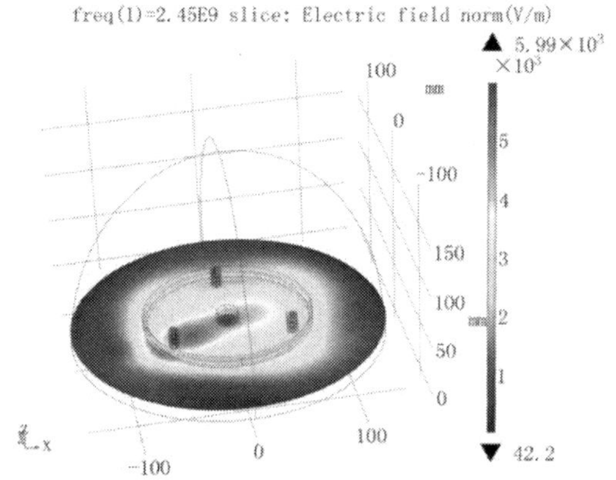

Figure 3. Electric field distribution cloud chart under load state(z=0.012m)

The distribution of electric field on the composite ring is an important factor affecting the curing effect of microwave.The more uniform the electric field distribution is, the more uniform the composites are heated during the microwave curing process, and the smaller the gradient of the curing degree is.Fig. 4 shows the cloud distribution of the electric field on the composite ring. It can be seen from the figure that the electric field distribution on the composite ring has some inhomogeneity. The results show that the strength of the electric field in the ring is mainly concentrated in the range of 3000-4000V / m, but in the local area close to the waveguide, the electric field intensity of the composite ring is slightly higher and the maximum electric field strength can reach 5827V/m. In the local edge region, the electric field strength of the composite ring is slightly lower, and the minimum electric field strength value is 1751V/m.

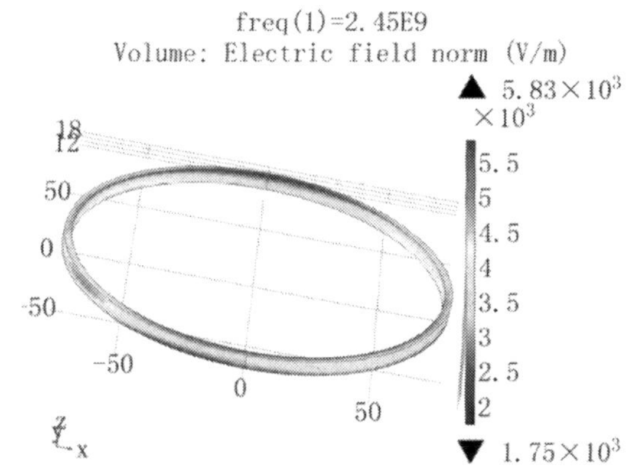

Figure 4. The cloud distribution of the electric field.

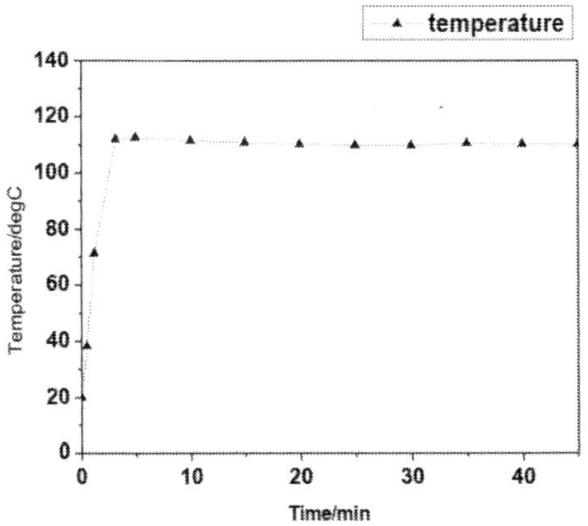

Figure 5. The curve of the temperature at the monitoring point with time.

4.2. temperature analysis

A point on the center line of the outer surface of the composite ring is arbitrarily selected as the temperature monitoring point, and the curve of the temperature at the monitoring point with time is shown in Figure 5. It can be seen from Figure 5 that under the microwave irradiation, the temperature of the composites increases rapidly, and the temperature at the monitoring point reaches the curing temperature 110℃ at 3.2 minutes. After 3.2min, the monitoring point temperature is controlled at about 110℃.

Figure 6 shows the temperature distribution of the composite ring at different time. At the initial stage, as the temperature of the composite ring gradually increases, the temperature difference between the inner and outer surfaces gradually increases. When heated to 3.2min, the monitoring point temperature reaches 110℃, at this time, the temperature difference between the inner and outer surface of the ring is about 5℃. It can be analyzed that the difference of the electric field distribution inside the composite ring is the main factor leading to the uneven temperature distribution. The electric field intensity near the inner edge of the ring is obviously higher than that in other regions, which makes the temperature rise in the region a little faster than that in the other regions, and the temperature is not controlled in the heating stage, which makes the temperature of the high electric field on the ring accumulates continuously, thus increasing the inhomogeneity of the temperature distribution. In addition, during the microwave curing process, the natural convection heat transfer between the outer surface and the air of the composite ring can also cause the loss of heat on the outer surface of the ring, which would increase the difference of the temperature distribution between the inner and outer surface of the ring.

When heated to 25min, the temperature difference between the inner and outer surfaces of the ring is only within 1℃, indicating that the temperature inside the composites has basically reached the same. It can be analyzed that the control of temperature makes the heat transfer from the high temperature part to the low temperature part and improves the uniformity of the temperature distribution in the composites. Secondly, with the progress of the curing process, the temperature of the air in the oven chamber increases continuously, which weakens the convective heat transfer between the outer surface of the ring and the air to a certain extent, thus reducing the heat loss on the outer surface of the ring.

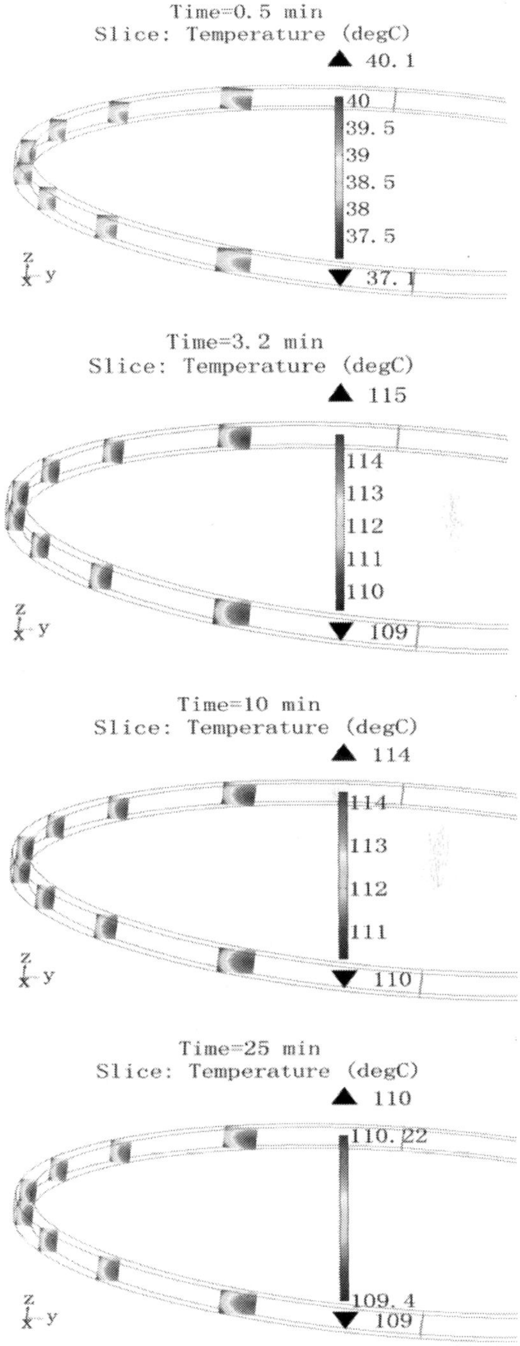

Figure 6. Temperature distribution of composite ring at different time.(YZ - plane slice)

5. Conclusions

Based on the theories of electromagnetics, heat transfer and solidification kinetics, a coupled model of electromagnetic field, temperature field, curing degree field for microwave curing process of glass fiber/epoxy composites with two internal heat sources is established. It can be found that the inhomogeneous electric field distribution leads to the difference of temperature distribution inside the composite ring during the initial stage of microwave curing, and the convection heat transfer between the outside of the ring and the air also increases the temperature difference. As the curing process carries on, the temperature difference will gradually decrease due to the control of the curing temperature. Therefore, it is necessary to take appropriate temperature control measures in the initial stage of microwave curing, in order to reduce the adverse effect of the difference of electric field intensity distribution on the heating rate and temperature distribution of the composites.

References

1. Li Y G, Li N Y and Gao J 2014 *J. International Journal of Advanced Manufacturing Technology*. **70** 591-606.
2. Li S J, Zhan L H, Chen R, et al 2014 *J. Fibers and Polymers*. **152** 404-09.
3. Joshi S C and Bhudolia S K 2014 *J. Journal of Composite Materials*. **48** 3035-48.
4. Hassan M R and Ganjeh B 2014 *J. Applied Mechanics and Materials*. **564** 310-14.
5. Durka M K, Dimitrakis G and Bogdal D 2015 *J. Chemical Engineering Journal*. **264** 633-44.
6. Nesbitt A, Navabpour P, Degamber B, et al 2004 *J. Measurement Science and Technology*. **15** 2313-2324.
7. Ma R L, Chang X L, Liao Y Q and Zhang X J 2016 *J. Polymer Material Science and Engineering*. **32** 96-101.
8. Ma R L, Chang X L, Liao Y Q and Zhang X J 2016 *J. Journal of Solid Rocket Technology*. **39** 809-14.

Microstructural Characteristics and Mechanical Properties of Water Cooling Bobbin-Tool Friction Stir Welded 6063-T6 Aluminum Alloy

Yunqiang Zhao, Chungui Wang and Chunlin Dong

Guangdong Provincial Key Laboratory of Advanced Welding Technology, Guangdong Welding Institute (China-Ukraine E.O.Paton Institute of Welding), Guangzhou, 510650, China

Abstract. In this study, a novel welding method called water cooling bobbin-tool friction stir welding (WBT-FSW) was developed. 4 mm-thick 6063-T6 aluminum alloy sheets were successfully jointed by WBT-FSW. Comparative studies on macro/microstructural characteristics and mechanical properties of the WBT-FSW and conventional bobbin-tool friction stir welding (BT-FSW) joints were carried out. The results indicated that the water mist cooling can significantly decrease the welding temperature and improve both the weld formation and the mechanical properties of the joint. The tensile strength of the WBT-FSW joint was 11.4% higher than that of BT-FSW joint.

1 Introduction

Friction stir welding (FSW) is a solid state joining technology, and it is quite suitable for joining aluminum alloys [1-3]. However, the curved and hollow sections cannot be joined by FSW, because a back plate is necessary to support the workpiece during welding [4]. In order to overcome this problem, bobbin-tool friction stir welding (BT-FSW) has been developed. For this bobbin tool, top and bottom shoulders jointed by the pin are applied, so the workpiece can be supported by this lower shoulder instead of the back plate.

Until now, some studies on the BT-FSW of aluminum alloys have been carried out in order to obtain high quality joint. For instance, Liu et al. [5] studied the BT-FSW of 6061-T6 aluminum alloy and they found that the precipitates deterioration caused by the welding heat input was the main softening mechanism of the joint. In addition, Al-Mg-Si alloy sheets were bobbin-tool friction stir welded using a specially designed tool by Zhou et al. [6]. The authors found that the lowest hardness location was in the heat affected zone of the joint and both the lowest hardness value and the tensile strength of the joint were increased with decreasing the welding heat input. Zhang et al. [7] conducted BT-FSW experiments on 2A14-T6 aluminum alloy sheets. The result also indicated that the strength of the joint can be increased by increasing the welding speed. It can be conclude that decreasing the welding heat input is an effective means for improving the mechanical properties of the BT-FSW joint. However, the effect by optimizing welding parameters is limited, because weld defect will be formed if the material cannot be sufficiently mixed at too high welding speed or too low tool rotation speed [8, 9]. In order to effectively decrease the welding heat input without changing the welding parameters, external water cooling was applied in the conventional FSW process in some studies [10]. The results indicated that the external water can take away the welding heat and then increase the mechanical properties of the FSW joints. For instance, Fratini et al. [11] applied water cooling on the top surface of the workpiece in the FSW experiment of 7075-T6 aluminum alloy, It was found that the tensile strength of the joint can be improved by the water cooling effect. Zhang et al [12-15] conducted the conventional FSW experiments on 2219-T6 aluminium alloy, during which the whole workpiece was immersed in the water environment. The results indicated that water cooling can be utilized to improve the mechanical properties of the FSW joint.

For increasing the joint strength, water cooling has been applied in the conventional FSW process. However, for the BT-FSW which suffers from high welding heat input and heat softening effect, the effects of the water cooling on the welding heat input and mechanical properties of the joint has not been studied. In this study, in order to effectively decrease the welding heat input and increase the strength of the BT-FSW joint, water spraying was applied during BT-FSW. This novel BT-FSW method was defined as water cooling bobbin-tool friction stir welding (WBT-FSW). 6063-T6 aluminum alloy which is widely applied in high-speed train structures was joined by both BT-FSW and WBT-FSW with the aim of highlighting the advantages of this novel welding method. The focus is placed on the weld formation, macro/microstructural characteristics, hardness distributions and tensile properties.

Table 1 Chemical compositions mechanical properties of 6063-T6 aluminum alloy

Chemical compositions, wt.%									Mechanical properties	
Mg	Si	Fe	Mn	Cr	Zn	Cu	Ti	Al	Tensile strength	Elongation
0.45~0.9	0.2~0.6	0.35	0.1	0.1	0.1	0.1	0.1	Bal.	260 Mpa	12%

2 Material and Experiment

4 mm-thick 6063-T6 aluminum alloy was taken as the base material whose chemical compositions and mechanical properties are listed in Table 1. For the bobbin tool, the diameters of both top and bottom shoulders were 16mm, and the shoulder surfaces were flat with spiral groove. A cylinder pin with two symmetric threads was applied and its diameter and length were 8 mm and 4mm respectively. During both BT-FSW and WBT-FSW, the tool rotation speed and welding speed were fixed at 1200 rpm and 200 mm/min respectively. The plunge depth of both top and bottom shoulders was set to 0.05 mm. In the WBT-FSW process, water mist with a constant flow rate of 0.1 L/min was sprayed to the leading edge of the top shoulder, as shown in Figure 1a. In order to measure the welding thermal cycles, two K-type thermocouples were embedded into the workpiece on the advancing side (AS) of the weld with a depth of 2 mm. Their distances from the weld center were 9 mm and 12 mm respectively, as illustrated in Figure 1b.

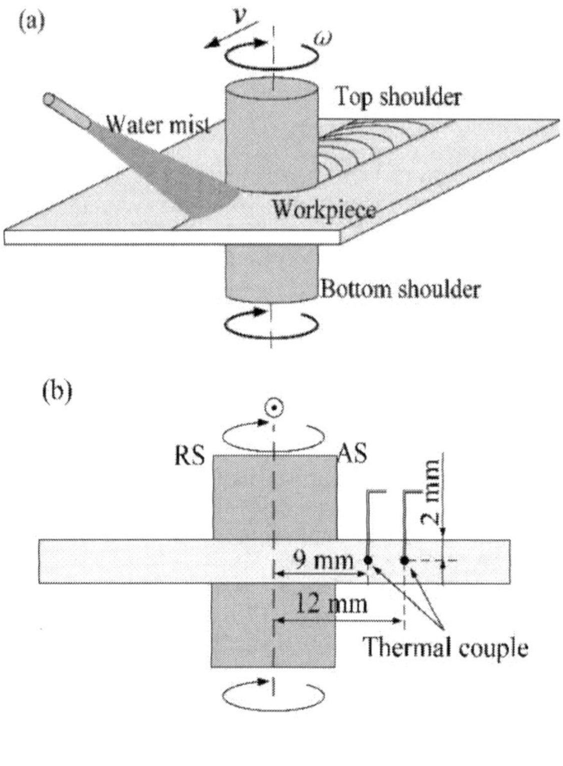

Figure 1. Schematic illustrations of (a) WBT-FSW process and (b) thermal couple positions

After welding, the metallographic specimens were successively sanded, polished and then etched using Keller's reagent. The metallographic analyses were carried out by optical microscopy (OM, Olympus-MPG3). Microhardness distributions on the cross-sections of the joints were measured along three lines whose distances from the workpiece top surface were 1mm, 2mm and 3mm respectively. The space between the adjacent indentations was 0.5 mm, and the testing load was 1.96 N for 10 s. The tensile tests were carried out using an electron universal testing machine (Instron 1186). The fracture morphologies of the failure specimens were observed using scanning electron microscope (SEM, Hitachi-S3400N).

3. Results and Discussions

3.1. Thermal cycles.

For the heat-treatable aluminum alloys, the welding thermal cycle can lead to precipitates deterioration and grain coarsening [16]. This is the main weakening mechanism for the BT-FSW or FSW joints [17]. For the 6063 aluminum alloy in this study, the main strengthening phase is β" which are coherent with the aluminum matrix. In addition, the β' phase which is semi-coherent with the matrix has weaker strengthening effect [18]. The artificial aging temperature of 6063 aluminum alloy is 175 ℃ at which the β" particles are accelerated to precipitate. When the temperature is in the range of 207-250 ℃, the small β" dissolution and the large β" growth will occur. When the temperature is in the range of 250–320 ℃, the β" precipitates will be coarsened and transform into β' precipitates. With the temperature further increasing in the range of 400–480 ℃, the β' precipitates will be dissolved into the matrix.

Figure 2 shows the welding thermal cycles at different positions during BT-FSW and WBT-FSW. The measured positions are respectively located in the thermo-mechanically affected zone (TMAZ) and the heat-affected zone (HAZ) of the weld. For the BT-FSW, the peak temperature and the dwelling time at high temperature (artificial aging temperature, 175 ℃) in the TMAZ are 419 ℃ and 10 s respectively. It can be concluded that the β" precipitates in the TMAZ experience dissolution and transformation into the β' precipitates. After that, the β' should be coarsened and partially dissolved into the matrix with the temperature increasing. In the HAZ of the BT-FSW joint, the peak temperatures and the dwelling time at high temperature are 309 ℃ and 8.7 s respectively. The β" precipitates in this zone should be partially dissolve and the large β" precipitates should transform into the β' precipitates. In

https://doi.org/10.1051/matecconf/201820603002

the WBT-FSW process, the measured peak temperatures and the dwelling time at high temperature were dramatically decreased. This can be attributed to the significant cooling effect of the water mist. In addition, the measured welding thermal cycles in the WBT-FSW process are more fluctuant, owing to the instability of the water mist cooling. For the WBT-FSW, the peak temperature and dwelling time at high temperature in the TMAZ are 278 ℃ and 2.8 s respectively. It can be concluded that the β'' precipitates should partially dissolve and transform into the β' precipitates. The peak temperature and dwelling time at high temperature in the HAZ are 197 ℃ and 0.5 s respectively. As a result, only β'' precipitates coarsening should occur in this zone.

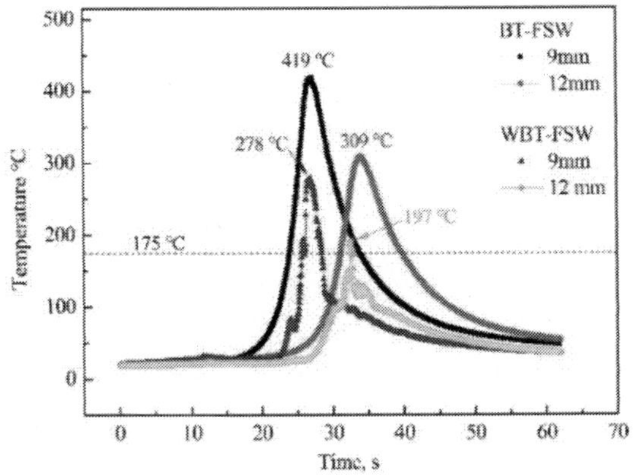

Figure 2. Welding thermal cycles with different distances from weld center in BT-FSW and WBT-FSW

3.2. Macro/microstructural characteristics.

As shown in Figure 3, defect-free joints can be obtained by both BT-FSW and WBT-FSW. For BT-FSW, the weld surface is rough and superfluous flash is formed at the retreating side (RS) of the weld (Figure 3a). This can be attributed to the softened material with low viscosity extruded out from the top and bottom shoulder at high temperature. By contrast, the weld surface obtained by WBT-FSW is smoother and the superfluous flash is vanished (Figure 3b), owing to the cooling effect of the water mist.

Figure 3c and d display the cross-section macrographs of BT-FSW and WBT-FSW joints. For both BT-FSW and WBT-FSW, the geometry of the SZ displays a dumbbell-shape. The width of the SZ at the mid-thickness is a little larger than the pin diameter, and the widths of the top and the bottom surfaces are almost equal to the shoulder diameter. The transition between the SZ and the TMAZ on the advancing side (AS) is sharper than that on the RS. Mishra et al. [19] ascribed this to the different flow behaviours and the degree of plastic deformation of the material on both sides of welding tool. For the WBT-FSW joint, the water mist cooling suppresses the flow of the plastic material, as a result, the sizes of the SZ, the TMAZ and the HAZ are obviously narrower than those in the BT-FSW joint.

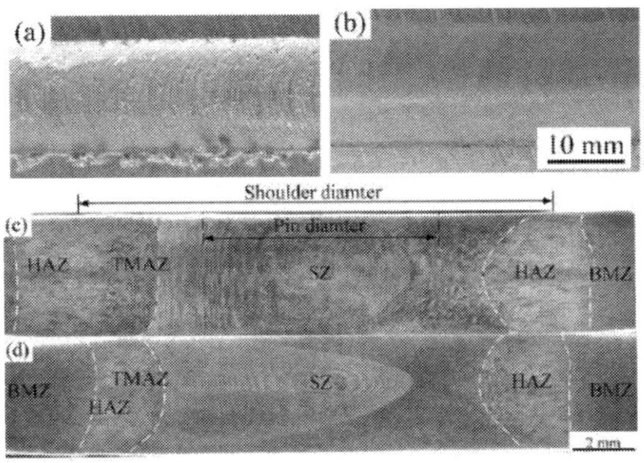

Figure 3. Macrostructural characteristics of BT-FSW and WBT-FSW joint: Surface appearances of (a) BT-FSW and (b) WBT-FSW joints, cross-sections of (a) BT-FSW and (d) WBT-FSW joints

Figure 4 displays the grain morphologies in the SZs of the BT-FSW and WBT-FSW joints. For both BT-FSW and WBT-FSW, dynamic recrystallization occurs in the SZ owing to high temperature and severe plastic deformation. As a result, fine and equiaxed grains are formed in this zone. The size of the recrystallized grain can be calculated according to the following equation [20]:

$$D^2\text{-}D_0^2=Aexp(\text{-}Q/RT)t_g \qquad (1)$$

where D_0 and D are the initial and final recrystallized grain size, A is a constant, Q is the activation energy for grain growth, R is the ideal gas constant, T is the absolute temperature, and tg is the time of the grain growth. Sato et al. [20] pointed that the D_0 in the FSW process was so small that the final grain size was mainly determined by the welding temperature and the dwelling time at high temperature. For the BT-FSW, the average size in the SZ is 22.5 μm. By contrast, the average grain size in the SZ of the WBT-FSW decreases to 16.1 μm, due to the lower welding temperature and shorter dwelling time at high temperature.

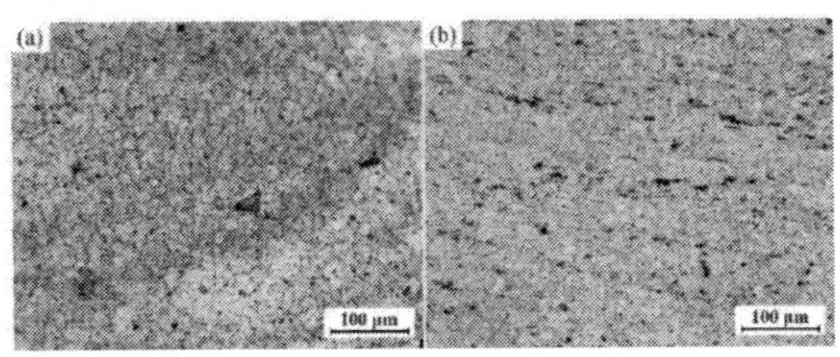

Figure 4. Grain morphologies in the SZs of the (a) BT-FSW and (b) WBT-FSW joints

3.3 Mechanical properties

Microhardness distributions on the cross-sections of the BT-FSW and the WBT-FSW joints are shown in Figure 5. Owing to the welding heat effect, precipitate deterioration occurs in the SZ, the TMAZ and the HAZ. As a result, a softening region is formed in both BT-FSW and WBT-FSW joints. However, the refinement strengthening caused by the dynamic recrystallization in the SZ and the dislocations strengthening caused by the severe plastic deformation can partly compensate the hardness loss in this two zones. As a result, the HAZ has lower hardness and the lowest hardness values locate in the HAZ close to the TMAZ of the joint. In addition, owing to the symmetrical thermal and mechanical effects of the top and the bottom shoulders, the hardness distributes symmetrically with respect to the mid-thickness of the joint. The hardness values near the mid-thickness of the join are lower than those near the top and the bottom surface. This can be attributed to the superimposed heat effects by the two shoulders and the poor cooling condition near the mid-thickness of the sheet. For the BT-FSW joint, the lowest hardness location is in the HAZ on the AS of the joint. On the contrary, the lowest hardness location for the WBT-FSW is in the HAZ on the RS of the joint. Comparing with the BT-FSW joint, the hardness values are increased and the softening region is significantly narrowed in the WBT-FSW joint. That is to say, the water mist cooling has significant positive effects on the mechanical strength of the joint.

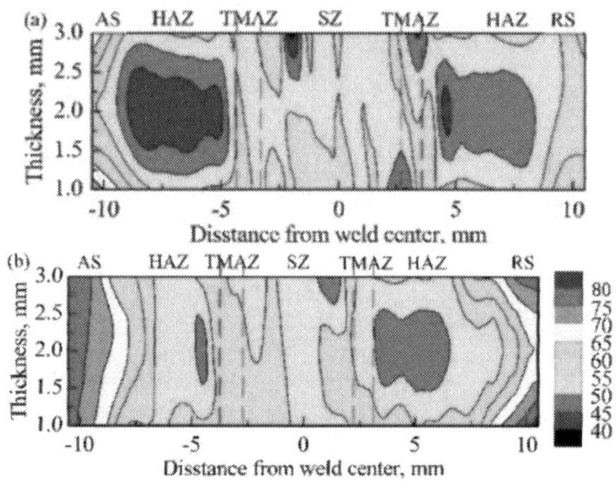

Figure 5. Microhardness distributions on the cross-sections of (a) BT-FSW and (b) WBT-FSW joints

The tensile properties of the BT-FSW and the WBT-FSW joints are displayed in Figure 6. For the BT-FSW joint, the tensile strength is 170 Mpa and the elongation is 10.1% (Figure 6a). The crack initiates and propagates in the HAZ on the AS of the joint (Figure 6b). The fracture location is in accordance with the lowest hardness location in the BT-FSW joint (Figure 5b). For the WBT-FSW, owing to the water mist cooling, the tensile strength of the joint increases to 178 Mpa and the elongation decreases to 8.7%. Unlike the BT-FSW, the fracture location for the failed WBT-FSW joint is in the HAZ on the RS of the joint Figure 6c), where the lowest hardness appears (Figure 5b). Figure 6d and e show the fracture surfaces characteristics of the BT-FSW and WBT-FSW joints. The fracture surfaces of both joints are characterized by large and deep dimples, and some second phase particles can be seen at the bottom of the dimples. That is to say, both the failed BT-FSW and WBT-FSW joints experience ductile fracture.

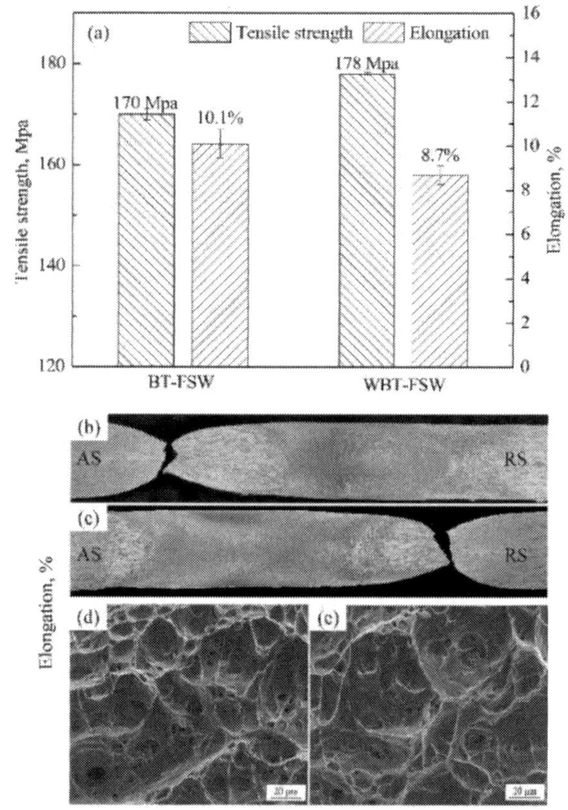

Figure 6. Tensile properties of the BT-FSW and WBT-FSW joints: (a) tensile strength and elongations of joints, fracture locations of (b) BT-FSW and (c) WBT-FSW joints, fracture surfaces of (d) BT-FSW and (e) WBT-FSW joints

4 Summary

In this study, a novel WBT-FSW method was developed and 6063-T6 aluminum alloy sheet was successfully jointed by WBT-FSW. Comparing with the conventional BT-FSW, the water mist cooling can significantly decrease the welding temperature in the WBT-FSW. As a result, for the WBT-FSW smoother weld surface and joint with narrower softening region can be obtained. The hardness of the WBT-FSW joint was obviously higher than that in the BT-FSW joint. The tensile strength of the WBT-FSW was 178 MPa which was 11.4% higher than that of the BT-FSW joint.

Acknowledgment

The authors are grateful to be supported by the Implementing Innovation-Driven Development Capacity Building Special Funds project of Guangdong Academy of Sciences (2017GDASCX-0847 and 2018GDASCX-0113), the Science and Technology Plan Project of Guangzhou City (201807010068 and 201704030056), and the Science and Technology Plan Project of Guangdong Province (2015B090922011).

References

1. G. Cam, S. Mistikoglu, J. Mater. Eng. Perform. 23, 1936-1953 (2014)
2. R.S. Mishra, Z.Y. Ma, Mater. Sci. Eng. R. 50, 1-78 (2005)
3. Z. Zhang, B.L. Xiao, Z.Y. Ma, Mater. Charact. 106, 255-265 (2015)
4. W.Y. Li, T. Fu, L. Huetsch, J. Hilgert, F.F. Wang, J.F. dos Santos, N. Huber, Mater. Des. 64, 714-720 (2014)
5. H.J. Liu, J.C. Hou, H. Guo, Mater. Des. 50, 872-878 (2013)
6. L. Zhou, G.H. Li, C.L. Liu, J. Wang, Y.X. Huang, J.C. Feng, F.X. Meng, Sci. Technol. Weld. Join. 22, 438-445 (2016)
7. H. Zhang, M. Wang, X. Zhang, G. Yang, Mater. Des. 65, 559-566 (2015)
8. T.-J. Yoon, J.-G. Yun, C.-Y. Kang, Mater. Des. 90, 568-578 (2016)
9. H. Shirazi, S. Kheirandish, M.A. Safarkhanian, Measurement. 76, 62-69 (2015)
10. M.A. Wahid, Z.A. Khan, A.N. Siddiquee, Trans. Nonferrous. Metal. Soc. Ch. 28, 193-219 (2018)
11. L. Fratini, G. Buffa, R. Shivpuri, Int. J. Adv. Manuf Technol. 43, 664-670 (2008)
12. H.J. Zhang, H.J. Liu, L. Yu, J. Mater. Eng. Perform. 21, 1182-1187 (2012)
13. H.J. Zhang, H.J. Liu, L. Yu, Sci. Technol. Weld. Joining.16, 459-464 (2011)
14. H. Zhang, H. Liu, Metallography, Microstructure, and Analysis.1, 269-281 (2012)
15. H.J. Zhang, H.J. Liu, L. Yu, Mater. Des. 32, 4402-4407 (2011)
16. J.Q. Su, T.W. Nelson, R. Mishra, M. Mahoney, Acta. Mater. 51, 713-729 (2003)
17. C.B. Fuller, M.W. Mahoney, M. Calabrese, L. Micona, Mater. Sci. Eng. A. 527, 2233-2240 (2010)
18. H.L. Peng Dong, Daqian Sun, Wenbiao Gong, Jie Liu, Mater. Des. 45, 524-531 (2013)
19. R.S. Mishra, Z.Y. Ma, Mater. Sci. Eng. R. 50, 1-78 (2005)
20. Y. Sato, M. Urata, H. Kokawa, Metall. Mater. Trans. A. 33, 625-635 (2002)

Quantifying the Difference of Uniformly Dispersed and Re-agglomerated Graphene Oxide-Based Cement Pastes on Rheological and Mechanical Properties

Haodao Li, Jingjie Wei, Wujian Long

Guangdong Provincial Key Laboratory of Durability for Marine Civil Engineering College of Civil Engineering, Shenzhen University, Shenzhen 518060, China.

Abstract. The state of GO dispersion is closely related to the properties of graphene oxide (GO)-based cement paste. This paper presents the effect of uniformly dispersed and re-agglomerated GO on the rheological and mechanical properties. The results showed that, compared to re-agglomerated GO cement paste, the yield stress and plastic viscosity of uniformly dispersed GO cement paste were higher. Moreover, the compressive and flexural strengths of uniformly dispersed GO pastes were higher than those of re-agglomerated GO pastes. Porosity analysis using mercury intrusion porosimetry showed that the well-dispersed GO can inhibit the formation of large-diameter pores and optimize the pore size distribution better than the re-agglomerated GO.

1 Introduction

Cement-based composites modified with graphene oxide (GO) additives have been increasingly used in recent years as engineering materials [1]. The use of GO is advantageous since GO is an advanced nanomaterial that exhibits superior performance of GO-incorporated cement pastes. It is clear that this additive does not function solely as a filler, its modifying effect is based on nucleation, followed by aggregation of cement hydration products and formation of a dense mesh structure, ultimately accelerating hydration and improving the early mechanical properties of cement pastes [2]. However, although GO is typically well-dispersed in aqueous solution, an equivalently uniform dispersion in a highly alkaline environment may not be obtained owing to the steric stabilization of GO with charged ions in a cement pore solution [3].

The agglomeration of GO monolithic phases critically affects the properties of GO incorporated cement pastes. It is clear that the abundance of calcium cations plays a significant role in the dispersion of GO in fresh cement pastes [4]. Many studies have been conducted to improve the dispersion of graphene, for example, surface modification of graphene during refining, introduction of foreign molecules and addition of surfactant molecules [5]. An appropriate way for consistent GO dispersion instead of re-agglomeration is mixing with PCE, which can reduce the van der Waals force between GO sheets and disperse them uniformly throughout cement matrix [6].

To compare the differences between uniformly dispersed and re-agglomerated GO, this study systematically investigates the effect of different dispersion states of GO on the rheological and mechanical properties of cement pastes using two distinct mixing sequences among GO, PCE, and cement.

2 Materials and experimental procedure

2.1 Raw materials

Type I ordinary Portland cement (OPC), conforming to the requirements of Chinese Standard GB 175[7], was used as the base material in this study. The chemical composition of OPC is provide in Table 1.

Table 1. Chemical composition of OPC (% by mass).

CaO	SiO_2	Al_2O_3	Fe_2O_3	SO_3	MgO	S_fO	Na_2O	Cl^-	LOI
64.65	21.88	4.49	3.45	2.44	2.36	0.90	0.51	0.010	1.31

Graphene oxide was purchased from Sixth Element Ltd., and used probe sonication, a suspension of GO (3 g/L) was prepared by dispersion the graphite oxide powder in water. The frequency and the time of ultrasound application were 25 Hz and 2 h, respectively. Polycarboxylate-ether was used as a surfactant to disperse GO in the cement pastes, conforming to the requirement of JG/T 223[8]. As a result, long molecules of the organic polymer wrap around the GO particles and prevent their aggregation by physically blocking them from approaching each other.

2.2 Test methods

2.2.1 Rheological measurements.

The rheological parameters of the cement paste were measured using an RM 100 touch device manufactured by Lamy Rheology Instruments. During the rheological test, the shear rate was increased from 3 to 240 s^{-1} over 15 speed intervals. The plastic viscosity (η_p) and the shear stress (τ_0) can be obtained from the slope and the intercept of the Modified Bingham (M-B) model relating the shear stress.

2.2.2 Mechanical measurements.

The strengths of the hardened cement pastes were tested according to the procedures of GB/T 17671[9]. The compressive and flexural loading speeds were 2.4 kN/s and 50 N/s, respectively.

2.3 Mix proportions

A PCE-to-GO mass ratio of 1:1 was chosen based on preliminary results and is consistent with the limited data reported in the literature [10]. The mix proportions of the GO-based cement pastes are given in Table 2 and the only difference in the mixing sequence of the two samples is the pre-mixing of GO with PCE before adding to the cement paste.

3 Results and discussion

3.1 Effect of dispersion state of GOs on the rheological properties

As can be seen in Figure 1(a), the fresh cement paste containing the re-agglomerated GO (Paste No.1) and uniformly dispersed GO (Paste No.2) show obvious differences in the curve shape but exhibit similar ascending trends prior to solidification, indicating that the two distinct dispersion states of GO in fresh cement pastes evaluated here have a significant effect on the shear stress under different shear rates. Moreover, the shear stress of Paste No.1 is smaller than that of Paste No.2 at the same shear rate, indicating that the re-agglomeration of GO may cause a decrease in the shear stress of paste. The possible reasons are the surface area of GO decreases after re-agglomeration, consequently decreasing water absorption and increasing the amount of free water in paste.

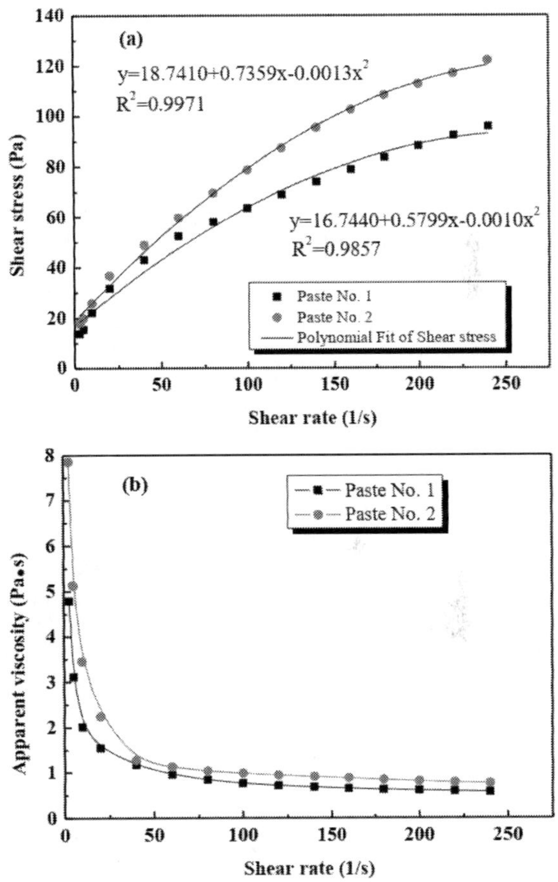

Figure 1. Rheological parameters of GO cement pastes: (a) shear rate vs. shear stress and (b) shear rate vs. apparent viscosity.

Table 2. Mix proportions of cement pastes.

Mix No.	Cement (g)	Water (g)	W/C	GO (g)	GO/Cement	PCE/GO
Paste No. 1	3600	1440	0.4	2.16	0.06%	1.0
Paste No. 2	3600	1440	0.4	2.16	0.06%	1.0

Table 3. Rheological parameters of cement pastes.

Mix No.	Fitting equation	τ_o (Pa)	η_p (Pa·s)	Correlation coefficient
Paste No. 1	$y = 16.7440 + 0.5799x - 0.0010x^2$	16.7440	0.5799	0.9857
Paste No. 2	$y = 18.7410 + 0.7359x - 0.0013x^2$	18.7410	0.7359	0.9971

Figure 1(b) shows that the apparent viscosity of Pastes No.1 and No.2 decrease rapidly with increase in the shear rate and eventually stabilize. Meanwhile, at the same shear rate, the apparent viscosity of Paste No.1 is less than that of Paste No.2 because the flocculated structure in Paste No.1 are destroyed by the applied shear stress, and the re-agglomerated GO is unable to connect any flocculated structure

The yield stress and plastic viscosity of the two cement pastes evaluated in this study are given in Table 3. The results were fitted by the Modified-Bingham (M-B) model, and the correlation coefficient is close to 0.99, indicating that this fitting model is feasible and appropriate.

3.2 Effect of dispersion state of GOs on mechanical properties

Table 4 shows the compressive and the flexural strengths of Pastes No. 1 and No. 2. Both the compressive and the flexural strengths of Paste No. 1 are lower than those of Paste No. 2. The possible reasons include uniform dispersion of GO in Paste No. 2, which facilitates the hydration process of the cement paste, and filling of the inner pores of the cement matrix with GO nanosheets, which increases the density of the cement matrix.

It can be seen from Table 4 that the compressive and flexural strengths of Paste No. 2 at 3, 7 and 28 days were increased by 8%, 5% and 4% and 27%, 26% and 19%, respectively, when compared to Paste No. 1. Therefore, it can be concluded that the homogeneous dispersion of GO increases the flexural strength of a paste at an early stage. Clearly, in order to ensure the strength of GO-cement composites, the incorporation of well-dispersed GO into the cement paste must be assured.

3.3 Analysis based on MIP

The total porosity and the ratio of different pore sizes to the total porosity in Paste No.1 and No.2 at 7 and 28 d of curing are shown in Fig. 2. According to the influence of different pore sizes on the properties of concrete, the pores can be divided into innocuous pores (less than 0.02 μm in diameter), less harmful pores (0.02–0.05 μm in diameter), harmful pores (0.05–0.2 μm in diameter) and more harmful pores (greater than 0.2 μm in diameter). Because both the pore size and pore ratio determine the performance of concrete, increasing the proportion of pores less than 0.05 μm in diameter, rather than solely decreasing the proportion of pores greater than 0.05 μm in diameter, is more favorable for improving the performance of a concrete, particularly in terms of the durability of the materials.

It can be seen from Figure 7 that the total porosity of Pastes No. 1 and No. 2 are 23.5%, 17% and 23%, 13% at 7 and 28 days, respectively. The total porosity of a sample trends to reduce as the age increases and the total porosities for different ages of Paste No. 1 are consistently greater than those of Paste No. 2, which fits well with the results of strength testing.

Prior to 28 d, the total porosity of Pastes No. 1 and No. 2 decreases from 26.2% and 23.9% to 17% and 13%, respectively, and the percentage of pores less than 0.05 μm in diameter increases from 63% and 83.3% to 84.4% and 85%, respectively. Therefore, between the ages of 7 and 28 d, the strength of Paste No. 1 increased because of decrease in both the total porosity and the percentage of pores less than 0.05 μm in diameter. However, the strength of Paste No. 2 remains higher because the percentage of pores less than 0.05 μm in diameter continue to be a significant proportion of the total porosity.

Table 4. Compressive and flexural strengths of cement pastes (in MPa).

Age (days)	Compressive strength (% increase)		Flexural strength (% increase)	
	Paste No. 1	Paste No. 2	Paste No. 1	Paste No. 2
3	43.1	46.7 (8%)	2.2	2.8 (27%)
7	58.5	62.0 (5%)	5.0	6.3 (26%)
28	62.3	64.8 (4%)	7.5	8.9 (19%)

 https://doi.org/10.1051/matecconf/201820603003

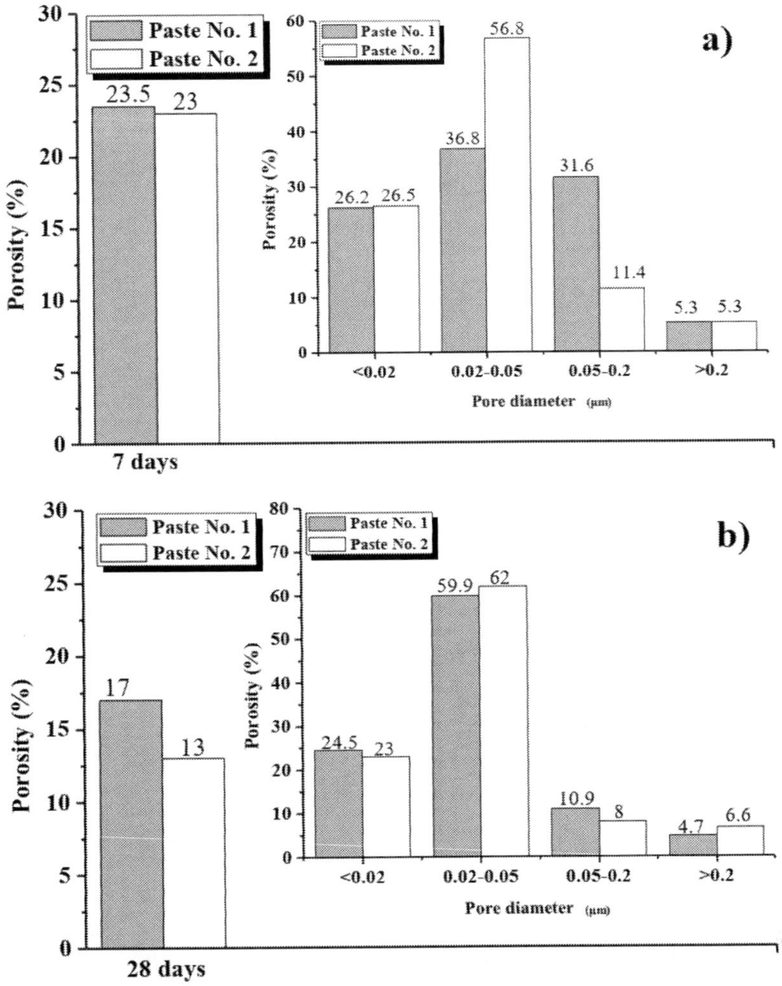

Figure 2. Total porosity of Paste No.1 and No.2 at (a) 7 and (b) 28 d of curing, and ratio of different pore sizes in range of <0.02 μm, 0.02–0.05 μm, 0.05–0.2 μm, and >0.2 μm to the total porosity.

In summary, hardened cement pastes containing a uniform dispersion of GO have a greater strength than those containing re-agglomerated GO. The hydration process of cement pastes containing uniformly dispersed GO is faster than that of cement pastes containing re-agglomerated GO, resulting in a lower cement-based porosity in uniformly distributed pastes at the same age. Additionally, the uniform dispersion of GO is able to increase the ratio of harmless and less-detrimental pores in cement pastes and optimizes the pore size distribution within the matrix, further enhancing the strength of the cement.

4 Conclusion

(1) The yield stress and plastic viscosity of Paste No. 2 were higher than those of Paste No. 1. It may conclude that the uniform dispersion of GO increases the water requirement of the paste and GO enhances connections between the flocculated structures of hydration products.

(2) The pores in the cement matrix were filled with GO, restricting the development of cracks. The mechanical properties of Paste No. 2, including compressive and flexural strengths, were higher than those of Paste No. 1.

Acknowledgements

The authors gratefully acknowledge the financial support provided by the National Natural Science Foundations of China (No. 51778368, No.51578341), and the Shenzhen Overseas High-Caliber Personnel Grant (No. 000095).

References

1. J. Silvestre, N. Silvestre, J.D. Brito, Eur. J. Environ. Civil Eng. **20**, 1-31 (2015)

2. M. Wang, R. Wang, H. Yao, Z. Wang, S. Zheng, Rsc Adv. **6**, 63365-72 (2016)
3. S. Rana, R. Fangueiro, J. Nanomater. **2013**, 1-19 (2013).
4. M. Saafi, L. Tang, J. Fung, M. Rahman, J. Liggat, Cem. Concr. Res. **67**, 292-99 (2015)
5. A.H. Korayem, N. Tourani, M. Zakertabrizi, A.M. Sabziparvar, W.H. Duan, Constr. Build. Mater. **153**, 346-57 (2017).
6. D.G. Gao, Y.J. Ma, Fine Chem. (2015)
7. GB/T175-2007 Chinese national standard (2007)
8. JG/T223-2007 Chinese construction industry standard (2007)
9. GB/T17671-2007 Chinese national standard (2007)
10. M.S. Konsta-Gdoutos, Z.S. Metaxa, S.P. Shah, Cem. Concr. Res. **40**, 1052-59 (2016)

Effect of Bonding Temperature on Microstructure Evolution during TLP Bonding of a Ni_3Al based Superalloy IC10

Xiong Yue[1,2], Fengmei Liu[1], Hexing Chen[1], Di Wan[1] and Hongbo Qin[1]

[1]*Guangdong Provincial Key Laboratory of Advanced Welding Technology, Guangdong Welding Institute (China-Ukraine E. O. Paton Institute of Welding), 510650 Guangzhou, China*
[2]*School of Materials Science and Engineering, Central South University, 410083 Changsha, China*

Abstract. Transient liquid phase (TLP) bonding of Ni3Al-based superalloy IC10 was carried out using the interlayer based on the base metal which added B and Hf as the melting point depressant elements. The effect of bonding temperature (1250 – 1270 °C) on the microstructure evolution of bonding joints were investigated. Microstructure of bonding joint composed of isothermally solidification zone (ISZ) formed γ' phase and athermally solidified zone (ASZ) which consists of newly formed γ+γ' reticular eutectic among with borides and carbides. Boride precipitates are not formed in diffusion affected zone (DAZ) and the boundary between ASZ and ISZ become not obvious. Isothermally solidification rate decreases as the increase of the bonding temperature.

1 Introduction

Nickel-based superalloy is an essential material for the key parts of aero-engine working blades, turbine disks and combustion chambers. With the advance of aero-engine technology, inlet temperature of the turbine is continuously increasing, which makes service temperature and service life of the turbine blades face a stern challenge[1]. As the newly developed Ni_3Al based superalloys, IC10 can be used as the material for advanced aero-engine fan with a service temperature up to 1100 °C[2]. Transient liquid phase (TLP) bonding is currently an optimized technology utilized for joining Ni-based superalloy, which combines the beneficial features of liquid phase joining and diffusion bonding techniques.

The directionally solidified superalloy IC10 contains 1.5 Wt.% Hf, which makes the alloy has excellent casting properties. In the late stage of solidification, Hf rich liquid show superior flow characteristics and good wettability. Hf also can strengthen the γ'(Ni_3Al) phase and improve its corrosion resistance. Basing on the Ni-Hf phase diagram, an appropriate content of Hf can reduce the liquidus temperatures of the alloy, and interlayers using Hf as the only melting point depressant element also had been successfully applied to TLP bonding of nickel-base superalloys[3]. However, due to the larger atomic radius of Hf, it takes a long time to reduce its concentration in the joint region. In this study, in order to decrease content of B in interlayer, B and Hf were both used as MPD elements in TLP bonding of the directionally solidified superalloy IC10. Appropriate bonding temperature have an important influence on the degree of isothermal solidification and the formation of phases and their morphology.

2 Experiment proceduce

The directionally solidified superalloy IC10 is mainly composed of mesh γ and γ' phase. Besides, some sunflower shaped γ+γ' eutectic together with some TaC and HfC carbides distribute among mesh γ and γ' phase[4]. The chemical compositions of the base metal and the interlayer numbered BH2 are shown in Table 1. The interlayer was melting with WK-I non-consumable vacuum arc melting furnace. Its melting temperature range was tested by Germany-resistant STA409PC synchronous thermal analyzer. The solidus and liquidus temperature of the interlayer are 1172.1°C and 1240.7 °C, respectively.

The TLP bonding operation was carried out in a United States CVI high vacuum diffusion furnace. According to DSC results of the interlayer alloy, the specimens were bonded at temperature of 1250, 1260, 1270 °C for holding time of 2 h. Then the 4, 6 and 8 h were respectively kept at 1250 °C. Samples were etched in a solution of HNO_3+HF+Glycerin (1:2:1). Microstructural examinations were carried out using Nova NanoSEM 430 ultra-high resolution field emissions scanning electron microscope equipped with a beryllium window energy dispersive spectrometer (EDS) system. X-ray diffraction (XRD) analysis of phases present in the joint area was carried out using a PANalytical X'Pert Power X diffractometer equipped with a Cu-Kα target. The wave length of Cu-Kα radiation was 0.15406 nm.

Table 1 IC10 and BH2 interlayer alloy chemical composition (wt.%)

Alloy	B	Hf	C	Al	Cr	Co	Mo	Ta	W	Ni
IC10	0.04-0.042	1.4-1.5	0.087-0.09	5.4-5.5	7.1-7.3	12.1-12.3	1.5-1.6	6.7-6.9	4.9-5.2	Bal.
BH2	2.4	2.6	0.087-0.09	5.4-5.5	7.1-7.3	12.1-12.3	1.5-1.6	6.7-6.9	4.9-5.2	Bal.

3 Result and discussion

3.1 Effect of temperature on microstructure evolution of TLP bonded joint

According to previous experiments, B and Hf were used as the MPD elements of TLP bonding IC10, the bonding joint mainly consisted of three zones: isothermally solidification zone (ISZ), athermally solidified zone (ASZ) and base metal (BM). Fig. 1 shows SEM images of the joint TLP bonded at 1250, 1260 and 1270 °C for 2 h with BH2 interlayer. Fig. 2 shows the analysis of X-ray diffraction patterns from tensile strength fracture surfaces of the joint, it revealed that the joint is composed of Ni_3Al, HfC, TaC, WB and $Cr_{1.8}W_{3.2}B_3$. Due to the complex composition of borides, it mostly called borides in this paper. It can be seen from Fig. 1 (a) and (b), the joint bonded at 1250 °C mainly consists of cellular γ' phase. Besides, newly formed γ+γ' reticular eutectic and a small amount of W rich borides distribute in the cellular γ' phase. As can be seen from Fig. 1 (c) and (d), the cellular γ' phase is larger at 1260 °C, which is related to the presence of a large amount of Al in liquid phase and un-restricted growth of γ' phase. As growth rate of cellular γ' phase increased, the boundary between ASZ and ISZ become not obvious. With the progress of isothermal solidification, solute elements whose partition coefficient less than one (k<1) discharged to the liquid front phase, resulting in the solute redistribution in the frontier enrichment of the solid-liquid interface. Although the latent heat of crystallization is released during the solidification of the Ni-based solid solution, the frontal solid-liquid interface is under a positive temperature gradient, and γ' phase interface is in a cell-like shape. Therefore, there is a definite component undercooling at the front of the interface. It can be seen from Fig. 1 (e) and (f) that the cellular γ' phase disappears at 1270 °C.

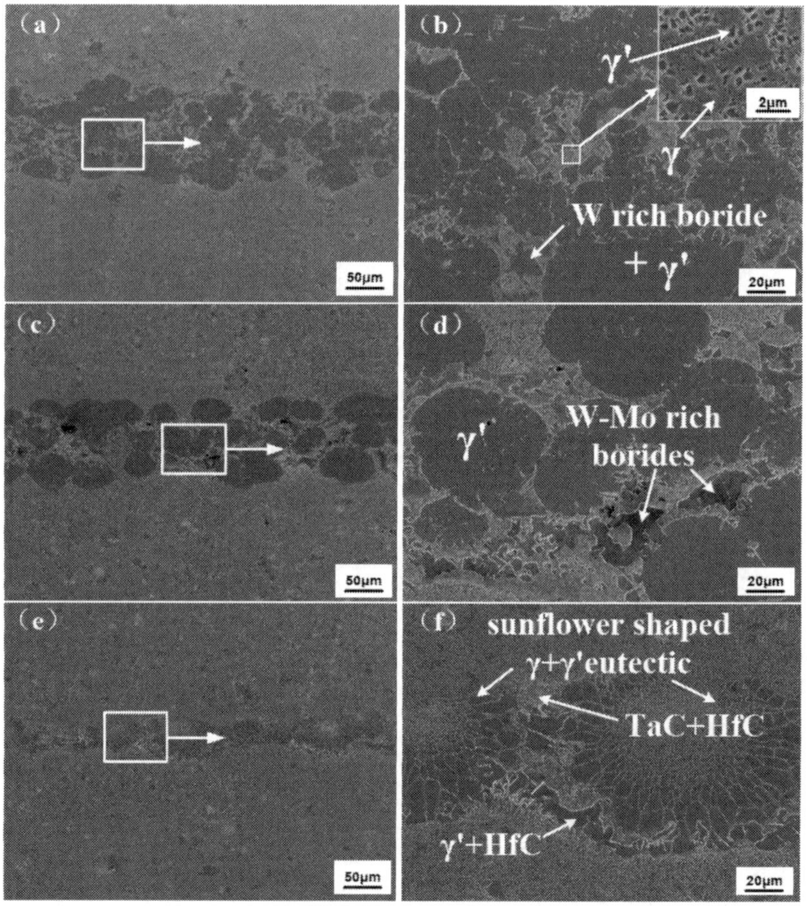

Fig. 1. SEM images of the joint TLP bonded at (a), (b) 1250 °C for 2 h; (c), (d) 1260 °C for 2 h; (e), (f) 1270 °C for 2 h.

The bonding joint mainly consists of sunflower shaped $\gamma+\gamma'$ eutectic. Besides, γ' and HfC symbiotic phases, HfC and TaC symbiotic phases distribute at the edge of sunflower shaped eutectic.

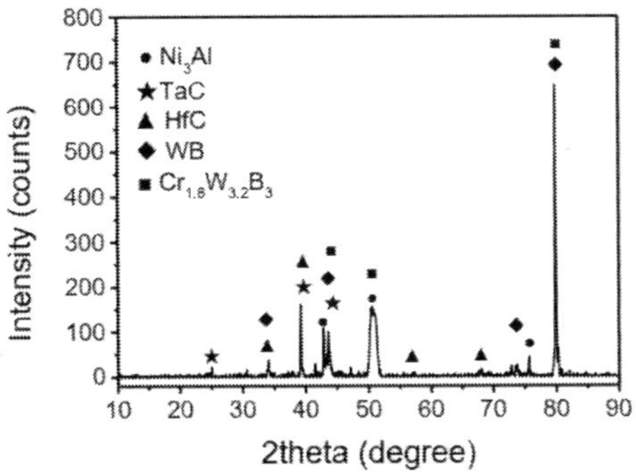

Fig. 2. XRD pattern of the tensile strength fracture surfaces of the TLP joint brazed at 1250 °C for 2 h.

It can be observed in Fig. 1, Boride precipitates are not not formed in diffusion affected zone (DAZ) like conventional TLP bonding. This phenomenon also had been confirmed by M. Pouranvari[5] and A. Schnell[6]. Isothermal solidification process is controlled by diffusion of B to base metal. And the diffusion of B from the ISZ to base metal is controlled by the concentration of B in the liquid and ISZ interfaces. If B content in ISZ is lower than B solubility in the base metal, boride precipitation in the base metal would not be formed. In addition, Cr, W and Mo from the interlayer and the base metal are easily combined with B which leads to the decrease of B content in ISZ. Therefore, it is more difficult to form boride precipitates.

3.2 Effect of temperature on isothermal solidification rate of TLP bonded joint

It can be seen from Fig. 1, the isothermal solidification rate decreases as the increase of the bonding temperature. Cellular γ' phase become obvious as the bonding temperature rises from 1250 to 1260 °C. When the bonding temperature increases to 1270 °C, the degree of solidification is low and the sunflower shaped $\gamma+\gamma'$ eutectic is formed. Jalilvand V et al.[7] believed that the low isothermal solidification rate is not only related to the enrichment of base metal alloy elements in the liquid phase, but also to the decrease of the solid solubility limit of B in base metal. Fick's second law ignores changes in D and concentration, and assumes a constant diffusion coefficient:

$$\frac{\partial c}{\partial t} = D\frac{\partial^2 c}{\partial x^2} \tag{1}$$

$\partial c/\partial t$ is the solid-liquid surface solid-state diffusion rate, D is the diffusion coefficient. $\partial^2 c/\partial x^2$ is concentration gradient of MPD elements as the diffusion distance changes and is determined by the solid solubility limit of MPD elements in the base metal. The concentration gradient of MPD elements in the base metal affects diffusion rate, controlling the isothermal solidification rate. According to the Arrhenius equation, an increase in temperature leads to an exponential increase in diffusion coefficient of MPD elements. Based on the Ni-B phase diagram, above the eutectic temperature, the solubility of B in Ni decreases with the increase of the bonding temperature. The decrease of the solubility of B in Ni leads to a decrease of the driving force for the diffusion of B at the solid-liquid interface, resulting in the enrichment of B in the liquid phase.

The rate of isothermal solidification also can be judged by the time required for isothermal solidification, which is controlled by the solid-phase diffusion of solute atoms. The diffusion problem can be treated as a diffusion matter in an infinite system with the width of $2h = W_0$ (W_0 is the thickness of the interlayer). The concentration of the diffusion element in the thickness of the interlayer is C_0, and the concentration of the other regions (the concentration of the diffusion element in the base metal) is C_m. From this, we can find the concentration distribution of diffusing elements in the system at a bonding temperature for a period of time. According to Fick's second law, we use the error function to solve[8, 9]:

$$C(x,t) = C_m + \frac{C_0 - C_m}{2}\left\{erf\frac{x+h}{2\sqrt{Dt}} - erf\frac{x-h}{2\sqrt{Dt}}\right\} \tag{2}$$

Where $C(x, t)$ is a function of distance x from the weld center and time t, and D is the diffusion coefficient of the diffusion element in the solid phase at base metal. From Eq. 2, the time t_f required for isothermal solidification can be obtained as the time when the diffusion element concentration in the interlayer decreases to the solidus concentration Cs at $x = 0$[10]:

$$C_s - C_m = C_0 - C_m\left\{erf\frac{h}{2\sqrt{Dt_f}}\right\} \tag{3}$$

the diffusion coefficient of MPD element in base metal (D) depends on bonding temperature through the Arrhenius relationship:

$$D = D_0\left(\frac{-Q}{RT}\right) \tag{4}$$

where R is the gas constant, Q is the apparent activation energy for diffusion of MPD element, D_0 is the frequency factor. If Hf as MPD element, the time of complete isothermal solidification (t_f) can be estimated from Eq. 2 and Eq. 4. Bergner indicated that the Q and D_0 of Hf is 287 KJ/mol and 1.8 × 10^{-4} m^2/s at the temperature range of 750 - 1150 °C; Kurtz reported that the Q and D_0 of Hf is 272.1 KJ/mol and 2 × 10^{-4} m^2/s at the temperature range of 750 – 1150 °C[11]. The diffusion activation energy of Hf is 272.1 - 287 KJ/mol. In this experiment, we take Q = 272.1 KJ/mol, D_0 = 2 × 10^{-4} m^2/s. When W_0 = 100 μm, C_0 = 2.6 wt.%, C_m = 1.4 wt.%, the time required for isothermal solidification (t_f) at 1250, 1260 and 1270 °C is 2.96, 3.55 and 4.72 h, respectively. If B as MPD element, the time of complete isothermal solidification (t_f) can be estimated from Eq. 3 and Eq. 4. The diffusion activation energy of B is 248 kJ/mol and D_0 is 0.81 m^2/s. When W_0= 100 μm, C_0 = 2.4 wt.%, C_m = 0, the time required for isothermal solidification at 1250, 1260 and 1270 °C is 1.99, 1.97 and 1.92 h, respectively. In view of the time required for the isothermal solidification of B diffusion from calculation, the increase of temperature has little effect on the time

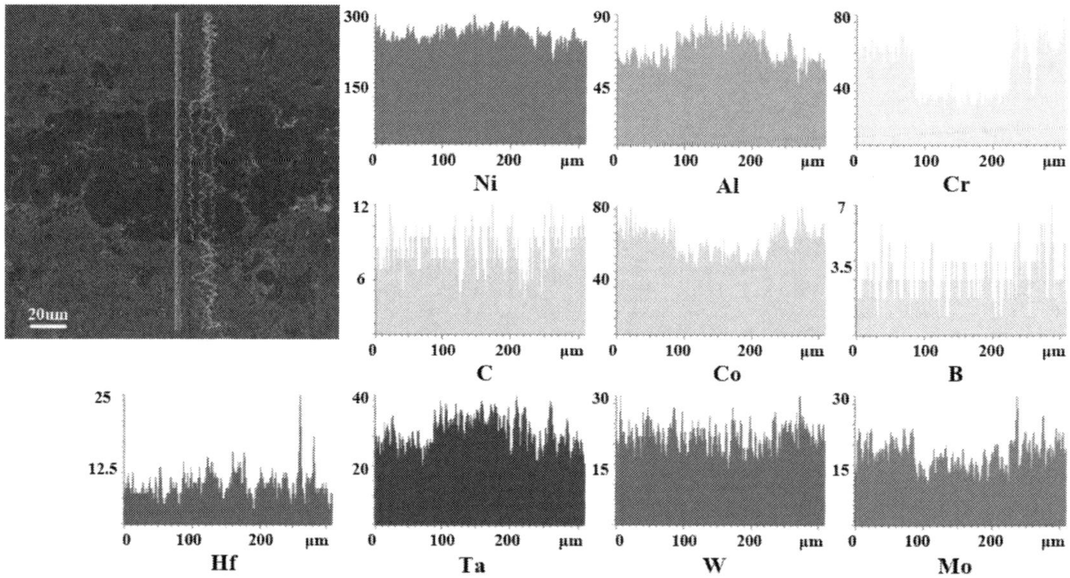

Fig. 3. SEM images in the TLP bonded sample brazed at 1250 °C for 4 h and it lines scan of Ni, Al, Cr, C, Co, B, Hf, Ta, W and Mo by SEM/EDS analysis.

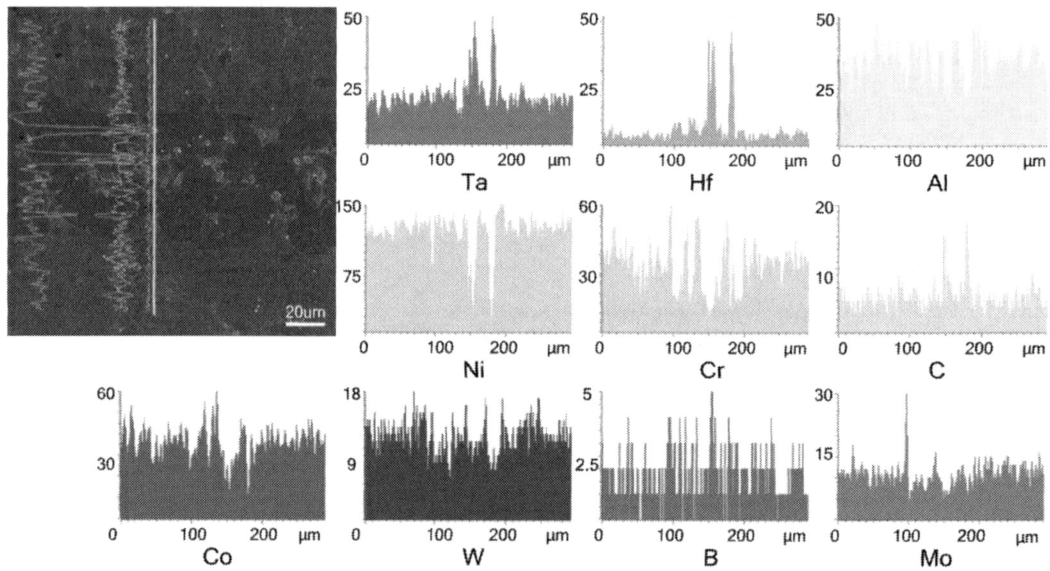

Fig. 4. SEM images in the TLP bonded sample brazed at 1250 °C for 8 h and it line scan of Ni, Al, Cr, C, Co, B, Hf, Ta, W and Mo by SEM/EDS analysis.

required to complete the isothermal solidification. However, the time required for the isothermal solidification of Hf diffusion depends strongly on bonding temperature. The time of complete isothermal solidification increased obviously, which indicates the decrease of isothermal solidification rate.

As can be seen from Fig. 3, the bonding joint almost completed the isothermal solidification at 1250 °C for 4 h. The distribution of Ni in bonded joint is uniform and content of Cr, Co, B, Mo and W is lower than that in base metal. It may be caused by Cr-W-Mo borides in the bonding region depleting Cr, Mo, W and B. However, Al, Ta and Hf in the bonding joint are higher than the base metal because Hf and Ta elements tend to form carbides and preferentially enter γ' phase, and Al is the constituent of Ni_3Al. Fig. 4 shows the lines scan of the sample bonded at 1250 °C for 8 h. The peaks of Hf, Ta and C are in accord with each other, while the peaks of Ni and Al are exactly the opposite. It can be judged that the lines scan passes through the carbides riched Hf and Ta. Increasing the bonding time to 8 h, carbides and borides don't decrease in the bonding joint. This phenomenon can be explained by the higher Hf content in the interlayer, which forms carbides and is not easily decomposed at the bonding temperature. This is in agreement with the previous works on TLP bonding of

SXG3 alloy using Hf-containing Ni-based interlayer by Z.R. Yu et al.[3] who reported that granular Hf rich carbide was observed in the joint when the bonding time is 15 min. But the SXG3 alloy which bonded at 1290 °C for 6 h has been reheated to 1290 °C and quenched, Hf rich carbide still can be observed. According to quench results, carbides precipitated in the liquid phase at the bonding temperature, whereas the eutectic is formed in the bonding joint during the furnace cooling.

4 Conclusions

(1) Microstructure of the TLP joint is composed of ISZ and ASZ. The ISZ formed γ' phase rather than the γ solid solution. The ASZ mainly consists of newly formed γ+γ' reticular eutectic, borides and carbides. Boride precipitates are not formed in DAZ like conventional TLP bonding.

(2) The isothermal solidification rate decreases as the increase of the bonding temperature. Cellular γ' phase become obvious as the bonding temperature rises from 1250 to 1260 °C. When the bonding temperature increases to 1270 °C, the sunflower shaped γ+γ' eutectic structure is formed. When the bonding temperature is 1250 °C, prolonging holding time does not eliminate borides and carbides in the joint.

Acknowledgments

This work was supported by the National International Science and Technology Cooperation Project of China (Grant No. 2015DFR50310), the Province Science and Technology Project of Guangdong (2017A070701026) and the 2018 GDAS' Special Project of Science and Technology Development (2018GDASCX-0113).

References

1. S. Steuer, R.F. Singer, Metall, Mater. Trans. A. **45**, 3545 (2014)
2. H. Zhang, W. Wen, H. Cui, Mater. Sci. Eng. A. **504**, 99 (2009)
3. Z.R. Yu, X.F. Ding, N.M. Cao, Y.R. Zheng, F. Q, Acta Metall Sin. **52**, 549 (2016).
4. H. Zhang, W. Wen, H. Cui, Y. Xu, Mater. Sci. Eng. A. **527**, 328 (2009)
5. M. Pouranvari, A. Ekrami, A. Kokabi, Mater. Sci. Eng. A. **568**, 76 (2013)
6. M. Pouranvari, A. Ekrami, A.H. Kokabi, J. Alloy Compd. **469**, 270 (2009)
7. W. Gale, E. Wallach, Mater. Trans. A. **22**, 2451 (1991)
8. V. Jalilvand, H. Omidvar, M.R. Rahimipour, H.R. Shakeri, Mater Des. **52**, 36 (2013)
9. O.A. Idowu, N.L. Richards, M.C. Chaturvedi, Mater. Sci. Eng. A. **397**, 98 (2005)
10. M.A. Arafin, M. Medraj, D.P. Turner, P. Bocher, Adv. Mater. Res. **15**, 882 (2007)
11. O.A. Ojo, N.L. Richards, M.C. Chaturvedi, Sci.Technol. Weld. Join. **9**, 532 (2004)

Effect of Mixed Rare Earths on the Wetting Behavior and Interfacial Reaction between Sn-0.70Cu-0.05Ni Solder and Amorphous $Fe_{84.3}Si_{10.3}B_{5.4}$ Alloy

Bin Hou[1], Fengmei Liu[1], Hongqin Wang[2], Yupeng Zhang[1], Jianglong Yi[1] and Hongbo Qin[1]

[1]*Guangdong Provincial Key Laboratory of Advanced Welding Technology, Guangdong Welding Institute (China-Ukraine E.O.Paton Institute of Welding), Guangzhou, 510650, China*
[2]*Reliability Research and Analysis Center, the Fifth Research Institute of MITT, P. R, Guangzhou, 510610, China*

Abstract. In order to explore the effect of addition of mixed rare earths (MRE) on the wetting behavior and interfacial reaction between Sn-0.70Cu-0.05Ni solder and amorphous $Fe_{84.3}Si_{10.3}B_{5.4}$ alloy, 0.25 wt.% percentage of the MRE, which are mainly elements La and Ce, were added into the solder. Results show it can refine the microstructure of the solder alloy, and there is limited change of melting temperature with the addition of MRE in the solder. The wettability of the solder on amorphous substrate is improved by adding 0.25 wt.% percentage of the MRE into Sn-0.70Cu-0.05Ni solder. Moreover, research results indicate that, with the increase of wetting temperature, the final equilibrium wetting angles of Sn-0.70Cu-0.05Ni and Sn-0.70Cu-0.05Ni-0.25MRE on amorphous substrate decrease gradually, indicating the better wettability at the higher wetting temperature. In addition, with the increase of temperature, the distribution of intermetallic compound (IMC) $FeSn_2$ formed at the interface between the two solders and amorphous substrate is changed from discontinuous state to continuous state. The thickness of the interfacial IMC layer between solder and amorphous substrates reduced with the addition of MRE, indicating that the presence of 0.25 wt.% percentage of the MRE is effective in suppressing the growth of IMC layer.

1 Introduction

Due to excellent mechanical, physical and chemical properties, amorphous alloys are ideal materials for aviation and spacecraft structures owing to the high specific strength, which is vital in the field of aerospace[1]. However, at present, the critical sizes of amorphous alloys with excellent deformability prepared by fast cooling method are limited within millimeters, constraining their wide applications as engineering materials[2]. In order to overcome the size limitation and spread their applications, it is necessary to explore appropriate joining techniques to obtain larger size amorphous alloys. As a low temperature joining process, soldering is an effective and reliable technique because of the advantage of little influence on the structure of amorphous alloys. In addition, the good wettability and interfacial reaction of solders on the amorphous alloys are prerequisites for the high soldering reliability. Therefore, the research on the wetting behavior and the interfacial reaction between solders and amorphous alloys are of great significance.

Owing to environmental and legal issues of using Pb in electronic packaging, large quantities of investigations have been carried out on lead-free solders for the substitutes of Sn-Pb solder[3]. Among the lead-free solders, SnCuNi solder is widely used in wave soldering process because of its cheapness. However, the physical properties and wetting ability of SnCuNi are lower than those of the traditional Sn-Pb solder[4]. Moreover, rare earth (RE) elements, which are called the "vitamin"of metals, can greatly enhance the mechanical and physical properties of metals. In addition, several researchers have reported that the addition of rare earth Ce or La+Ce into lead-free solders can obviously refine microstructures and improve mechanical properties[5,6]. Further, the wettability and interfacial reaction of solders with rare earth addition on the copper substrate have been widely studied[4,5]. However, thus far, the effects of the addition of RE elements on the wettability and interfacial reaction between the lead-free solders and the Fe-based amorphous alloy substrates are seldom discussed.

In this study, the mixed rare earth (MRE) elements La and Ce were selected as additives added into Sn-0.70Cu-0.05Ni lead-free solder. The purpose of the current research is to investigate the effects of trace amounts of MRE on microstructure evolution, melting property, wetting behavior and interfacial microstructure of Sn-0.7Cu-0.05Ni solder on amorphous $Fe_{84.3}Si_{10.3}B_{5.4}$ substrate.

2 Experimental procedure

 https://doi.org/10.1051/matecconf/201820603005

In this study, the raw materials used for this study are Sn, Cu, Ni (purity 99.9%) and MRE (with about 60% Ce and about 40% La). First, Cu-RE master alloy was melted in an electronic resistance furnace at 500 °C for two hours. Second, the master alloy and suitable weights of other metals (Sn, Cu and Ni) were mixed together and melted under the same conditions to prepare Sn-0.70Cu-0.05Ni and Sn-0.70Cu-0.05Ni-0.25MRE solders. A stainless steel rod was used to stir the liquid solder every 10 min to ensure the uniformity of solder. In order to prevent oxidation during the melting, KCl + LiCl molten salt with the mass ratio of 1:1.3 was used to cover the surface of the liquid solder. In order to measure and analyze thermodynamic parameters of the amorphous $Fe_{84.3}Si_{10.3}B_{5.4}$ alloy and the two solders, DSC (NETZSCH STA 449C) measurements were carried out at the heating rate of 20 K min^{-1} in the Ar atmosphere. The structures of the amorphous samples were examined by X-ray diffraction (D8 ADVANCE, Germany) with Cu Kα radiation. The XRD patterns and DSC curves of amorphous substrate are shown in Fig. 1. Fig. 1(a) presents that no crystalline peak is visible, which indicates that the substrate was in the amorphous state. There are two successive exothermic events in the crystallization process with peak temperatures about 515 °C (T_{p1}) and 555 °C (T_{p2}). Meanwhile, the glass transition temperature is about 488 °C (T_g), and the starting temperature of the crystallization is about 511 °C (T_x), as shown in Figure 1(b).

The wetting experiments of solders on amorphous $Fe_{84.3}Si_{10.3}B_{5.4}$ substrates were carried out by the sessile drop method using Dataphysics OCA-20 in the N_2 atmosphere. The solder samples were made into small balls and cleaned ultrasonically together with amorphous $Fe_{84.3}Si_{10.3}B_{5.4}$ substrates in the acetone. The solders with an activated flux were placed on the center of the amorphous substrates placed into stainless-steel chamber. Then, the chamber was heated to 250 °C, 300 °C, 350 °C and 400 °C, respectively, to investigate the wetting behavior between the solders and amorphous substrates. The drop profiles were recorded by a high resolution charge-coupled- device camera and analyzed by a drop-analysis program to record the variation of contact angles with time and calculate the final equilibrium contact angle. The wetting process was held for 30 min and then the samples were air-cooled by taking them out from chamber. The solidified samples were sectioned to examine the interfaces between solders and amorphous substrates. The interfacial microstructure and the chemical composition of interface compounds were identified using the scanning electron microscope (SEM, SUPRA 55V) equipped with energy diffraction spectrum (EDS).

3 Results and Discussion

3.1. Melting Temperature of Solders.

The initial point of the DSC heating curve is recognized as the solidus temperature while the peak point is considered to be the liquidus temperature of solders[7]. Fig.2 shows the solidus temperature and liquidus temperature of Sn-0.70Cu-0.05Ni and Sn-0.70Cu-0.05Ni-MRE solders. Clearly, the solidus temperature and liquidus temperature decrease slightly with the addition of 0.25 wt.% MRE. The liquidus temperature ranges from 230.66 to 231.29 °C, while the solidus temperature is between 228.85 and 227.56 °C. It can be concluded that the addition of MRE has a slight influence on the melting temperature of Sn-0.70Cu-0.05Ni solder.

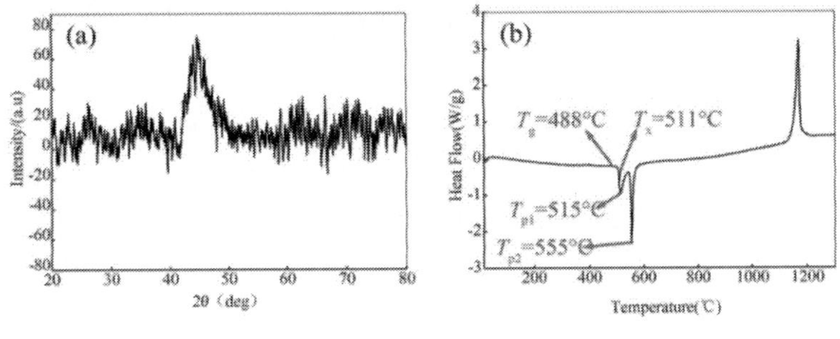

Figure 1. XRD patterns (a) and DSC curve (b) of amorphous substrate.

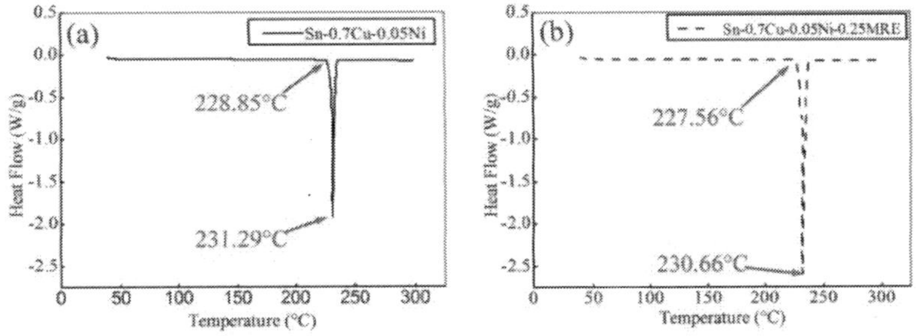

Figure 2. DSC curves of Sn-0.70Cu-0.05Ni (a) and Sn-0.70Cu-0.05Ni-MRE (b).

3.2 Microstructure of the Solders

Figure 3 is the typical optical microscope (OM) photograph of the as-solidified Sn-0.70Cu-0.05Ni and Sn-0.70Cu-0.05Ni-MRE solders, and there are three regions: the light-colored regions, which are pure β-Sn grains; the gray regions, which is eutectic network band containing both β-Sn and $(Cu,Ni)_6Sn_5$ phases and surround the light-colored regions, and the dark-colored regions, which is mainly $(Cu,Ni)_6Sn_5$. With the addition of 0.25 % MRE elements, intermetallic compounds $(Cu,Ni)_6Sn_5$ grains have been transformed into smaller grains, and refine the microstructure of the solder alloy[8].

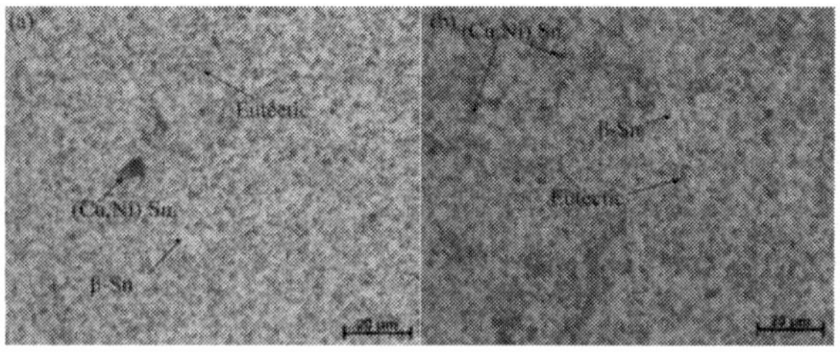

Figure 3. The typical OM photograph of the as-solidified Sn-0.70Cu-0.05Ni (a) and Sn-0.70Cu-0.05Ni-MRE(b) solders.

3.3 Wetting Behavior

Figure 4 shows the variations of contact angles with time and the final equilibrium contact angles for Sn-0.70Cu-0.05Ni solder on amorphous $Fe_{84.3}Si_{10.3}B_{5.4}$ substrate at 250 °C, 300 °C, 350 °C and 400 °C, respectively. Apparently, the initial contact angle is the largest when the solder is melted completely (i.e., 0 s), then it decreases gradually and finally keeps equilibrium state. The final equilibrium contact angle for Sn-0.70Cu-0.05Ni solder on substrate gradually decreases with the increase of wetting temperature, see Fig. 4(b). Similarly, the wettability of Sn-0.70Cu-0.05Ni-MRE solder on amorphous $Fe_{84.3}Si_{10.3}B_{5.4}$ substrate also follows the rules. The wettability can be measured by the contact angle determined by the Young equation as follows:

$$\cos\theta = \left(\sigma_{sg} - \sigma_{sl}\right) / \sigma_{lg} \quad (1)$$

where σ_{sg}, σ_{lg} , and σ_{sl} refer to the surface tensions of the solid-gas, liquid-gas and solid-liquid interfaces, respectively. The smaller contact angle (i.e. smaller σ_{lg}, σ_{sl} and larger σ_{sg}), the better the wettability. On the one hand, with the increase of wetting temperature, the surface tension of solder (σ_{lg}) is gradually decreased, resulting in the decrease of contact angle (θ), and the wettability becomes better. On the other hand, the activity and diffusion capacity of Fe atoms in amorphous alloy substrate and Sn atoms in solder gradually increased, inducing the formation of interfacial intermetallic compound, thus further promoting wetting process and improving wettability.

Figure 5 shows the curves of variations of contact angles with different time at 250 °C and the final equilibrium contact angles at different temperatures for Sn-0.70Cu-0.05Ni and Sn-0.70Cu-0.05Ni-MRE solders on amorphous $Fe_{84.3}Si_{10.3}B_{5.4}$ substrate. Clearly, the final equilibrium contact angle of Sn-0.70Cu-0.05Ni solder on amorphous substrate is larger than that of Sn-0.70Cu-0.05Ni-MRE solder at the same wetting temperature. Meanwhile, the final equilibrium contact angle can be reduced obviously with the addition of MRE. It can be concluded that the wettability of Sn-0.70Cu-0.05Ni are improved by adding 0.25 wt.% MRE elements.

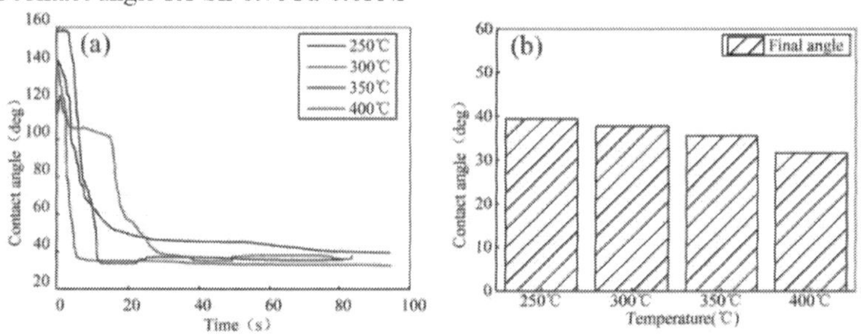

Figure 4. Wetting curves for Sn-0.70Cu-0.05Ni solder on amorphous $Fe_{84.3}Si_{10.3}B_{5.4}$ substrate at different temperatures: (a) variations of contact angles vs time; (b) final equilibrium contact angles vs temperatures.

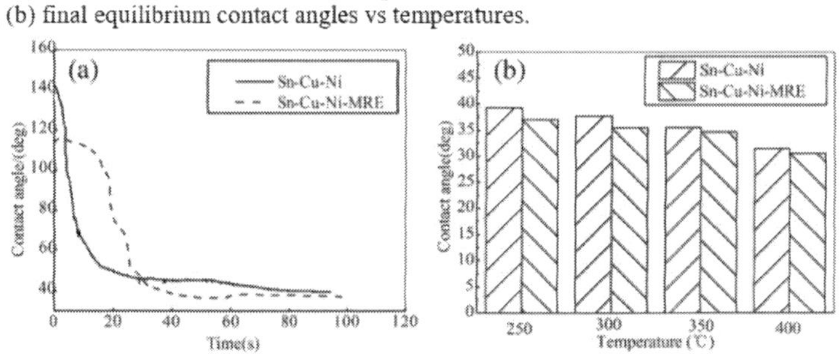

Figure 5. Wetting curves for Sn-0.70Cu-0.05Ni and Sn-0.70Cu-0.05Ni-MRE solders on amorphous $Fe_{84.3}Si_{10.3}B_{5.4}$ substrate: (a) contact angles vs time at 250 °C; (b) final equilibrium contact angles vs temperatures.

The improvement of wetting obtained by adding 0.25 wt.% MRE elements can be explained by the effects of MRE on the interfacial tensions. It is easier for MRE to accumulate at the solder interface in the molten state due to the high reactivity. Thus, MRE reduce the interfacial tension of molten solder effectively. According to Eq. (1), the decrease of σ_{lg} is beneficial for obtaining a smaller contact angle, resulting in improving wetting quality after the addition of MRE[9].

3.4 Interfacial Characteristics

Figure 6 presents SEM micrographs at the center of interface between Sn-0.7Cu-0.05Ni solder and amorphous $Fe_{84.3}Si_{10.3}B_{5.4}$ substrate after wetting at different temperatures for 30 min. Clearly, thin-layer intermetallic compounds discontinuously distribute at the interface between Sn-0.7Cu-0.05Ni solder and amorphous substrate wetting at 250 °C and 300 °C. However, a continuous compound layer is formed at the interface when the wetting temperature rises to 350 °C and 400 °C, and there are intermetallic compounds in block state inside the solder close to the interface. Accordingly, with the temperature is increased from 250 °C to 450 °C, the intermetallic compound distributed at the interface between Sn-0.7Cu-0.05Ni solder and amorphous substrate is changed from discontinuous to continuous distribution. It can be concluded that the interfacial reaction between Sn-0.7Cu-0.05Ni solder and amorphous substrate becomes more and more intense with the increase of wetting temperature. EDS analysis shows that the big block of intermetallic compounds containing Sn and Fe formed at the interface and in the solder near the interface, and the ratio of atomic percentage of Sn and Fe is about 2:1, as shown in the Table 1. According to the Sn-Fe phase diagram[10], it is believed that $FeSn_2$ is formed at the interface for the wetting tests performed at 400 °C[11].

For Sn-0.7Cu-0.05Ni-MRE, a similar phenomenon can also be clearly observed that big block of intermetallic compounds $FeSn_2$ are formed forms at the interface and in the solder closing to the interface when wetting at 400 °C, both consist of Sn and Fe, and the atomic percentage is about 2:1, as shown in the Table 1, and intermetallic compounds $FeSn_2$ at the interface are changed from discontinuous to continuous state with the increase of wetting temperature, see Fig. 7.

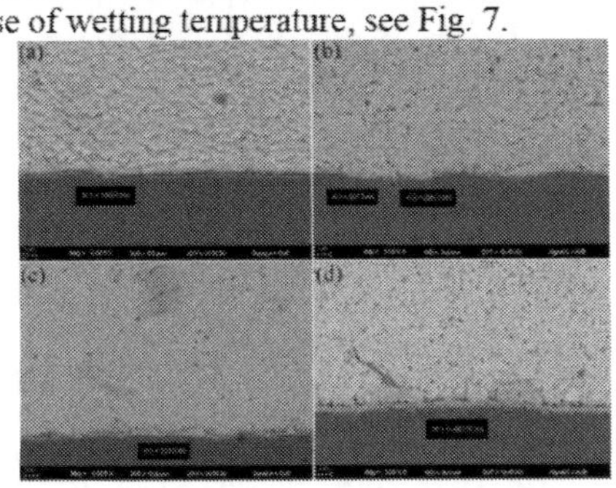

Figure 6. SEM micrographs at the center of interface between molten Sn-0.7Cu-0.05Ni and amorphous $Fe_{84.3}Si_{10.3}B_{5.4}$ substrate after wetting at different temperatures for 30 min: (a) 250 °C; (b) 300 °C; (c) 350 °C; and (d) 400 °C.

It is found that the thickness of the interfacial intermetallic compounds is reduced by adding MRE. Especially when wetting at 350 °C, the intermetallic compounds are distributed continuously at the interface between Sn-0.7Cu-0.05Ni solder and amorphous substrate, while the interfacial intermetallic compounds are in discontinuous state with the addition of MRE. It is concluded that the growth of interfacial intermetallic compounds is inhibited with the presence of MRE.

The thickness of the interfacial intermetallic compounds $FeSn_2$ reduced by adding MRE elements is attributable to the fact that La and Ce have higher affinity for Sn[12]. It will lead to a reduction in the activity of Fe with Sn due to the formation of Sn-MRE compound, thus depressing the growth of the $FeSn_2$.

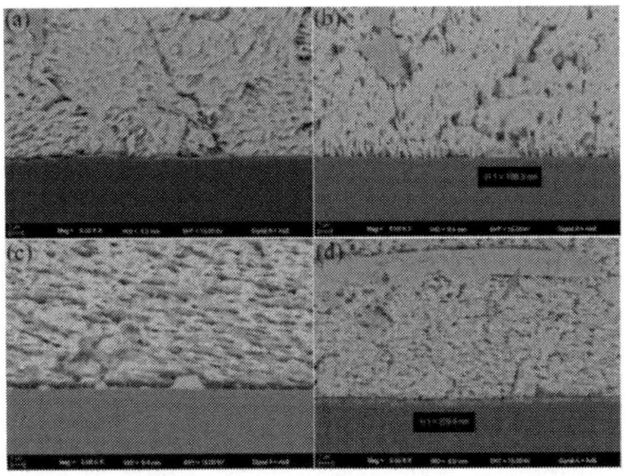

Figure 7. SEM micrographs at the center of interface between Sn-0.7Cu-0.05Ni-MRE solder and amorphous $Fe_{84.3}Si_{10.3}B_{5.4}$ substrate after wetting at different temperatures for 30 min: (a) 250 °C; (b) 300 °C; (c) 350 °C; (d) 400 °C.

Table 1. EDS analysis of points marked in Fig.6 and Fig.7

	Sn (%)	Fe (%)	Si (%)
A	63.60	33.41	2.99
B	68.44	31.56	-

4 Summary

The solidus temperature and liquidus temperature decrease slightly with the addition of 0.25 wt.% MRE. The final equilibrium contact angle for both Sn-0.70Cu-0.05Ni and Sn-0.70Cu-0.05Ni-0.25MRE solder on amorphous $Fe_{84.3}Si_{10.3}B_{5.4}$ substrate gradually decreases with the increase of wetting temperature. With the increase of wetting temperature, the intermetallic compounds $FeSn_2$ distributed at the interface are changed from discontinuous to continuous state. The wettability of Sn-0.70Cu-0.05Ni on amorphous $Fe_{84.3}Si_{10.3}B_{5.4}$ substrate is improved by adding 0.25 wt.% MRE elements. The presence of 0.25 wt.% percentage of the MRE is effective in preventing the growth of the intermetallic compounds $FeSn_2$ formed at the interface between solders and amorphous $Fe_{84.3}Si_{10.3}B_{5.4}$ substrate, results in a decrease in the thickness of the interfacial intermetallic compounds by adding MRE.

Acknowledgements

This work is presently funded by National Key Research and Development Plan of China (2017YFB0305700), the National Natural Science Foundation of China (Grant No.51304176 and No.51405092), the Science and Technology Cooperation Plan Project of Guangzhou City (201807010028) and Implementing Innovation-Driven Development Capacity Building Special Funds project of Guangdong Academy of Sciences (2018GDASCX-0113).

References

1. C.A.C. Souza, D.V. Ribeiro, C.S. Kiminami, J. Non-Cryst. Solids. **442**, 56 (2016)
2. W.H. Wang, Adv Mater. **21**, 4524 (2009)
3. A. Pietriková, J. Bednarčík, J. Ďurišin, J. Alloy. Compd. **509**, 1550 (2011)
4. G. Zeng, S. Xue, L. Gao, L. Zhang, Y. Hu, Z. Lai, J. Alloy. Compd. **509,** 7152 (2011)
5. J.X. Wang, S.B. Xue, Z.J. Han, S.L. Yu, Y. Chen, Y.P. Shi, H. Wang, J. Alloy. Compd. **467**, 219 (2009)
6. C.M.T. Law, C.M.L. Wu, D.Q. Yu, L. Wang, J.K.L. Lai, J. Electron. Mater. **35**, 89 (2006)
7. L. Zhang, L. Tian, Y.H. Guo, L. Sun, Y. Min, J. Rare. Earth. **32**, 1184 (2014)
8. M. Zhou, X. Ma, X. Zhang, Rare. Metal. Mat. Eng. **40**, 25 (2011)
9. C.M.L Wu, D.Q. Yu, C.M.T. Law, L. Wang, J. Electron. Mater. **31**, 928 (2002)
10. Y.N. Dai, *Binary Alloy Phase Digrams* (Science Press, Beijing, 2009).
11. Q.Q. Lai, L. Zhang, N. Eustathopoulos, Acta. Mate. **61**, 4127 (2013)
12. X. Ma, Y. Qian, F. Yoshida, J. Alloy. Compd. **334**, 224 (2002)

Effect of Time and Temperature on the Aging of 791-H SBS Modifier

Christian Magloire Ndjegwes[1], Mouhamed Bayane Bouraima[1], Xiao-hua Zhang[2], Martin Bitolog[1] and Yanjun Qiu[1]

[1]School of Civil Engineering, Southwest Jiaotong Univ., Highway Engineering Key Laboratory of Sichuan Province, Chengdu China.
[2]Sichuan Provincial Transport Department Highway Planning, Survey, Design and Research Institute, Chengdu China.

Abstract: The effects of time and temperature level on the aging of the 791-H SBS modifier have been investigated through experimental study. A series of Thin Film Oven (TFO) Test was conducted at the temperature of 100℃ and 163℃ and time of 1, 2, 5 and 10h. The changes in the chemical composition due to oxidative aging are detected by Fourier Transform Infrared Spectroscopy (FTIR) test. The results show that the visual observation of SBS after aging within 10h is almost unchanged when the aging temperature is 100℃. At 163℃ within 1-2h aging, the SBS particles gradually softened and become slightly bonded to each other; at the same temperature after 5h of aging, the particles of SBS became yellow. The result from FTIR test shows that temperature is an important factor in the aging of SBS modifier and its effect on the aging of SBS modifier is much greater than that of aging time.

1 Introduction

The term aging has been related to multiple mechanisms in asphalt mixtures. For some researchers, it indicates the overall deterioration of bitumen pavements, thus seriously affecting the performance and service life of asphalt pavement [1-6]. While in others cases, it is defined as a series of changes in asphalt concrete such as evaporation, oxidation, polymerization, and even the variations of the internal structure of asphalt [7]. Aging is also due to thermal aging and ultraviolet radiation aging [8, 9], oxidative aging and moisture related damage [10]. Others researchers [11-12] showed that a series of physical and chemical variations took place under the action of natural conditions such as wind and rain, change of temperature during the storing, processing, construction, use of asphalt and because of exposure to the air. In addition to the increase of functional groups having oxygen, the other chemical composition of asphalt also changed which leads to its hardening, crisping and cracking. However, it couldn't keep on exerting the intrinsic cohesive and seal functions [13]. According to the study in [14], aging is defined as the irreversible changes of colloid structure, physical and chemical properties, or mechanical performance that asphalt had exhibited. The effect of extraction residue on the properties of asphalt binders has been investigated and results showed that ageing asphalt binder is related to extraction residue even is less sensitive [15]. The aging research of SBS modified asphalt has been increased and gradually aroused people's recognition with its increasing application [16, 17]. The road surface's span life is influenced by the quality (good or bad) of the anti-aging performance of SBS modified asphalt. More study on aging has been done using basic asphalt in the past, but that of SBS modified asphalt has been less investigated. In 2011, the Sichuan Provincial Department of Communications and Transportation issued the notice on improving the performance (rutting resistance) of expressway asphalt pavement (Letter of Sichuan [2011] No. 37). Modified asphalt has been widely used in the surface and middle layers of asphalt pavement in the local highway and the regeneration of modified asphalt mixture as it is the main material for the regeneration of the asphalt pavement in the province. In this study, the aging test of SBS modifier agent under different time and temperature is investigated in the thin film oven test. The effects of time and temperature on the aging of the commonly 791-H SBS modifier produced by Yueyang Petrochemical Company are also detected by means of infrared spectrum analysis.

2 Experimental program

2.1 Materials

In the experiment, the 791-H SBS that Yueyang petrochemical company produced as the modifier is used and shown in Figure 1. The general properties are shown in Table 1.

Figure 1. 791-H SBS modifier

 https://doi.org/10.1051/matecconf/201820603006

Table 1 Property of 791-H SBS

Composition	SBS
Molecular structure	Linear
Block ratio- (S/B)	30/70
Oil-extended content (%)	0
Volatile content (%)	0.7
Ash (%)	0.5
300% tensile modulus (≥MPa)	2
Tensile strength	18
Elongation at break (≤%) %)	750
Residual deformation (≤ %)	30
Shore hardness (A)	75
Melt flow rate (g/10min)	0.50

2.2 Methods

The effect of temperature level and exposure time of 791-H SBS modifier has been investigated through experimental study. A series of Rolling Thin Film Oven (TFO) Test was conducted at the temperature for 100℃ and 163℃ and times for 1, 2, 5 and 10h.

3 Test results and discussion

3.1 SBS aging test

The effects of aging temperature and aging time on SBS were investigated. The visual observation of SBS after aging is shown in Figure 2.

(a) Aging temperature: 100 ℃; aging time: 1h

(b) Aging temperature: 100℃; aging time: 2h

(c) Aging temperature: 100℃; aging time: 5h

(d) Aging temperature: 100℃; aging time: 10h

(e) Aging temperature: 163℃; aging time: 1h

(f) Aging temperature: 163℃; aging time: 2h

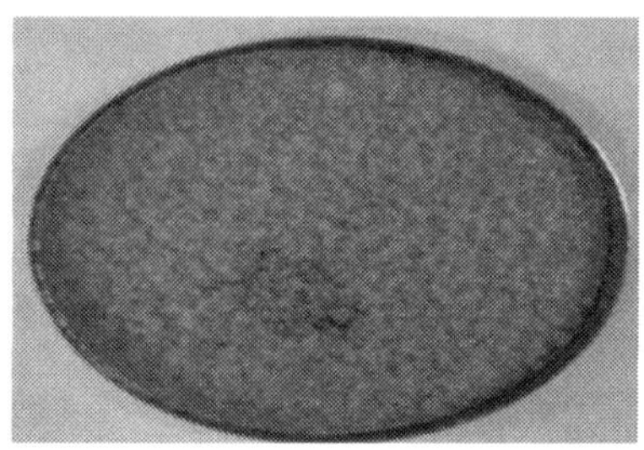

(g) Aging temperature: 163℃; aging time: 5h

(h) Aging temperature: 163℃; aging time: 10h

Figure 2. Visual observation of SBS after aging under different conditions

As can be seen from Figure 2, when the aging temperature is 100℃, the visual observation of SBS after aging within 10h is almost unchanged. When the temperature of the test was increased to 163℃, the SBS particles gradually softened after 1～2h aging and the particles were no longer in the dispersed state but were slightly bonded to each other. After aging for 5h, the particles of SBS became yellow, the particles are tightly bonded, and the SBS elasticity is gradually lost and converted into rigid plastic bodies. The change in mass of SBS at different aging times with temperature is shown in Table 2.

3.2 Infrared spectrum analysis of SBS after aging

The SBS was analyzed by Fourier transform infrared spectroscopy before and after aging, and the change of SBS functional groups during aging was observed. The test instrument uses the Agilent Cary 630 FTIR model, the test results shown in Figure 4.

The descriptive analysis of the infrared spectrum results of the aged SBS at 100°C at 10h from figure 5 (a) showed that there is no change with that of the original SBS and there is not the production of new functional groups during the process of 10h aging at 100°C. From figure 5 (b), it can be seen that there is still no change in the infrared spectrum results of SBS aging at 163°C within 1~2 h compared with the original SBS. After 5h aging (163°C), the infrared spectrum of SBS varies greatly and there are new absorption peaks at 3370cm^{-1} and 1715cm^{-1}. The characteristic peak at 3370cm^{-1} belongs to the characteristic absorption peak of OH while the characteristic peak at 1715cm^{-1} belongs to the characteristic absorption peak of C = O. This indicated that after aging at 163°C for 5 h, SBS had an oxygen reaction, and OH and C = O functional groups appeared in the molecule. The result shows that temperature is an important factor in the aging of SBS modifier. The effect of temperature on the aging of SBS modifier is much greater than that of aging time. At 100°C under the conditions aging of 1h to 10h, it can be shown that the performance was basically unchanged from the visual observation and infrared spectroscopy analysis. However, the performance at 163°C is more obvious, with the visual observation ce of yellow color at first and gradually deepened with time. The infrared spectrum analysis showed that oxygen reaction occurred in SBS.

Table 2 Change in mass of SBS quality different aging times

Aging temperature	Mass changes at different aging times (%)			
°C	1h	2h	5h	10h
100	0.2	0.2	0	-0.4
163	-0.3	-0.3	1.5	2.2

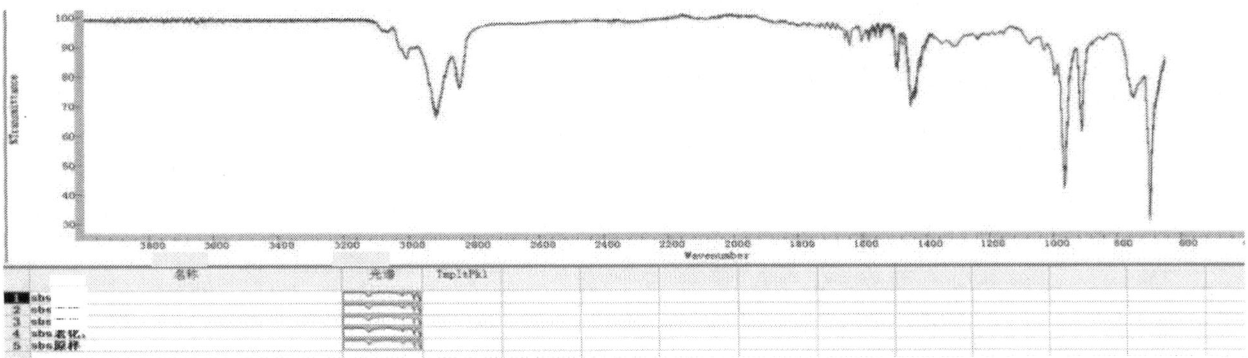

(a)Comparison of spectra of aged SBS at 100°C with original SBS at different time 1-100°C, 1h; 2-100°C, 2h; 3-100°C, 5h; 4-100°C, 10h; 5-original SBS

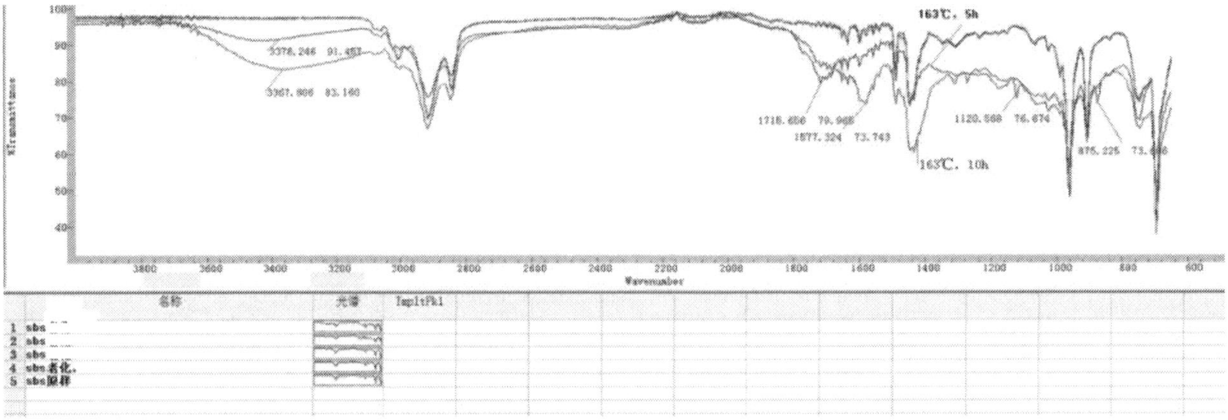

(b)Comparison of spectra of aged SBS at 163°C with original SBS at different time 1-163°C, 1h; 2-163°C, 2h; 3-163°C, 5h; 4-163°C, 10h; 5-original SBS.

Figure 3. Infrared spectrogram of original SBS and SBS after aging.

4 Conclusion

The aim of this study is to investigate the effect of time and temperature on the aging of the 791-H SBS modifier by means of visual observation after Thin Film Oven (TFO) test and infrared spectrum analysis. The main conclusions drawn from the current research work are presented as follows:

- The visual observation of SBS after aging within 10h is almost unchanged when the aging temperature is 100°C
- At 163°C within 1-2h aging, the SBS particles gradually softened and become slightly bonded to each other; at the same temperature degree after 5h of aging, the particles of SBS became yellow.
- The temperature is an important factor in the aging of SBS modifier and its effect on the aging of SBS modifier is much greater than that of aging time.

References

1. Siddiqui, M. N., and Ali, M. F. "Studies on the aging behavior of the Arabian asphalts." Fuel (1999); 78(9), 1005-1015.
2. Mouillet, V., Farcas, F., and Besson, S. "Ageing by UV radiation of an elastomer modified bitumen." Fuel (2008); 87(12), 2408-2419.
3. Dehouche, N., Kaci, M., and Mokhtar, K. A. "Influence of thermo-oxidative aging on chemical composition and physical properties of polymer modified bitumens." Constr. Build. Mater.(2012); 26(1), 350-356.
4. Xiao, F., Punith, V., Amirkhanian, S., and Thodesen, C. "Improved resistance of long-term aged warm-mix asphalt to moisture damage containing moist aggregates." J. Mater. Civ. Eng., (2013a); 10.1061/(ASCE) MT.1943-5533.0000567, 913-922.
5. Chen, J., Chen, S., Liao, M., and Huang, S. "Laboratory evaluation of asphalt blends of recycling agents mixed with aged binders." J. Mater. Civ. Eng., (2015); 10.1061(ASCE) MT.1943-5533.0001081, 04014143.6.
6. Cheolmin Baek, Benjamin Shane Underwood, Yongbaek Richard Kim. Effects of oxidative aging on asphalt mixtures properties. Transportation Research Record. 2012
7. Huang S C, Grimes W. Influence of aging temperature on rheological and chemical properties of asphalt binders [J]. Transportation Research Record, 2010, 2179(1): 39-48
8. Lu XH, Isacsson U. Effect of ageing on bitumen chemistry and rheology.Constr Build Mater (2002); 16:15-22.
9. Durrieu F, Farcas F, Mouillet V. The influence of UV aging of a styrene/butadiene/styrene modified bitumen: comparison between laboratory and on site aging. Fuel(2007); 86:1446-51.
10. Wright, J.R. Weathering: Theoretical and Practical Aspects of Asphalt Durability. Chapter 8 in Bituminous Materials: Asphalts, Tars and Pitches Volume II, Ed. A.J. Hoiberg, Inter-science Publishers, New York, 1965.
11. EngineYener, Sinan Hınıslıoğlu. Effects of exposure time and temperature in aging test on asphalt binder properties. International Journal of Civil and Structural Engineering, volume 5, issue 2, 2014 Pages: 112 -124.
12. Engine Yener, Sinan Hınıslıoğlu. The influence of the temperature and time variations on the aging of bituminous binders. 9th International Congress on Advances in Civil Engineering, 27-30 September 2010.
13. Yuan, J., and Zhao, K. (2000). The dose of modifier and the property of modified asphalt. Petrol. Asphalt 13:10.
14. Pan, Jielin. "A study of Asphalt Aging Behavior using Molecular Dynamics Simulations." (2016). http://digitalrepository.unm.edu/ ce_etds/12.
15. Bouraima, Mouhamed Bayane, Xiao-hua Zhang, Shui-wen Zhou, and Yanjun Qiu. "Effect of Extraction Residue on the Properties of Asphalt Binders." The Global Journal of Researches in Engineering. Vol 17, No 2-E (2017).
16. Y. Gao, F. Gu, L.Y. Zhang. Thermal oxidative aging characterization of SBS modified asphalt. J. Wuhan Univ. Technol., 2 (2013), pp. 88-91
17. F. Zhang, J. Y. Yu, S. P. Wu. Effect on ageing on rheological properties of storage-stable SBS/sulfur-modified asphalts. J. Hazard. Mater. 507-517 (2010).

MATEC Web of Conferences **206**, 04001 (2018) https://doi.org/10.1051/matecconf/201820604001
ICCEMS 2018

Modeling and Simulation Technology of High Frequency Vibratory Stress Relief Treatment for Complex Thin - Walled Workpiece

Xian-yang ZHAO[1], Nian-song ZHANG[1] and Ai-min WANG[2]

[1]*School of Mechanical Engineering, Nanjing University of Science and Technology, Nanjing 210000, China*
[2]*School of Mechanical Engineering, Beijing Institute of Technology, Beijing 100081, China*

Abstract. The current research based on vibratory stress relief (VSR) is focused on low frequency excitation, but it is seldom mentioned for complex thin-walled workpiece with higher natural frequency. In this paper, the mechanism of high frequency VSR is studied by microdynamic theory. Aiming at the rotating frame of complex thin-walled workpiece in aerospace equipment, the high frequency VSR numerical simulation is carried out by finite element software Abaqus, the modal analysis technology and harmonic response analysis technology are proposed, and the exciting force, excitation frequency and other important parameters are determined. The simulation results are analyzed to verify the superiority of high frequency VSR to eliminate residual stresses of complex thin-walled workpieces, and provide a theoretical basis for optimal selection of excitation parameters for high frequency VSR.

1 Introduction

Complex thin-wall workpiece has the characteristics of light weight, high strength and so on, and is widely used in aerospace and military industry. In the process of machining, it is easy to produce uneven plastic deformation, resulting in residual stress, the existence of residual stress is the main cause for the cracking or deformation of the workpiece. According to the literature [1], compared with the traditional way of eliminating residual stress such as natural aging and thermal aging, in twenty-first century, the technology of VSR has become an effective energy-saving and environmental protection technology. Foreign scholars, Kelso and Walker C A, have made a comprehensive elaboration on the mechanism of low frequency VSR [2-4]. The domestic professor Song Tianmin explained the mechanism of VSR by experiment from the angle of energy transformation in dislocation motion [5]. But these researches are mainly focused on the traditional low frequency VSR, whose excitation frequency is less than 200Hz, which is seldom involved in high frequency VSR of complex thin-walled workpieces with high natural frequencies.

Based on the theoretical study of the mechanism of high frequency VSR, aiming at the key parts of aerospace equipment, a modeling and simulation technology of high frequency VSR for complex thin-walled workpiece is proposed, through modal analysis and harmonic response analysis of finite element model of workpiece, the maximum dynamic stress of structure is calculated, the relationship curve between the maximum dynamic stress of structure and excitation frequency and amplitude of excitation force is established, and the high frequency VSR parameters are determined, which can be used to guide the practical engineering application.

2 Mechanism of high frequency VSR

According to lattice dynamics, the relationship between crystal particles in the crystal is regarded as a simple mass spring system, as shown in Figure 1, the motion equation of material particles with multiple degrees of freedom damped vibration system can be expressed as:

$$M\ddot{x}+C\dot{x}+Kx=F(t) \qquad (1)$$

In the formula (1), M is the mass matrix, C is the damping matrix, K is the stiffness matrix, and $F(t)$ is the force vector.

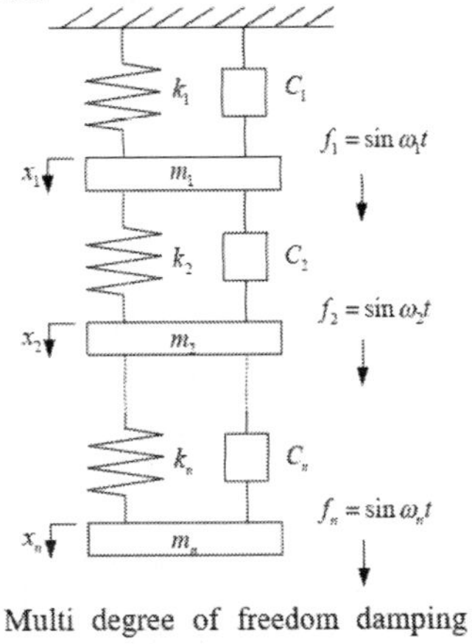

Figure 1. Multi degree of freedom damping vibration system diagram.

According to the modal analysis theory, the displacement admittance (frequency response function) [6] of any mass element m_l under the excitation of mass element m_p can be expressed as:

$$H_{lp}(\omega)=\sum_{i=1}^{N}\frac{\varphi_{li}\varphi_{pi}}{K_i-\omega^2 M_i+j\omega C_i} \quad (2)$$

In the formula (2), K_i、M_i、C_i are the i-th order damped modal stiffness, modal mass and modal damping. φ_{li}、φ_{pi} are the i-th order damped natural mode coefficient of m_l、m_p.

As can be seen from equation (2), for a system with multiple degree of freedom and damping, that system has multiple natural frequency, when the excitation force frequency is $\omega=\sqrt{1-2\xi_i^2}\omega_{ni}$ (ξ_i is damping ratio, and $\xi_i=C_i/(2\sqrt{M_iK_i})$; ω_{ni} is the i-th order natural frequency of the undamped mode, and $\omega_{ni}=\sqrt{K_i/M_i}$), m_l has the largest response and generate resonance. Figure 2 is a displacement frequency response diagram of multi-degree-of-freedom damped vibration system.

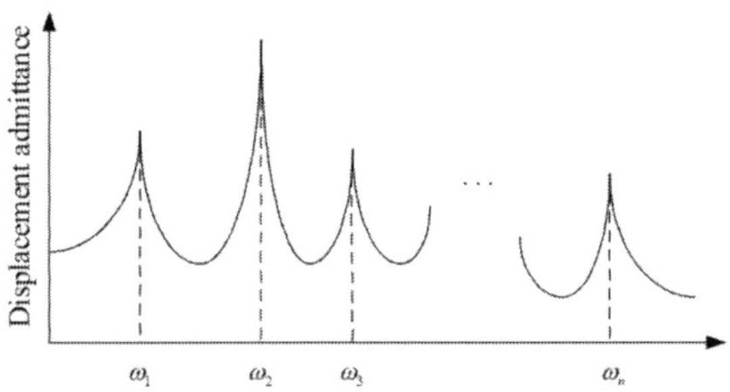

Figure 2. Displacement admittance graph of a multi degree of freedom damping vibration system.

When a high frequency vibration is applied to the crystal, the crystal system can be seen as a mass spring vibration system with multiple degrees of freedom. The system has multiple resonance frequencies, which are high because of the small mass of crystal particles and the large spring stiffness. However, under the condition of external high frequency excitation, when the external vibration frequency is consistent with a certain order resonance frequency of the system, the crystal grains in the crystal will move violently, so that the high-energy atoms in a stable state in the crystal material can obtain enough energy, thus getting rid of the bondage of the surrounding atoms and quickly returning to the original low-energy stable position. For crystal materials, dislocations and internal stresses disappear, so that we can eliminate the residual stress in the material.

The mechanism of common low frequency VSR is to cause dislocation proliferation, accumulation and entanglement in the material. Different from the mechanism of low frequency VSR. The mechanism of high frequency vibration aging is to make the dislocation density decrease or even disappear, which makes the material performance closer to the original stress free state. The residual stress distribution in the component is more uniform than the low frequency VSR, and the dimensional stability of the component is also better. At the same time, because of the high frequency excitation, the macroscopic deformation of the specimen is very small, so it is not easy to cause the fatigue damage of the specimen, which will lay the foundation for the popularization and application of the technology.

3 Numerical simulation analysis of high frequency VSR of rotating frame

3.1 Establishment of finite element model of rotating frame

The rotating frame is the main supporting structure of an aerospace equipment servo mechanism inner frame combination, which belongs to a typical complex thin-wall workpiece, as shown in Figure 3, and its material is ZL114a. In order to facilitate the subsequent analysis of the structure and reduce the calculation, the key structure of the rotating frame is extracted on the basis of not affecting the simulation results, only the main features are retained, the parts are simplified and the structures such as guide angle, guide edge, fillet and reinforcement are removed, as shown in Figure 4.

The rotating frame material is ZL114A, and the material mechanics performance parameters, such as yield limit, modulus of elasticity, Poisson's ratio and density, are required to determine in the stress field analysis process. According to the manual of metal materials, the physical and mechanical properties of ZL114A material below 4000℃ were found. The density was $2850kg/m^3$, the elastic modulus was 71GPa, and the Poisson's ratio was 0.34. When the initial residual stress is applied to the workpiece, the expansion coefficient of the temperature field analysis is 2.38e-5 (1/ ℃).

Figure 3. Original model.

Figure 4. Simplified model.

After a series of pre-processing such as simplifying the rotating frame model, defining the performance parameters and so on, the model is meshed and the finite element model is established. Figure 5 is the finite element model of a rotating frame.

Figure 5. Finite element model.

3.2 Modal analysis technology of rotating frame

3.2.1 Free mode analysis

First of all, the free modal analysis of the rotating frame is carried out to get its natural mode, and the reasonable vibration mode is selected as the initial mode of high frequency vibration aging treatment, and the excitation point, support point and excitation frequency are determined according to the vibration mode.

Based on the finite element model established above, the frequency extraction analysis step is created, and the free modal analysis of the workpiece is carried out. The resonant frequencies and mode shapes of the first eight orders of the natural frequencies of the workpiece are extracted, and the corresponding mode shapes are described. The results of free mode analysis are shown in table 1.

Table 1. Free mode analysis results.

Order	Natural frequency (Hz)	Mode description
1	0	First order rigid body displacement
2	0	Second order rigid body displacement
3	0	Third order rigid body displacement
4	1.70E-03	Fourth order bending
5	2.60E-03	Fifth order bending
6	3.10E-03	Sixth order bending
7	1151.7	Seventh order bending
8	1511.2	Eighth order bending

Figure 6 is the free mode analysis nephogram of first to eighth order of a three-dimensional model of the rotating frame workpiece. It can be seen from the figure that the first three natural frequencies are all zero, corresponding to the rigid body displacement of the workpiece in three degrees of freedom. The mode shapes of the fourth, fifth and sixth order modes show a small bending state, and the natural frequency is very small, which is close to the displacement of the rigid body, so it is not suitable for VSR. In the seventh order modal shape, the displacement deformation near the circular hole in the middle of the workpiece is zero, and the four end parts of the workpiece are relatively large in displacement deformation. Therefore, the middle part of the workpiece with zero displacement deformation can be selected as a supporting point, a three-point support method is adopted, and the supporting points are arranged near the circular hole in the middle part, and the excitation point is selected at the right end part of the workpiece with relatively large displacement deformation. The eighth order modal shape is complex, and the vibration mode is ignored here.

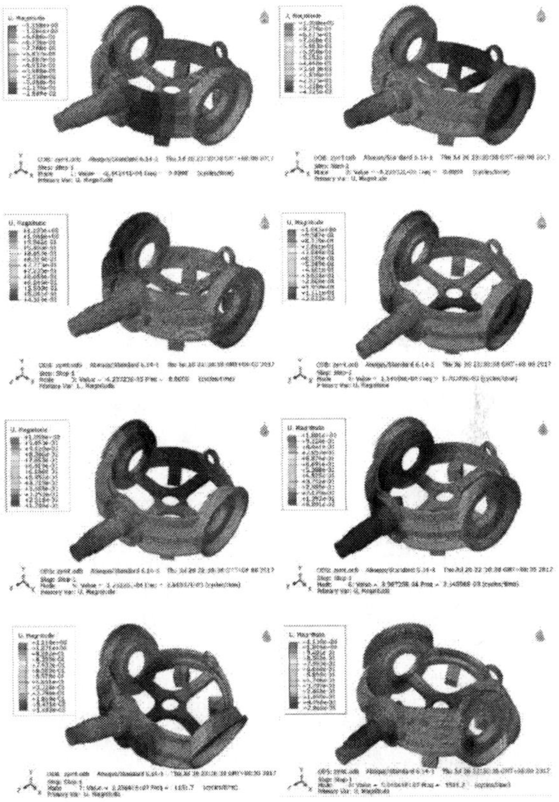

Figure 6. One to eight order free modal analysis nephogram of rotating frame.

The analysis shows that the bending vibration mode of the seventh order is selected as the initial vibration mode of the high frequency VSR, the natural frequency value is 1151.7Hz, three-point support is adopted, the supporting point of the VSR excitation is arranged in three directions of zero displacement near the circular hole. The larger displacement value, that is, the right end of the workpiece is selected as the excitation point of the high frequency VSR.

3.2.2 Constrained mode analysis

Next, constraint modal analysis of rotating frame is carried out. The simulation of the constrained modal analysis continues to follow the finite element analysis model of free mode analysis, and select the constraint points according to its seventh order mode, specify the number of modes to be analyzed, add linear boundary conditions. Then, analyze and solve the inherent frequency and corresponding mode of vibration, and enter the post processing to look at the analysis results.

On the basis of the free mode analysis, the seventh order mode of free vibration is taken as the prototype. According to the actual conditions of the experiment, the middle part of the workpiece is selected as the supporting point, and the three point support is adopted to add the three full degree of freedom constraint boundary conditions in the three directions of the middle hole of the model, and the constrained modal analysis of the rotating frame is carried out. The natural frequencies and modes of the first eight orders of the constrained modal analysis of the rotating frame are extracted, and the corresponding mode shapes are described, as shown in table 2.

Table 2. Constrained mode analysis results.

Order	Natural frequency (Hz)	Mode description
1	348.95	First order bending
2	594.58	Second order bending
3	914.94	Third order bending
4	929.57	Fourth order bending
5	1182.4	Fifth order bending
6	1573	Sixth order bending
7	2295.4	Seventh order bending
8	2404.4	Eighth order bending

The first eight natural frequencies and mode shapes are still extracted, and the fifth order mode is selected with a natural frequency of 1182.4Hz. The vibration frequency is selected range from 1120Hz to 1180Hz by equation (3) [7].

$$\omega = \omega_1 + 1/3(\omega_0 - \omega_1) \sim \omega_1 + 2/3(\omega_0 - \omega_1) \quad (3)$$

In the formula (3): $\omega_0 - \omega_1 = \zeta\omega_0$, Where ζ represent that damping ratio of the material.

3.3 Harmonic response analysis technology of rotating frame

The finite element analysis model is consistent with the modal analysis process, and harmonic response analysis is carried out by using the same grid division method. This paper mainly introduces the following four problems that should be solved in harmonic response analysis.

(a) Determine the value of the exciting force

The strength limit, yield limit and fatigue limit of cast aluminum alloy 114A can be known from the metal material handbook. As shown below:

$$\sigma_b = 360MPa$$

$$\sigma_s = 325MPa$$

$$\sigma_{-1} = \sigma_b / 3$$

According to the standard formula of excitation force (4), the value of exciting force can be calculated.

$$\sigma_s - \sigma_r < \sigma_d < \sigma_{-1} \quad (4)$$

According to the calculation formula of excitation force (4), the range of dynamic stress can be obtained.

(b) Determine the excitation frequency

According to the results of free modal analysis and constrained modal analysis, the range of excitation frequency is determined. It is known from the previous article the range of excitation frequency is $1120Hz < F_0 < 1200Hz$.

(c) Applied residual stress field

In the harmonic analysis of the rotating frame, the residual stress field in the process of the material should be taken into consideration. Therefore, the initial residual stress should be applied to the rotating frame work. The residual stress field of the workpiece is simulated by induction hardening. The initial temperature field is defined as 20 ℃ In the analysis step, the temperature of the hardened zone is raised to a certain temperature of T_{high}, and the rest temperature remains 20 ℃ . This temperature difference causes compressive stress in the high temperature region, which is equivalent to the simulated residual compressive stress. The maximum residual compressive stress of the workpiece can be obtained from the analysis of the residual stress field simulated by 282MPa.

(d) Harmonic response analysis

Different excitation force amplitudes F_0 are taken as 20N, 40N, 60N, 80N, 100N for analysis and solution in the same frequency range $1120Hz < F_0 < 1200Hz$, and the results are plotted as a graph of the relationship between the maximum stress value and the excitation frequency of the rotating frame at different excitation force amplitudes as shown in Figure 7.

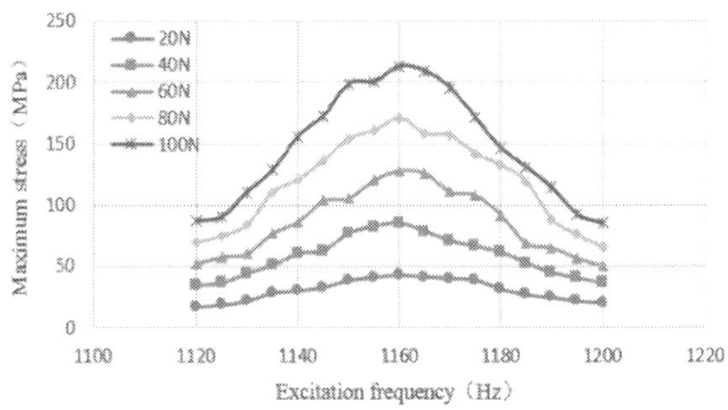

Figure 7. Frequency-maximum stress curve under different excitation force amplitudes.

As can be seen from that graph of frequency and maximum stress in Figure 7, under the action of different excitation force amplitudes, the rotary frame workpiece can generate the maximum equivalent stress value at the excitation frequency of 1160Hz. Its value reflects the effect of vibration stress relief to eliminate residual stress, so the excitation frequency can be chosen as 1160Hz. The relation between the exciting force and the maximum dynamic stress under the excitation frequency is shown in table 3.

The relation curve between the maximum dynamic stress and the exciting force of the rotating frame structure at 1160Hz is made, as shown in Figure 8. According to the mechanism of VSR, the sum of dynamic stress and residual stress should be greater than the yield limit of the vibrating material, and the

https://doi.org/10.1051/matecconf/201820604001

dynamic stress is less than the fatigue limit of the material. Therefore according to the dynamic stress standard calculation formula $\sigma_s-\sigma_r < \sigma_d < \sigma_{-1}$, the range of dynamic stress that can be obtained is 43MPa $< \sigma_d <$ 120MPa. According to the relationship between the maximum dynamic stress and the excitation force shown in Figure 8, we can obtain the amplitude range of the required harmonic excitation force is 21.2N $< F_0 <$ 56N, and the F_0 = 25N is preferable for experiment.

Table 3. The relationship between the exciting force and the maximum dynamic stress when frequency is 1160Hz.

Exciting force/N	20	40	60	80	100
Maximum dynamic stress/Mpa	42.6	85.2	127.8	170.4	213

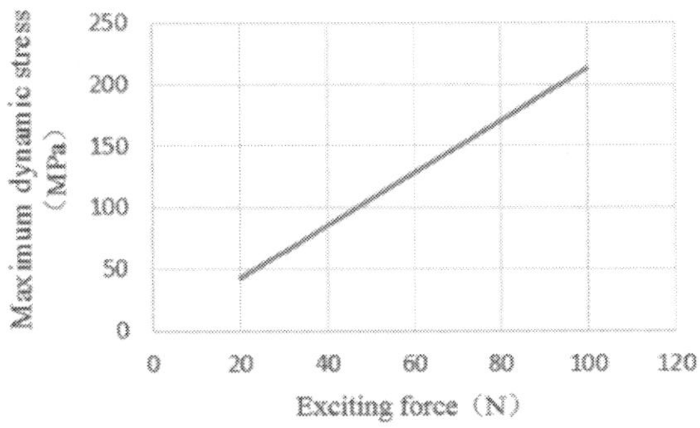

Figure 8. Curve of excitation force and maximum dynamic stress at excitation frequency of 1160Hz.

4 Analysis of high frequency VSR simulation results of rotating frame

The high frequency excitation simulation is carried out at 25N excitation force and 1160Hz excitation frequency, the stress distribution of the workpiece before and after excitation is shown in Figure 9. Ten nodes with obvious structural characteristics are selected, and the residual stress values before and after high frequency VSR treatment (harmonic response analysis) are recorded respectively, the results are plotted as a graph as shown in Figure 10.

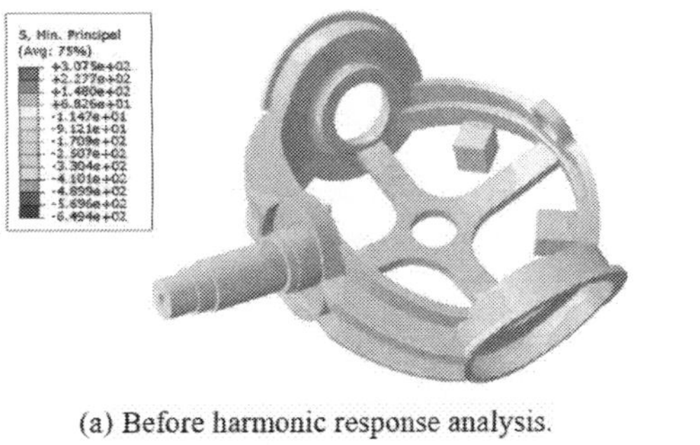

(a) Before harmonic response analysis.

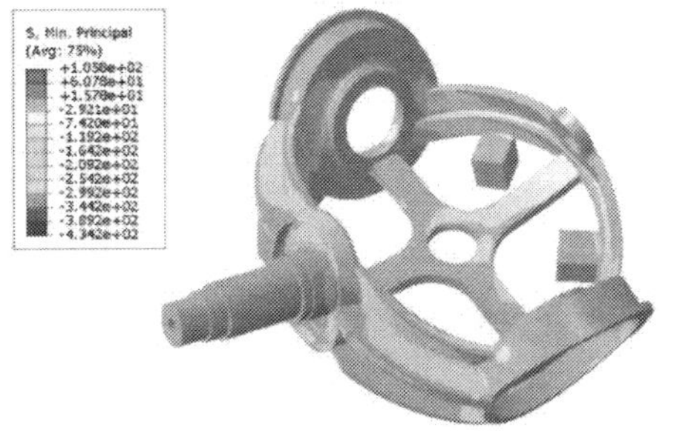

(b) After harmonic response analysis.

Figure 9. Stress distribution of workpiece before and after harmonic response analysis.

Comparing and analyzing the stress data before and after the high frequency VSR treatment in Figure 10, the excitation force has different influence on the residual stress field in different regions of the rotating frame. Where the initial residual stress value is larger, the reduction of residual stress is larger, and the homogenization degree is also higher.

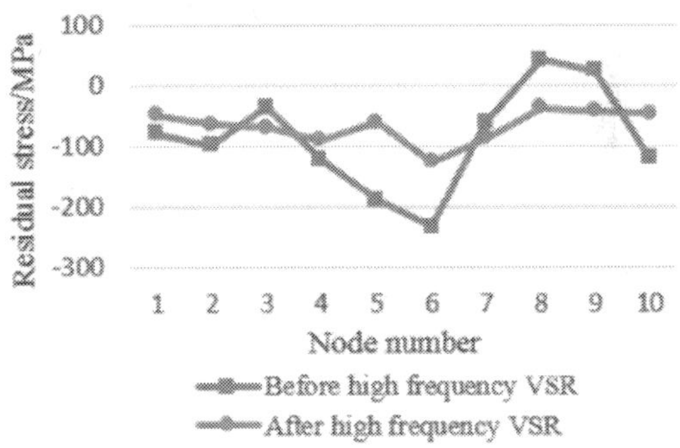

Figure 10. Stress change before and after high frequency VSR simulation analysis.

5 Conclusion

In this paper, the mechanism of high frequency VSR is studied by microdynamic theory, and the relationship between material particles in the specimen is considered as a damping vibration system with multiple degrees of freedom. Under the condition of vibration excitation, when the external vibration frequency is consistent with a certain resonance frequency, the particles will move violently and quickly return to the original stable equilibrium position, so as to reduce and homogenize the internal residual stress of the material, and the material performance is closer to the original stress-free state.

Taking the analysis of typical complex thin-walled workpiece rotating frame as an example, the finite element model is established, and the high frequency VSR parameters are determined by combining modal analysis and harmonic response analysis technology. The simulation results are analyzed to verify the effect of reducing and homogenizing residual stress, which provides a theoretical basis for the optimization of high frequency VSR excitation parameters. Further technical development and implementation will be carried out in combination with the actual high frequency VSR experiment.

References

1. Zhang Yong, Chin. J. Mech. Eng **13**(19) pp 1640–1642 (2002)
2. C.A. Walker, A.J. Waddell, D.J. Johnston, Proc. Inst. Mech. Eng, Part E: J. Proc. Mech. Eng **209**(1) pp 51–58 (1995)
3. G. Gnirss, Weld. World 26(11) pp 284–291 (1988)
4. A.S. Munsi, A.J. Waddell, C.A. Walker, J. Stra. Anal. Eng. Des 36(5) pp 453–464 (2001)
5. Song Tianmin, Research on the mechanism of VSR (J. Jilin. U: Techno. Ed (1) pp 53–56, 1995)
6. Chen Yaodong, Mechanical vibration (Hangzhou: Zhejiang University Press) (1988)
7. Li Quanhua, Heat treatment technology 400 questions and analysis (Beijing: Machinery Industry Press) pp 552–553 (2002)

Influence of Main Road Alignment on Handling Stability of Runaway Vehicle on Truck Escape Ramp

Pinpin Qin[1], Yaqi Han[1], Heyu Hu[2], Guangyi QIN[1] and Jie WANG[1]

[1]*College of Mechanical Engineering, Guangxi University, Nanning, Guangxi, China*
[2]*RWS Group, Mechanics Department, No. 76 of West Yanta Road, Xi'an, Shaanxi, China*

Abstract. Main road alignment exerts an influence on handling stability of a runaway vehicle to a certain extent, as rational designs of the main road alignment play an important role in ensuring that a runaway vehicle enters a truck escape ramp safely. Firstly, forces acting on the runaway vehicle on the main road are analysed, and the minimum curve radius value and the corresponding approach angle that can prevent lateral rollover and lateral slip are discussed. Secondly, a simulation software, TruckSim, is used to establish a simulation model of the runaway vehicle and the road, wherein the minimum curve radius value and the corresponding approach angle that can prevent lateral rollover and slip are taken as input parameters of the road model. Finally, the established simulation model of the vehicle and road is used to simulate the influence that three types of main road alignment, i.e. a straight line, a right curve and a left curve, respectively have on the handling stability of the runaway vehicle entering a truck escape ramp.

1 Introduction

A truck escape ramp is usually arranged on one side of a long, steep downhill grade section of a road, and is used to guide a runaway vehicle or trailer with a brake failure from the main road into the truck escape ramp. In the United States, studies of truck escape ramps are mainly conducted in the 1980s, during which a series of research results were achieved to guide the design and application of truck escape ramps [1-4]. Both the research and application of truck escape ramps in China lag far behind those at abroad; moreover, engineering norms or design guidelines have not been introduced yet. There are still some deficiencies in the design of current truck escape ramps, which have led to a variety of accidents of runaway vehicles on the truck escape ramps, such as a vehicle rushing out of a truck escape ramp, a rollover of a vehicle on an arrester bed, goods squeezing into a cab, backward sliding of a vehicle on an arrester bed, and a vehicle crashing into a guardrail or the other vehicles on a truck escape ramp [5-6].

Research triggered by these accidents is directed to the safety of truck escape ramps, and studies on truck escape ramps in China mainly focus on the design of the location, the cross-section width, the approach angle and approach roads, the deflection angle, the vertical curve radius at the entrance, etc. of truck escape ramps [7-13]. Few studies have investigated the influence of design parameters of the truck escape ramp on the safety performance of a runaway vehicle; however, the safety performance of a truck escape ramp should further include the handling stability of a runaway vehicle when entering the truck escape ramp. This research expects to find out the main vehicle dynamics indicators that reflect the handling stability of a vehicle, so as to provide the basis for the design of main road alignment.

2 Research method

2.1 Force analysis of runaway vehicle on main road

The vehicle running on a horizontal curve produces a centrifugal force, which is directed away from the centre of the circle, and the magnitude of which is directly proportional to the square of the travelling speed and inversely proportional to the radius of the horizontal curve. The calculation formula is as follows:

$$F = \frac{Gv^2}{gR} \tag{1}$$

wherein F is the centrifugal force; R is the radius of a horizontal curve; v is the travelling speed of a runaway vehicle; and G is the gravity.

The lateral force analysis of a runaway vehicle at a horizontal curve is shown in Fig. 1, in which, F is the centrifugal force, G is the gravity of the runaway vehicle, and ih is the superelevation slope. The tread of the runaway vehicle is b, and the height of the centre of gravity thereof from the road surface is h.

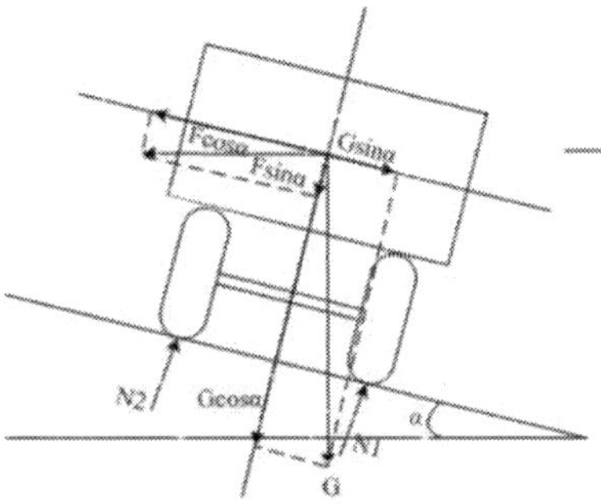

Figure. 1 Lateral force analysis of runaway vehicle at horizontal curve

The centrifugal force F and the runaway vehicle gravity G are decomposed into a horizontal force parallel to the road surface and a vertical force perpendicular to the road surface, and thus the lateral stability balance formula is as follows:

$$F \cdot \cos\alpha - G \cdot \sin\alpha = F_s \quad (2)$$

As lateral inclination angle of the road surface is generally small, then

$$\frac{Gv^2}{gR} = G \cdot \varphi + G \cdot i_h = G(\varphi + i_h) \quad (3)$$

wherein φ is the lateral adhesion coefficient of the road surface.

When the vehicle speed is greater than a certain value, the runaway vehicle may slip outward and cause an accident. In order to prevent the runaway vehicle from sideslipping, the travelling speed must satisfy the following equation:

$$v \le \sqrt{(\varphi + i_h) gR} \quad (4)$$

When runaway vehicles are driving on a horizontal curve with superelevation, there is also a risk of lateral rollover around the point of contact between the outer wheel and the road. In order to prevent runaway vehicle from overturning, the roll moment must be less than or equal to the stabilizing moment, i.e.

$$(F \cdot \cos\alpha - G \cdot \sin\alpha) \cdot h \le (F \cdot \sin\alpha + G \cdot \cos\alpha) \cdot \frac{b}{2} \quad (5)$$

Substituting $\sin\alpha = \tan\alpha = i_h, \cos\alpha \approx 1$ into the above equation, since Fi_h is much smaller than G and thus negligible, the following equation is obtained:

$$\frac{Gv^2}{gR} \le G(i_h + \frac{b}{2h}) \quad (6)$$

Therefore, in order to prevent the lateral rollover of a runaway vehicle on a horizontal curve, the travelling speed thereof must satisfy the following equation:

$$v \le \sqrt{(i_h + \frac{b}{2h}) gR} \quad (7)$$

According to the steering characteristics of the runaway vehicle driving from the main road into the truck escape ramp, the trajectory can be approximately considered as a horizontal curve, as shown in Fig. 2 [5]. According to geometric principles, the relationship between the approach angle θ of the truck escape ramps, the length L of the transition section and the turning radius R are as follows:

$$\theta = \cos^{-1}(\frac{R - 3.75/2}{R + B/2}) \quad (8)$$

$$L = (R + \frac{B}{2}) \sin\theta \quad (9)$$

wherein B is the width of the arrester bed.

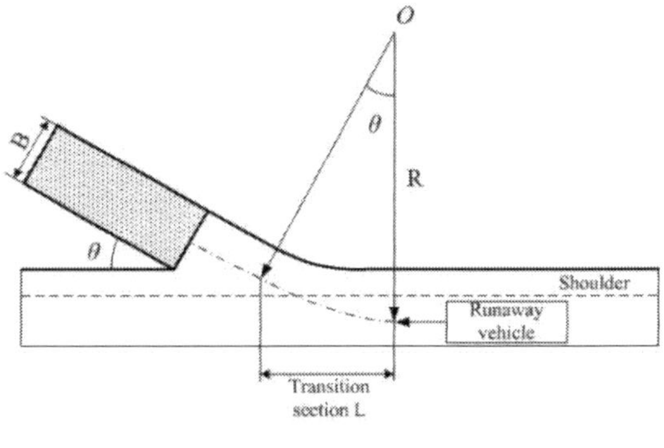

Figure. 2 Travelling trajectory of runaway vehicle entering truck escape ramp

2.2 Determination of minimum curve radius and approach angle

The maximum speeds that a runaway vehicle can reach in the case of different long, steep downhill grade sections are not the same. For the sake of safety, it is assumed that the speeds of the runaway vehicle entering truck escape ramps are 90km/h, 108km/h, 120km/h and 125km/h respectively. It is assumed that the tread of the runaway vehicle is 1.815 m, and the height of the centre of gravity thereof from the road surface is 1.25 m. The width of the road surface is assumed as 3.75 m. The coefficient of road adhesion is related to the condition of a road surface, tyres and vehicle speeds, and considering the most unfavourable situation, the lateral adhesion coefficient of the road surface is assumed as 0.24. Since it is provided that the superelevation is no more than 10%, superelevation rates of 2%, 4%, 6% and 8% are selected herein, based on which the minimum curve radius that can prevent a lateral rollover and lateral slip of a runaway vehicle under different combinations are calculated, as shown in Tab. 1.

Table. 1 Minimum curve radius that can prevent lateral rollover and lateral slip of vehicle

Approach speed (km/h)	90	108	120	125
Superelevation 8%	79	114	141	153
	199	287	354	384
superelevation 6%	81	117	144	157
	212	306	377	410
superelevation 4%	83	120	148	161
	227	328	404	439
superelevation 2%	85	123	152	165
	245	353	436	473

According to equations (1)-(9), the minimum trajectory circle radius, i.e. R, of the runaway vehicle entering the truck escape ramp is calculated, so as to determine the maximum approach angle of the truck escape ramp corresponding to different situations (wherein the incoming speed is 120km/h) for calculation, and the results are shown in Tab. 2.

Table. 2 Approach angle of truck escape ramp

(incoming speed: 120 km/h)

Approach angle/ (°)	Length of transition /(m)	Turning radius /(m)
2.5	254	5768
5	249	2915
7.5	105	797
10	75	426
12.5	65	296
15	55	207

3 Simulation experiment

3.1 Establishment of vehicle simulation model

In order to make the test result closer to the actual situation, a Balong Heavy Truck of Dongfeng Liuzhou Motor, which is a 385-horsepower 6x4 dump truck having a large market share, is selected as the simulation vehicle. Users can set and adjust vehicle model parameters in TruckSim as needed.

3.2 Establishment of road simulation model

With reference to the relevant parameters in Tab.3. TruckSim is used to establish truck escape ramp models. Users can set and adjust the road parameters of the truck escape ramp, such as the length of a transition section, the approach road, horizontal alignment, deflection and grade, in TruckSim according to requirements.

Table. 3 Truck escape ramp parameter

Width of approach/(m)	3.7～5.5
Width of shoulder/(m)	0.75
Width of arrester bed /(m)	7.5～12

3.3 Experimental design

In this paper, the main road alignment is divided into three different types of alignment, i.e. a straight line, a right horizontal curve and a left horizontal curve, to simulate the process of a runaway vehicle entering a truck escape ramp. On the premise that the runaway vehicle meets the stability requirement of the minimum radius when entering the truck escape ramp from a main road, the approach angle is set as 5°, 10° and 15°, and the vehicle speed is 120 km/h. In order to study the stability of the vehicle during the incoming process, four vehicle dynamics responses, i.e. the lateral acceleration, the yaw rate, the tyre slip angle and the roll angle, are taken as the indicators for evaluation.

4 Analysis of simulation result

4.1 Main road alignment being straight line

In the case of the main road alignment being a straight line, the simulation results of the lateral acceleration, yaw rate, tyre slip angle and roll angle of the runaway vehicle entering the truck escape ramp are shown in Fig. 3.

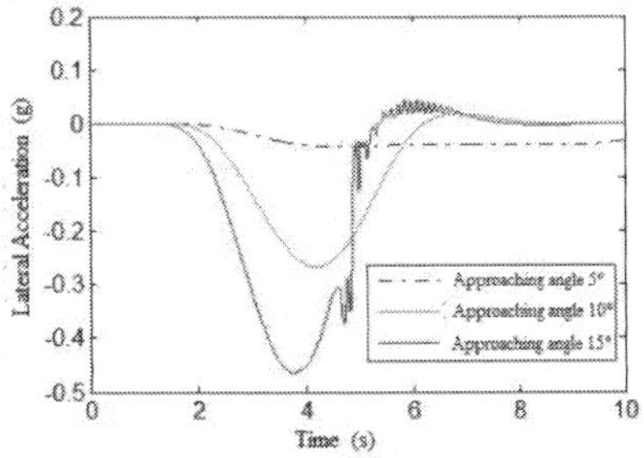

(a) Vehicle lateral acceleration over time

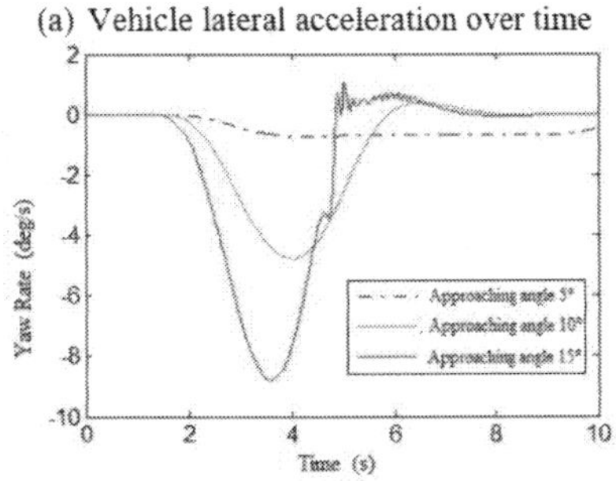

(b) Vehicle yaw rate over time

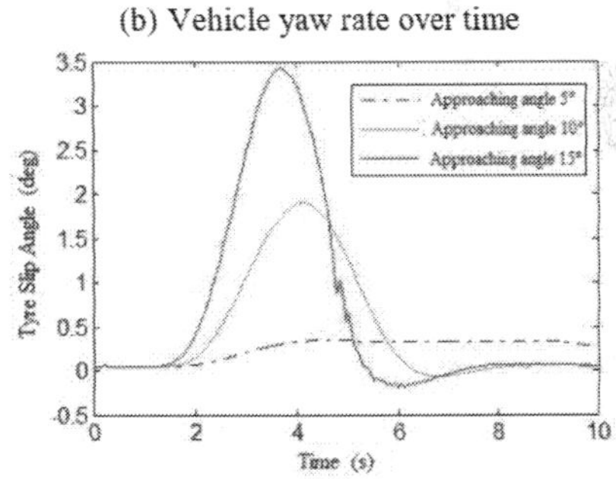

(c) Tyre slip angle over time

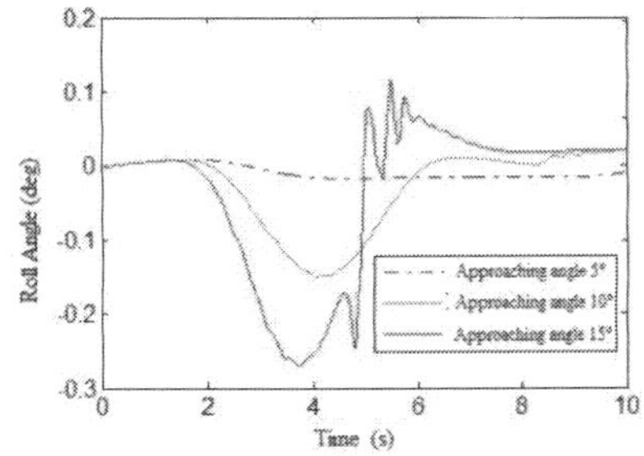

(d) Vehicle roll angle over time

Figure. 3 Simulation result for straight line main road

From the simulation results in Fig. 3(a), it can be seen that when the runaway vehicle enters the truck escape ramp from the straight main road, the lateral acceleration increases gradually as the runaway vehicle approaches the entrance of the truck escape ramp. Moreover, as the approach angle increases, the increase of lateral acceleration is more obvious. When the approach angle is 15°, the lateral acceleration of the runaway vehicle is of a large magnitude that is close to 0.5g, in which case the vehicle is in a dangerous situation and the vehicle is close to a rollover state. After the runaway vehicle enters the approach road of the truck escape ramp, the lateral acceleration of the vehicle gradually decreases. However, the vehicle in the whole incoming process is not stable, and a rollover may occur.

From the simulation results in Fig. 3(b), it can be seen that the yaw rate of the vehicle increases with the increase of the approach angle of truck escape ramps, and the regularity is obvious. The yaw rate of the vehicle does not change much when the approach angle is 5°, whereas the yaw rate thereof reaches 9 deg/s when the approach angle is 15° and the vehicle enters the approach road.

From the simulation results in Fig. 3(c), it can be seen that the tyre slip angle is constantly increasing in the process of the runaway vehicle entering the truck escape ramp. When the runaway vehicle enters the entrance of the truck escape ramp with an approach angle of 15°, the tyre slip angle is large. Although the tyre slip angle does not exceed the determination threshold, it becomes more difficult for the driver to handle the vehicle.

From the simulation results in Fig. 3(d), it can be seen that the roll angle of the vehicle increases gradually as the runaway vehicle enters the truck escape ramp from the straight main road. Moreover, as the approach angle increases, the increase of the roll angle of the vehicle is more obvious. When the approach angle is 15°, the roll angle of the vehicle is close to 0.3 deg, in which case the vehicle is in a dangerous situation, since the vehicle body tilts to one side, and the driver is unable to control the posture of the vehicle body. After the runaway vehicle enters the approach road of the truck escape ramp, the roll angle of the vehicle gradually decreases. However, the vehicle in the whole incoming process is not stable, and a rollover may occur.

4.2 Main road alignment being right curve

In the case of the main road alignment being a right curve, the simulation results of the lateral acceleration, yaw rate, tyre slip angle and roll angle of the runaway vehicle entering the truck escape ramp are shown in Fig. 4.

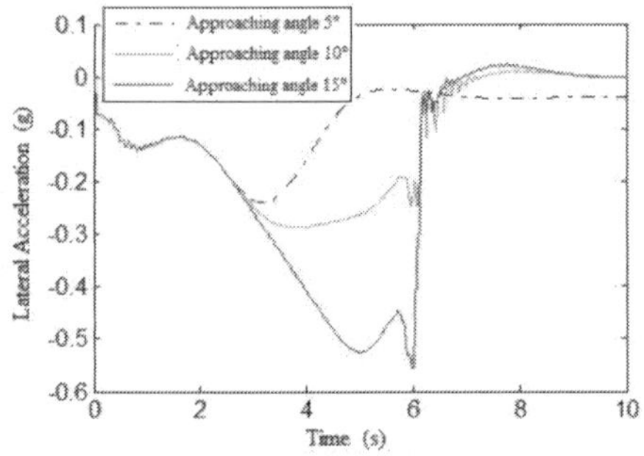

(a) Vehicle lateral acceleration over time

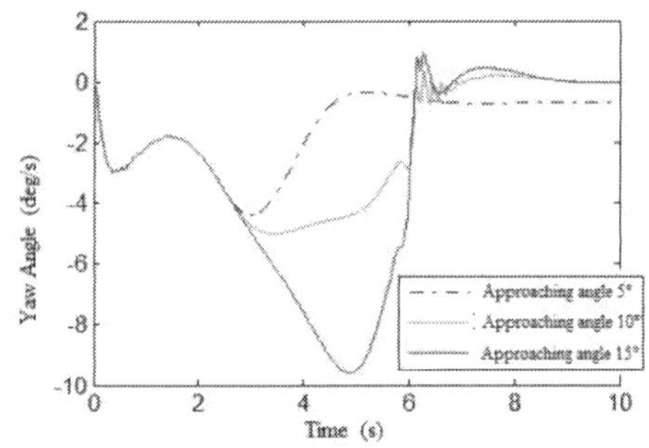

(b) Vehicle yaw rate over time

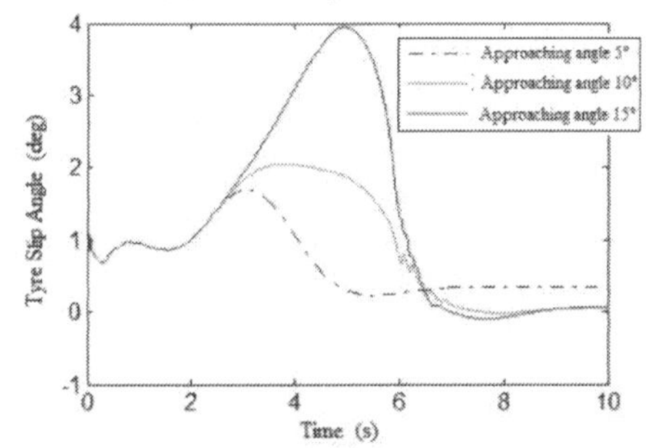

(c) Tyre slip angle over time

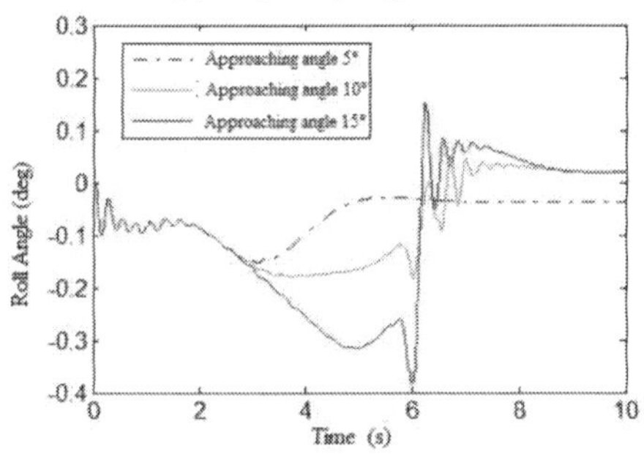

(d) Vehicle roll angle over time

Figure. 4 Simulation result for right curve main road

From the simulation results in Fig. 4(a), the lateral acceleration increases gradually as the runaway vehicle approaches the entrance of the truck escape ramp, and the lateral acceleration is already of a large magnitude when the runaway vehicle reaches the entrance, where the vehicle is in a dangerous situation, and although the instability state of the vehicle does not occur, it is already very difficult for the driver to control the vehicle at this time. After the runaway vehicle enters the approach road of the truck escape ramp, the lateral acceleration of the vehicle gradually decreases. However, the vehicle in the whole incoming process is not stable, and a rollover is very likely to occur.

From the simulation results of Fig. 4(b), it can be seen that as the runaway vehicle enters the truck escape ramp, the yaw rate increases to a certain extent, then decreases gradually, and finally becomes stable.

From the simulation results in Fig. 4(c), it can be seen that the tyre slip angle is constantly increasing in the process of the runaway vehicle entering the truck escape ramp. When the runaway vehicle reaches the entrance of the truck escape ramp, the value of the tyre slip angle is large, which indicates that the driver has some difficulties in handling the vehicle at this time; however, the tyre slip angle does not exceed the determination threshold of 5°, the state of instability of the vehicle does not occur.

From the simulation results in Fig. 4(d), it can be seen that the roll angle of the vehicle increases gradually as the runaway vehicle enters the truck escape ramp from the straight main road. Moreover, as the approach angle increases, the increase of the roll angle of the vehicle is more obvious. When the approach angle is 15°, the roll angle of the vehicle is close to 0.4 deg, in which case the vehicle body obviously tilts to one side, the vehicle is in a dangerous situation, and the posture of the vehicle is out of control. After the runaway vehicle enters the approach road of the truck escape ramp, the roll angle of the vehicle gradually decreases. However, the vehicle in the whole incoming process is very unstable, and a rollover of the vehicle is most likely to occur in this case.

The analysis of the above simulation data shows that the runaway vehicle did not lose its stability when the runaway vehicle enters the truck escape ramp with the limit radius of the right curve at the speed of 120km/h. Although the tyre slip angle and vehicle roll angle are relatively large when the vehicle reaches the entrance of the truck escape ramp, the tyre slip angle and vehicle roll angle do not exceed the determination threshold, and experienced drivers can generally control the vehicle. To some extent, the simulation experiment data support the fact that the value of the limit radius of the right turn curve is rational in theory.

4.3 Main road alignment being left curve

When the main road alignment is a left curve and the truck escape ramp is arranged in a tangent direction of the main road, the runaway vehicle drives into the truck escape ramp along a straight line, in which case the lateral acceleration, yaw rate, tyre slip angle and roll angle are very small, so that the handling stability of runaway vehicle is equivalent to that of the vehicle travelling along a straight line. It is known from the engineering experience that it is the safest to arrange the truck escape ramp in the tangent direction of a left curve.

5 Conclusion

The influence of main road alignment on the handling stability of a runaway vehicle is well reflected by the four vehicle dynamics responses, i.e. the lateral acceleration, the yaw rate, the tyre slip angle and the roll angle. When the minimum curve radius that can prevent a lateral rollover and lateral slip of a runaway vehicle is used as an input parameter of the road simulation model, the runaway vehicle in the entering process is extremely unstable in the case of the main road alignment being a right curve and the approach angle being 15°. Therefore, the minimum curve radius that theoretically does not cause a lateral rollover and lateral slip is underestimated in the actual travelling process. The response indicators of vehicle dynamics can also be used as the basis to determine whether the design of main road alignment is rational and feasible. Only if the safety and stability of runaway vehicles are ensured during the process of entering a truck escape ramp from a main road, can the truck escape ramp achieve its design purpose.

References

1. Eck, Ronald W. Development of warrants for the use and location of truck escape ramps. Volume **I** - final report [R].1980.
2. Eck, Ronald W. Development of warrants for the use and location of truck escape ramps. Volume **II** - appendices [R].1980.
3. J.C. Wambold, L.A. Rivera-Ortiz and M.C. Wang and Wambold, James C. A field and laboratory study to establish truck escape ramp design methodology. Volume **I**, Executive summary [R].1988.
4. J.C. Wambold, L.A. Rivera-Ortiz and M.C. Wang and Wambold, James C. A field and laboratory study to establish truck escape ramp design methodology. Volume **II**, Final report [R].1988.
5. Zhang Jianjun, Zhang Gaoqiang. Analysis of incident in emergency escape ramps and legal risk aversion [J].ICTE,2013:706-715.
6. MA Zhuanglin, ZHANG Honglu, TAN Xiaowei. Temporal-spatial distribution characteristics of traffic accidents on freeway continuous downgrade segment[J]. China Safety Science Journal, 2014, (05): 85-91.
7. GUO Xin, XU Yang, LIU Mingwen, et al. Research on safety of truck escape ramps on continuous long steep downgrades[J]. China Safety Science Journal, 2014, 24(4):152-158.
8. HUANG Xiucheng, TIAN Xidong, SUN Zhiyong. Safety evaluation of truck escape ramps of Guangdong Jiangxi Expressway[J]. Highway, 2005(12):102-104.
9. NIU Jianfeng, WANG Junhua, DONG Xianyuan. Research on the angle to truck escape ramps of mountain roads[J]. Journal of Chongqing Jiaotong University (Natural Science), 2010, (04):604-608.
10. PAN Binghong. Minimum radius of vertical curve at the entrance of truck escape ramps [J]. Highway, 2011, (06):35-38.
11. SONG Cancan, GUO Zhongyin, LIN Lin. Departure angle and approach length of truck escape ramp[J].

Journal of Tongji University, 2016, 4(44)587-592.

12. SONG Cancan, GUO Zhongying. Width of cross section on truck escape ramp based on driving simulation[J]. Journal of Tongji University (Natural Science), 2017, (01):46-52.

13. LI Hai, LIU Xiaoyong, SHI Hongxing. Preliminary analysis on the method of installation for drift angle on truck escape ramp[J]. Highway Engineering. 2013(6):84-86.

MATEC Web of Conferences **206**, 04003 (2018) https://doi.org/10.1051/matecconf/201820604003
ICCEMS 2018

Application of Thermal Images for Determining the Vegetation Coverage Ratio at Free-Frame Shotcrete Grid Beam Protected Slope

Chien-Yuan Chen[1] and Zheng-Sheng Jiang[2]

[1] *National Chiayi University, Taiwan*
[2] *Fifth District Maintenance Engineering Department, Highway Administration, Taiwan*

Abstract. In this study, infrared thermography was used to determine the vegetation coverage ratio (VCR) at a free-frame shotcrete grid beam-protected slope. An infrared thermal imager can detect the changes in surface radiation temperature on naked and vegetation-covered slope areas. Regional temperature analysis was performed through thermography, and the average temperature derived from the captured thermal images was used as the index for the vegetation area; the VCR was also evaluated. This method can be used in engineering practice after construction to reduce the errors in subjective visual judgment and can also serve as an effective and scientific evaluation methodology. A long-distance noncontact detection method for VCR evaluation can increase the engineering applicability of the proposed method after construction.

1. Introduction

Mountainous roads pose a challenge in maintaining traffic safety during the rainy season and typhoon events. The focus of expressway maintenance in mountainous areas is on engineering management at the upper and lower slopes. In the face of climate change, regular maintenance and postdisaster reconstruction of mountainous roadways is emphasized for improving the road slope for disaster resistance. With the increasing awareness of natural environment protection and landscaping for scenery enhancement, shotcrete grid beam-protected slopes are a commonly used engineering measure in Taiwan, with advantages such as low cost, short construction time, less earthwork, and high landscaping potential, compared with large-scale engineering measures.

Free-frame shotcrete grid beams are commonly used to protect unstable slopes or for recovery from naked landslides near expressways in mountainous areas. Such beams are suitable for areas with varying levels of topographic steepness. Free-frame shotcrete can protect slopes with vegetation against soil erosion caused by rainfall and landscape planting. In such protection systems, drainage engineering includes a catchwater drain for collecting surficial water to a hillside ditch to reduce the regolith erosion rate.

The construction flowchart for a free-frame shotcrete grid beam-protected slope includes the following steps (Figure 1):

(a).Designing the material compounding ratio and audit,
(b).Executing slope cutting to remove unstable soils and rocks,
(c).Constructing a galvanized wire mesh, establishing anchor bolts, and assembling reinforcement,
(d).Paving sheet PVC at the vegetation area,
(e).Executing shotcrete construction,
(f).Spraying a plant growth medium.

https://doi.org/10.1051/matecconf/201820604003

Figure 1. The constructure flowchart for the shotcrete grid beam-free frame protected slope.

The free frame is fixed in size and the shape is flexible to changes in topography. The commonly used free-frame size is 130 cm × 130 cm along the expressway, and the frame is fixed using anchor bolts. The mix design proportion of the shotcrete material is compounded by cement, sand, and water (Table 1).

Table 1 Mix design proportion of shotcrete material (Directorate General of Highways, MOTC).

Material	Unit	Quantity per 100 m^2
Cement	kg	3800
Sand	m^3	10.5
water	Liter	1900

The key aspect of a shotcrete grid beam-protected slope is the vegetation cover. In general, the area for spraying the plant growth medium is within the free frame with a size of 95 cm × 95 cm. The size and ruggedness are affected by the steepness and aspect of the slope. During construction, the vegetation coverage ratio (VCR) at the shotcrete site has a major influence on the protection of the slope. The vegetation is typically designed to grow naturally, and the determination of the VCR at the shotcrete site is affected by the topography; thus, visual inspection is commonly performed without any effective quantification methodology.

Thermal images have been applied to various fields for inspection and monitoring. Infrared thermography has been applied to nondestructive testing of concrete and masonry bridges [1] and to anomaly detection in adhesive wall tiling systems [2]. In recent years, thermal images have been widely applied for slope monitoring. Infrared thermography has been used to examine potentially unstable slopes [3,4]. It has also been commonly used for monitoring landslides [5–7] and earth dam failure modes [8]. Similar to a camera, using a thermal imager for monitoring has advantages such as noncontact and long-distance monitoring, and a thermal imager is suitable for both long-term monitoring and short-term inspection of sites with surficial radiation temperature changes.

2. Material and method

The monitoring sites in this study were located on Provincial Highway 18, with mileage of 73K + 550 and 73K + 980 (Figure 2). Table 2 lists the coordinates and corresponding area of the protected slope.

Table 2 Locations of the monitoring sites.

Milage (start date) (finish date)	Coordinate (TWD97)		Free frame area (m^2)	Grass species
	X (m)	Y (m)		
72K+550 (105.11.29) (106.02.16)	22311 0.383	25977 03.256	75.16	Festuca arundinacea , Cynodon dactylon, Eragrostis curvula,
73K+980 (101.04.11) (101.05.20)	22411 1.783	25975 92.561	473.80	

Landslides occurred at the monitoring sites during Typhoon Kalmegi in 2016 and a rainstorm in 2017. Both sides of the slope along Provincial Highway 18 in the mountainous region were naked with a steep gradient and weathered regolith, and were prone to landslides during torrential rainfall. The slope protection engineering process at mileage 72K + 550 was commenced on November 29, 2016, and completed on February 16, 2017, and that at mileage 73K + 980 commenced on April 11, 2012, and completed on May 20, 2012. The vegetation period was 3 months at 72K + 550 and 5 years at 73K + 980 to the date of monitoring. The acceptable standard for visual inspection is for a VCR of up to 90% in Taiwan. The site at mileage 73K + 980 was completed with a VCR of 100%, whereas that at mileage 72K + 550 was naked. The two sites were selected for comparison and further monitoring and analysis. Monitoring was performed using a thermal imager on different dates, and images captured on April 16, 2017, were used for analysis.

https://doi.org/10.1051/matecconf/201820604003

Figure 2. Monitoring site along Provincial Highway 18 at mileage (a) 72K + 550 (b) and 73K + 980 (April 16, 2017).

In this study, the captured thermal images were divided into 36 square grids for the monitoring of radiation temperature changes. The average radiation temperature (T in °C) was measured for each grid corresponding to a different VCR. The thermal imager was used for monitoring the selected sites, and images were captured at different times. A direct linkage between T and VCR was required to avoid visual adjudgement.

The thermal imager NEC G100D, with an accuracy of ±2 °C (below 100 °C) or ±2% (above 100 °C) and sensitivity of 0.08 °C, was used for the monitoring process. The detector comprises an 8–14 m microbolometer for long waves and 3–5 μm for short waves. The thermal imager can also be used for capturing visible-light digital images with a resolution of 200 megapixels.

3. Results and analysis

3.1 Thermal images for 0% VCR

The site at mileage 72K + 550 with a VCR of 0% was selected for monitoring. The monitoring area could be divided into parts for the plant growth medium and reinforced concrete made of the free frame. Thermal images revealed that the changes in infrared temperature in the area with the plant growth medium were larger than those in the area with free-frame reinforced concerete (Figure 3).

The temperature changes abruptly increased at 11:02 because of sunlight. In general, the average temperature was 42 ℃ in the area fully covered from sunlight (shadow area), 43–48 ℃ in the area partially covered from sunlight, and more than 50 ℃ in the area fully exposed to sunlight. The area partially covered from sunlight had larger infrared temperature changes than the other areas. The infrared temperature was in the range of 40–43 ℃ at 13:57 (Figure 4). The overall average infrared temperature for the site with a VCR of 0% at different monitoring times was 43.39 ℃ (Figure 5).

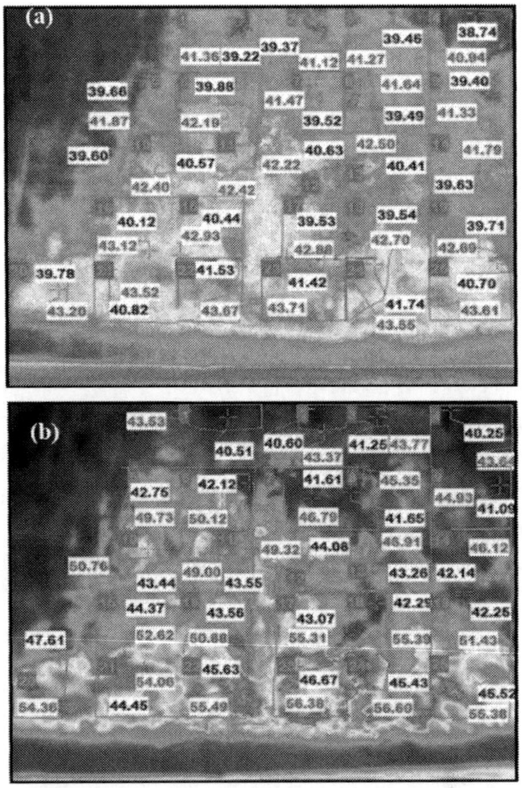

Figure 3. Thermal images for 0% VCR captured at (a) 11:02 and (b) 13:57 (captured on April 16, 2017).

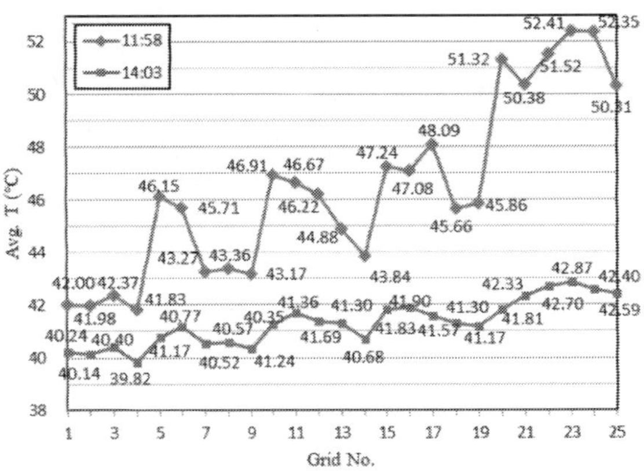

Figure 4. Average infrared temperature changes for 0% VCR at different grids at 11:02 and 13:57.

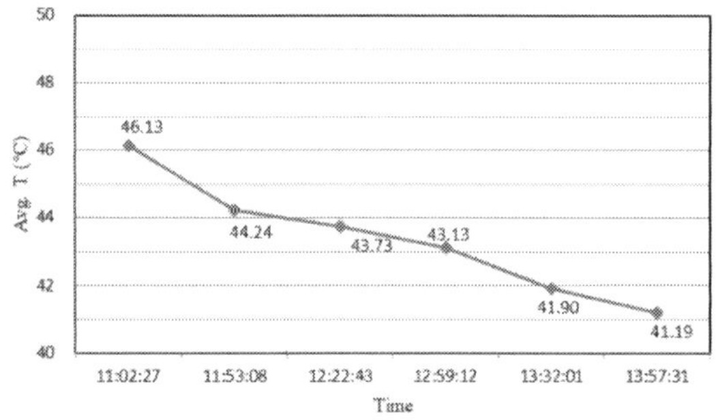

Figure 5. Average infrared temperature changes for 0% VCR at different times.

3.2 Thermal images for 100% VCR

The site at mileage 73K + 980 with 100% VCR was selected for monitoring at the shotcrete-protected slope. The well-vegetated area showed the lowest infrared temperature, followed by the area with free-frame; the area with a VCR of 0% had the highest infrared temperature (Figure 6). The overall average infrared temperature was between 43 and 45 ℃ at 11:58 and between 40 ℃ and 41 ℃ at 14:03 (Figure 7). The overall average infrared temperature for the area with a VCR of 0% at different monitoring times was 46.3 ℃ (Figure 8). Two additional sites along the expressway with VCRs of 10% and 50% were monitored, and their average infrared temperatures were recorded. Figure 9 shows the regression line for determining the VCR by using thermal images. The data can be regressed as follows:

$$T\,(°C) = 0.286(VCR) + 43.25\ (r^2 = 0.99) \qquad (1)$$

The VCR can be determined by the equation and the average infrared temperature can be monitored using the thermal imager.

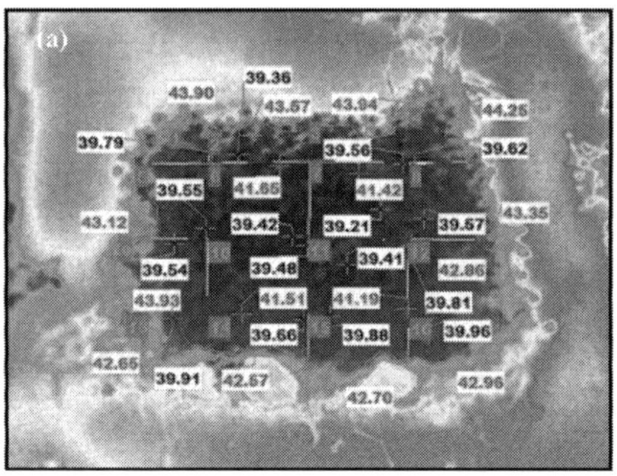

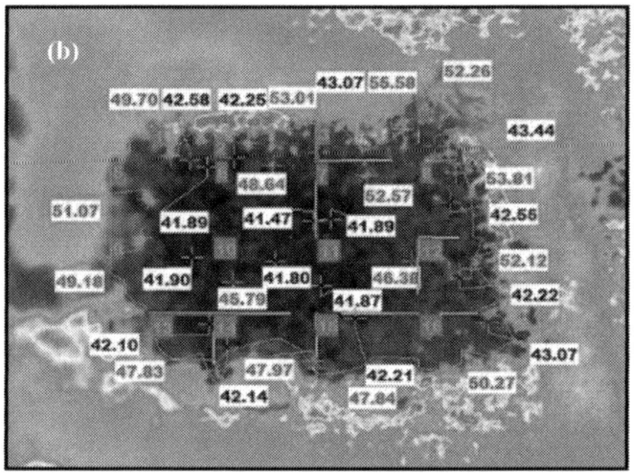

Figure 6. Thermal images for 100% VCR on (a) 11:58 and (b) 14:03 (captured on April 16, 2017).

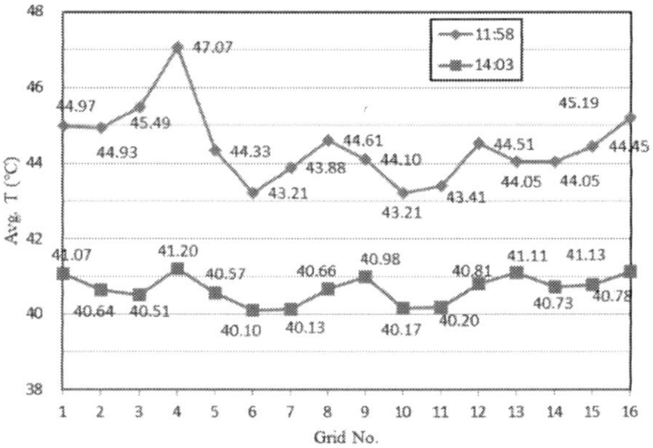

Figure 7. Average infrared temperature changes for 100% VCR at different grids at 11:58 and 14:03.

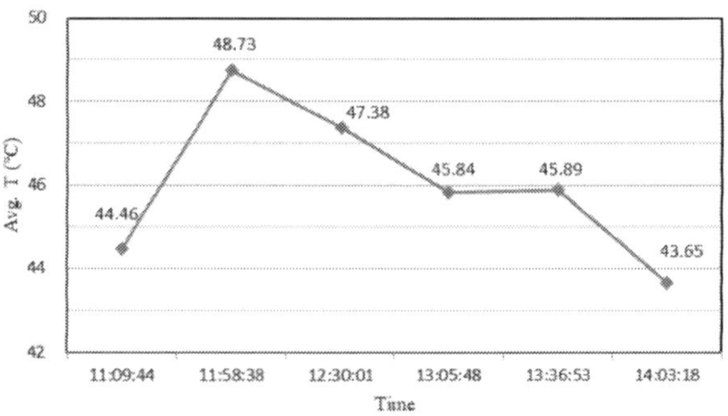

Figure 8. Average infrared temperature changes for 100% VCR at different time.

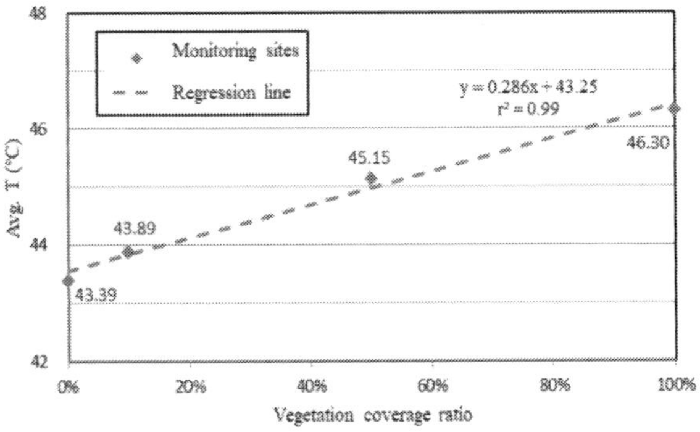

Figure 9. Regression equation for the VCR and average infrared temperature changes at the monitored slope.

4. Conclusions

In this study, a thermal imager was used to monitor the VCR at a free-frame shotcrete grid beam-protected slope. The infrared thermal imager could detect the changes in surficial infrared temperature at naked and vegetation-covered areas of the slope. Shotcrete-protected slope areas with VCRs of 100% and 0% (naked) were selected for monitoring and analysis. The regional temperature was analyzed through thermography, and the average temperature derived from the thermal images was used as the index for evaluating the VCR. The overall average infrared temperature was used to establish the relationship between the infrared temperature and VCR. An interpolation method was then used to determine the VCR by using the thermal imager. The method can be used in engineering practice after construction to reduce the errors in subjective visual judgment and can also serve as an effective and scientific evaluation methodology.

Acknowledgments

Authors wishing to acknowledge financial support from Ministry of Science and Technology in Taiwan under contract No. MOST 106-2625-M-415-003.

References

1. Clark M R, McCann D M and Forde M C 2003 Application of infrared thermography to the non-destructive testing of concrete and masonry bridges *NDT International* **36** 265–275
2. Lourenço T, Matias L and Faria P 2017 Anomalies detection in adhesive wall tiling systems by infrared thermography *Construction and Building Materials* **148** 419–428
3. Baroň I, Bečkovský D and Míča L 2012 Application of infrared thermography for mapping open fractures in deep-seated rockslides and unstable cliffs *Landslides* **36** 265–275
4. Nolesini T, Frodella W, Bianchini S and Casagli N 2016 Detecting slope and urban potential unstable areas by means of multi-platform remote sensing techniques: the Volterra (Italy) case study Remote Sensing **8**, 746, doi:10.3390/rs8090746
5. Wu J H, Lin H M, Lee D H and Fang S C 2005 Integrity assessment of rock mass behind the shotcreted slope using thermography *Engineering Geology* **80** 164–173
6. Martino S and Mazzanti P 2014 Integrating geomechanical surveys and remote sensing for sea cliff slope stability analysis: the Mt. Pucci case study (Italy) *Natural Hazards and Earth System Sciences* **14** 831–848
7. Mineo S, Pappalardo G, Rapisarda F, Cubito A and Di Maria G 2015 Integrated geostructural, seismic and infrared thermography surveys for the study of an unstable rock slope in the Peloritani Chain (NE Sicily) *Engineering Geology* **195** 225–235
8. Chen C Y, Chen S C and Chen K H 2016 Earth dam monitoring by using infrared thermography detection *Proceedings of Engineering Technology and Innovation* **4** 40–42

https://doi.org/10.1051/matecconf/201820604004

Evaluation of Measurement Uncertainty Based on Monte Carlo Method

X M Wang, J L Xiong and J Z Xie

School of Intelligent Science, National University of Defense Technology, Changsha, China

Abstract. The developments of scientific technology are inseparable from various measurement methods. Measurement uncertainty is an indication that evaluates the credibility of the measurement results directly, and it affects the development of technology and economy indirectly. Monte Carlo method (MCM) is an effective method to evaluate the measurement uncertainty, because it can evaluate the measurement uncertainty in the complex model and environment. The application range of MCM is large than the traditional method that recommended in the "Guide to the Uncertainty in Measurement (GUM)". Based on the study of Monte Carlo method, this paper establishes a model for MCM to evaluate the uncertainty of measurement. Finally, as an example, we use the MCM to evaluate the measurement uncertainty of the six-and-a-half digital multimeter (DMM), which verifies the validity of MCM for evaluating the measurement uncertainty.

1 Introduction

The measurement uncertainty is the uncertainty and change of measurement result. It is an estimation that the true value of measured value is in a certain range. And it is a parameter included in the measurement result to indicate the dispersion of the measured value. The definition of measurement uncertainty shows that a complete measurement result should contain both the measured value and the measurement uncertainty[1]. The measurement uncertainty is usually assessed by the mathematical model of the measurement process and the uncertainty transmission rate. The method used in the field of measurement uncertainty assessment is GUM, which is mainly derived from ISO / IEC GUIDE 98-3: 2008 "Measurement Uncertainty-Part 3: Measurement Uncertainty Representation Guide"[2]. On the basis of this international standard and JJF 1059-1999, China issued the new edition JJF 1059.1-2012 "Evaluation and Expression of Measurement in Uncertainty" on December 3th, 2012, and it implemented on June 3th, 2013[3]. The GUM obtains the measurement uncertainty of measurand through the relationship between input and output, and using the uncertainty transmission rate of input optimal estimation and its associated standard uncertainty. However, the application of GUM has some limitations and assumptions[4]. When the probability distribution of input should be distributed symmetrically, such as normal distribution and t-distribution, and the measurement model is a linear model or can be converted into a linear model, the GUM can be used to assess the measurement uncertainty.

MCM is a numerical method to achieve distributed distribution by repeated sampling. MCM obtains the discrete sampling values of measurand Y through sampling the input X_i according to its probability density function (PDF) and selecting the transmission model of the input distribution according to the measurement model. Finally, obtain the optimal estimation, the standard uncertainty and the inclusion interval of measurand Y through the calculated discrete sampling values Y[5]. MCM can be applied in a wide range, in addition to the scope of the traditional GUM. What's more, the application range suitable for MCM: the scale of the uncertainty component is not approximate; the deviation of measurement model is difficult to obtain; the PDF of measurand is not a normal distribution or a t-distribution; the estimated value of measurand and its standard uncertainty are equivalent; the measurement model is complex or non-linear; the PDF of input is asymmetric[6]. When the conditions required by the GUM are not or not sure satisfied, and the GUM is difficult to calculate measurement uncertainty due to the complexity of the model, the MCM is all suitable to assess the measurement uncertainty. In this paper, we use the MCM to evaluate the measurement uncertainty of the six-and-a-half DMM based on the MATLAB platform to get the measurement uncertainty result of DMM.

2 Measurement uncertainty evaluation using MCM

2.1 Measurement uncertainty

The measurement uncertainty is defined as the parameter that is reasonably assigned to the dispersion of the measured value, which is associated with the measurement result[7][8]. Measurement uncertainty is usually assessed by the mathematical model of measurement process and the uncertainty transmission rate. If the relationship between measurand Y and input X_1, X_2, …, X_n is $Y=f(X_1, X_2, …, X_n)$, the measurement uncertainty can be obtained according to the uncertainty transfer diagram shown in Fig. 1, and the synthetic standard uncertainty of Y can be calculated according to formula (1).

$$u_c(y)=\sqrt{\sum_{i=1}^{N}(\frac{\partial f}{\partial x_i})^2 u^2(x_i)+2\sum_{i=1}^{N-1}\sum_{j=1,j\neq i}^{N}\frac{\partial f}{\partial x_i}\frac{\partial f}{\partial x_j}r(x_i,x_j)u(x_i)u(x_j)} \quad (1)$$

Where $u(x_i)$ is the standard uncertainty of the input X_i, which can be type A standard uncertainty or type B standard uncertainty. Where $u_c(y)$ is the synthetic standard uncertainty of measurand Y, which indicates the dispersion of measurand Y. Where $\partial f/\partial x_i$ is the partial derivative of the function $Y=f(X_1, X_2, …, X_n)$, $r(x_i, x_j)$ is the correlation coefficient of the uncertainty, and N is the total number of the input quantity. If each standard uncertainty of the input X_i can be regarded as unrelated, i.e. $r(x_i, x_j)=0$, then the synthetic standard uncertainty can be reduced to the formula (2)

$$u_c(y)=\sqrt{\sum_{i=1}^{N}(\frac{\partial f}{\partial x_i})^2 u^2(x_i)} \quad (2)$$

If the measurand Y is measured by the measuring device directly, the relationship between measurand and input cannot be expressed by a fixed function. Through the uncertainty analysis, there are N obvious uncertain components u_i in the measurement result, and each uncertain component is not correlated. Suppose that each uncertainty component has the same sensitivity degree to the measurement result, the synthetic standard uncertainty can be calculated according to formula (3).

$$u_c(y)=\sqrt{\sum_{i=1}^{N}(u_i)^2} \quad (3)$$

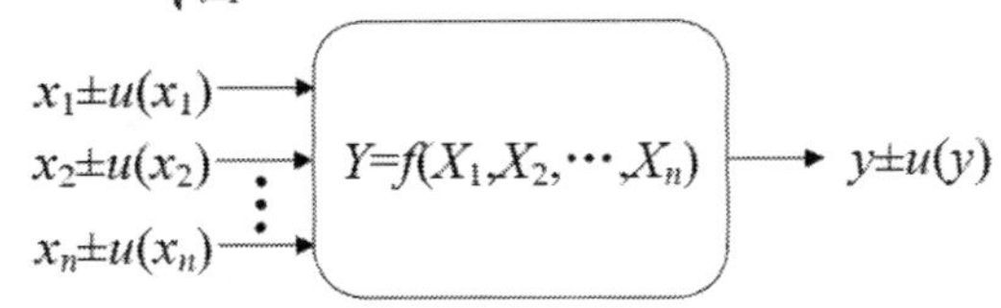

Figure 1. Uncertainty transfer diagram

2.2 Monte Carlo method

MCM is a statistical test method, which be used to solve the mathematical physics problem, that is difficult to determine the formula, through statistical sampling theory. For a statistic with uncertainty PDF, MCM is easier to solve the synthetic uncertainty which approaches the real solution[4][9]. Unlike the GUM method using the linearized model propagation uncertainty, MCM evaluates measurement uncertainty by a numerical method to achieve distributed propagation.

When the measurement model is linear and the output PDF is a normal distribution, the GUM method can provide accurate uncertainty and inclusion intervals[10]. Fig. 2 depicts the independent input distributions transmission, and the PDF of output quantity Y is obtained through three PDFs of input quantities X_1, X_2, X_3 with the model propagation. The three PDFs of input quantities X_1, X_2, X_3 in Fig. 2 are normal distributions, triangular distributions and normal distributions respectively, which are independent, and the PDF of output quantity Y is an asymmetry distribution. If the PDF of output quantity Y is asymmetry like in Fig. 2, then the measurement uncertainty evaluation. MCM carries out the discrete sampling of the input quantity PDF, and then calculates the discrete sampling of output quantity Y by the measurement model. Next, the best estimate, the standard uncertainty, and the inclusion interval of the output quantity Y are calculated by the discrete distribution value of the output quantity, which are improved as the number of sample M of the input X increases.

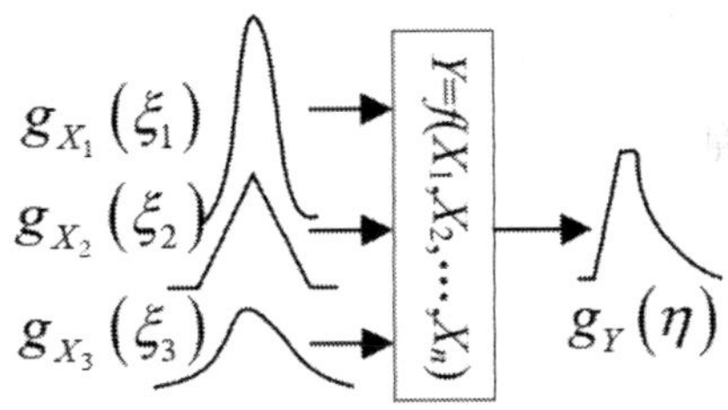

Figure 2. Independent input distributions transmission

2.3 Measurement uncertainty evaluation using MCM

The parameter estimation is mainly divided into two kinds, one is that estimate unknown parameters according to the samples when the overall distribution is known and one or several of the parameters is unknown; the other is that the overall distribution and the parameters are unknown. If the overall distribution is known, the commonly methods include moment estimation, maximum likelihood estimation, Bayesian estimation and least squares estimation. The intelligent algorithm is more and more important in parameter estimation, whose principle is to simulate the biological population activities, the information transmission between individuals and individuals.The parameter estimation is mainly divided into two kinds, one is that estimate unknown parameters according to the samples when the overall distribution is known and one or several of the parameters is unknown; the other is that the overall distribution and the parameters are unknown. If the overall distribution is known, the commonly methods include moment estimation, maximum likelihood estimation, Bayesian estimation and least squares estimation. The intelligent algorithm is more and more important in parameter estimation, whose principle is to simulate the biological population activities, the information transmission between individuals and individuals.

In the measurement uncertainty evaluation using MCM, the mathematical model $Y=f(X_1, X_2, \ldots, X_n)$ between the input $(X_1, X_2, \ldots, X_n)$ and the output Y is chosen first according to the physical model, mathematical formula or experience of the measurement. Then determine the distribution and its parameters of the input $(X_1, X_2, \ldots, X_n)$ quantities uncertainty, which according to generate the various distributed random variables $\xi_1, \xi_2 \ldots, \xi_n$. Generally, the various distributed random variables are usually generated by the inverse function of the random variable obeyed uniformly distribution in the interval (0, 1). The output value $y_i=f(\xi_1, \xi_2 \ldots, \xi_n)$ is obtained by putting the random variables $\xi_1, \xi_2 \ldots, \xi_n$ into the mathematical model $Y=f(X_1, X_2, \ldots, X_n)$. Since the range of each random variable is distributed in entire domain, the obtained y_i may cover all possible values of the output quantity Y. The mean value E_Y and the standard deviation S_Y of the output quantity Y are calculated by the formula (4) and (5) using the M values $y_1, y_2 \ldots, y_M$ generated by the MCM.

$$E_Y = \frac{1}{M}\sum_{i=1}^{M} y_i \tag{4}$$

$$S_Y = \frac{1}{n}\sum_{i=1}^{n}(y_i - Y)^2 / (n-1)]^{1/2} \tag{5}$$

The specific implementation steps of measurement uncertainty evaluation using MCM are shown in Figure 3, which mainly including MCM input, MCM transmission, MCM output and result report. The MCM input contains the following five aspects: defining the output of Y, which is the measurand, determining the input quantity $X=(X_1, X_2, \ldots, X_n)$ associated with the output Y, establishing the model $Y=f(X)$ between the output Y and the input X, determining the PDF of the each input X_i with the information obtained, and selecting the MCM sampling size M which is taken 10^6 generally. The more common distributions of input X are normal distribution, the uniform distribution and so on. The MCM transmission is calculating the corresponding output value Y_j for each sample $(x_{1j}, x_{2j}, \ldots, x_{nj})$ according to the model $Y=f(X)$. The samples $(x_{1j}, x_{2j}, \ldots, x_{nj})$ are extracted from the PDF of input X_i ,where $i=1, 2, \ldots, n$ and $j=1, 2, \ldots, M$. The MCM output is mainly sorting these M values of output Y by the strictly ascending order, and getting the discrete representation G of the PDF of the output Y according to the sorted sample values $(y_1, y_2, \ldots, y_n)$. Finally, according to the discrete representation G, the result report is mainly calculating the estimated value y and the standard uncertainty $u(y)$ of the output Y, and the inclusion interval $[y_{low}, y_{high}]$ of Y with given inclusion probability p.

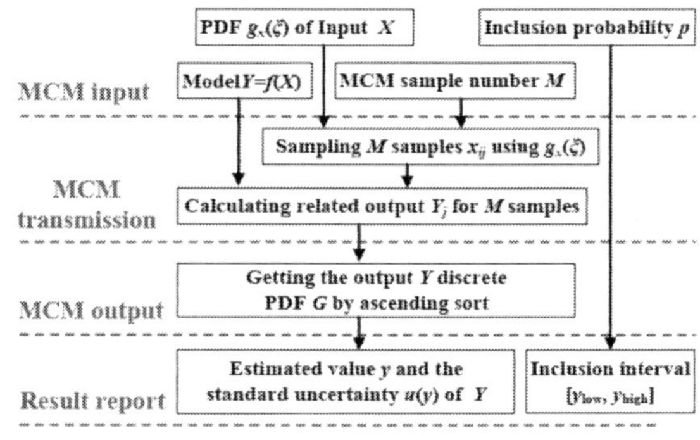

Figure 3. Specific implementation steps of measurement uncertainty evaluation using MCM

3 MCM for evaluating digital multimeter uncertainty

Uncertainty evaluation of digital multimeter by MCM, first of all, we need to determine the mathematical model of input and output. The uncertainty source of digital multimeter mainly include: the voltage drift of the multifunction calibrator since the last calibration, the effect of bias, nonlinearity and gain change on the voltage value of the multifunction calibrator instrument, the influence of ambient temperature on the voltage value of the calibrator, the influence of the power supply voltage, and the load effect caused by the limited input impedance of the calibrated DMM. According to the reference [11], the measurement model of DMM is established as shown in formula (6).

$$u_c = V_i - V_s + \Delta V_i - \Delta V_s \tag{6}$$

Where V_i is the voltage value measured by the DMM, V_s is the output voltage of the multifunction calibrator, and these two variables V_i and V_s are subject to the normal distribution. Where ΔV_i is the effect of the DMM resolution on the measurement result, ΔV_s is the comprehensive effect on the output voltage value of multifunction calibrator due to many kinds of reasons, and these two variables ΔV_i and ΔV_s generally obey the uniform distribution.

We use the MCM to evaluate the measurement uncertainty of the six-and-a-half digital multimeter, and verify the validity of MCM for evaluating the measurement uncertainty. Fig. 4 shows the experimental environment of DMM measurement uncertainty evaluation. It can be assumed that the reading itself of DMM doesn't introduce errors because the DMM is a digital display, so the standard deviation is 0. The uncertainty of the multifunction calibrator is given by the instructions or calibration certificate, and when set multifunction calibrator voltage 100V, the standard deviation of multifunction calibrator is 0.00027. The PDF of each input and related information are shown in Table 1. In the method of MCM evaluating the DMM uncertainty, the selected sample size is $M=10^6$. The measurement result of DMM from 1 to 1000 is shown in

Fig. 5, and the PDF of the DMM uncertainty is shown in Fig. 6. The mean value of DMM measurement uncertainty is E_Y=5.13×10^{-7}, and the standard deviation of DMM measurement uncertainty is S_Y=2.72×10^{-4}. 95% of the inclusion interval is [-0.53×10^{-3}, 0.53×10^{-3}]. The PDF of the uncertainty of DMM measurement uncertainty is similar to the normal distribution, which can DMM measurement uncertainty N(5.13×10^{-7},0.000272^2), so its 95% of the inclusion interval is symmetric interval. From the evaluation result, the uncertainty of the DMM is small and its measurement result is credible. At the same time, the uncertainty of the DMM measurement result mainly comes from the uncertainty of multifunction calibrator, while the uncertainty of other aspects has little effect on the uncertainty of the measurement result of DMM.

Table 1. PDF of DMM measurement model input and related information

Variable	Distribution type	PDF	Sensitivity coefficient
V_i	Normal distribution	N(100,0)	1
V_s	Normal distribution	N(100,0.000272)	-1
ΔV_i	Uniform distribution	R(-0.00005,0.00005)	1
ΔV_s	Uniform distribution	R(-0.000035,0.000035)	-1

Figure 4. The experimental environment of DMM measurement uncertainty evaluation

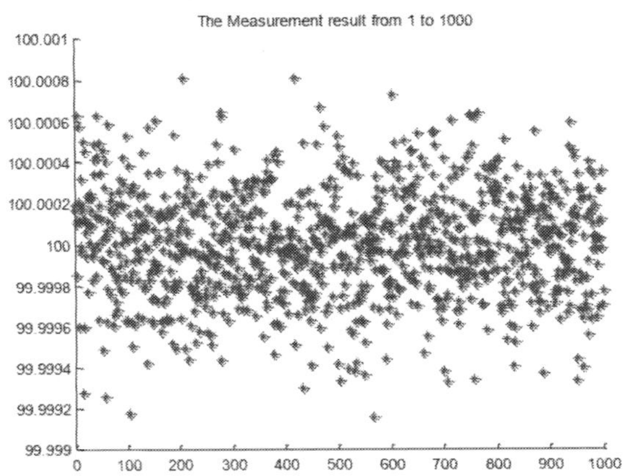

Figure 5. The measurement result of DMM from 1 to 1000

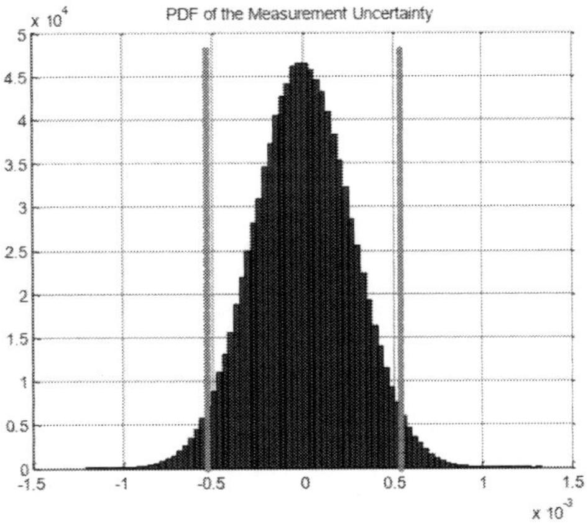

Figure 6. PDF of the DMM uncertainty

4 Conclusions

This paper mainly studies the measurement uncertainty evaluation of the digital multimeter by Monte Carlo method. Monte Carlo method has a strong applicability in evaluating measurement uncertainty. The advantages of Monte Carlo method for evaluating measurement uncertainty are analyzed theoretically, and the steps and precautions of Monte Carlo method for evaluating measurement uncertainty are introduced in detail. Finally, the six-and-a-half digital multimeter uncertainty is evaluated by the Monte Carlo method, and the probability density function of digital multimeter uncertainty obeys the normal distribution, which has high precision. The Monte Carlo method used to evaluate the measurement uncertainty can obtain the probability density function of output directly. Therefore, it has a very prominent advantage and operability in the complex nonlinear model.

Acknowledgments

This work is supported by grants from The Chinese People's Liberation Army General Armament Department. The department of instrument offers the guidance of MCM on DMM.

References

1. Gen W 2013 *Evaluation of measurement error and uncertainty (2th edition)* (Beijing: China zhijian publishing house, Standards press of China) pp 77-90
2. BIPM, IEC, IFCC, ISO, IUPAC, IUPAP, OIML 1993 *Guide to the expression of uncertainty in measurement* (Geneva, Switzerland.
3. General Administration of Quality Supervision, Inspection and Quarantine of the People's Republic of China 2013 *Evaluation and expression of uncertainty in measurement* (Beijing: China zhijian publishing house)
4. Chen X, Bo X, Wang H 2005 The evaluation of combined uncertainty based on the theory of Monte Carlo *Chinese Journal of Scientific Instrument* vol 26(z1) pp 759-761

5. Ge L 2006 *Uncertainty evaluation for virtual instrument measurements based on Monte Carlo method* (Hefei: HeFei University of Technology) pp 25-37
6. Chen H, Cao Y, Han J 2011 Evaluation of uncertainty in measurement based on a Monte Carlo method *Journal of Electronic Measurement and Instrumentation* vol 25(4) pp 301-308
7. Wang W, Song M, Chen Y, Gu L, Tao J 2008 Application of Monte-Carlo method in measurement uncertainty evaluation of complicated model *Chinese Journal of Scientific Instrument* vol 29 pp 1446-1449
8. Wang J, Liu Q, Yang H 2016 Moisture detector standard uncertainty evaluation based on Monte Carlo method *World Sci-Tech R & D* vol 2 pp 297-300
9. Ni Y 2015 *Practical evaluation of uncertainty in measurement (5th edition)* (Beijing: China zhijian publishing house, Standards press of China) pp 35-54
10. Cao Y, Chen W 2012 Analysis of uncertainty using MATLAB *Metrology & Measurement technique* vol39(2) pp 55-56
11. Huang F, Zhou X, Chen G 2016 Uncertainty evaluation of digital multimeter measurement and implementation based on LabVIEW *Industrial Metrology* vol s2 pp 43-45

AUTHOR INDEX

AUTHOR INDEX

EDP Sciences – Web of Conferences
17, Avenue du Hoggar
Parc d'Activité de Courtabouf
BP 112
F-91944 Les Ulis Cedex A
France

ISBN 978-1-5108-7179-3

9 781510 871793

CURRAN ASSOCIATES INC.
proceedings.com